# Salinity Monitoring and Modelling at Different Scales

Guest Editors

**Maria da Conceição Gonçalves**
**Mohammad Farzamian**
**Tiago Brito Ramos**

MDPI

Basel • Beijing • Wuhan • Barcelona • Belgrade • Novi Sad • Cluj • Manchester

*Guest Editors*

Maria da Conceição Gonçalves
Instituto Nacional de Investigação Agrária e Veterinária
Oeiras
Portugal

Mohammad Farzamian
Instituto Nacional de Investigação Agrária e Veterinária (INIAV)
Oeiras
Portugal

Tiago Brito Ramos
Instituto Superior Técnico, Universidade de Lisboa
Lisbon
Portugal

*Editorial Office*
MDPI AG
Grosspeteranlage 5
4052 Basel, Switzerland

This is a reprint of the Special Issue, published open access by the journal *Land* (ISSN 2073-445X), freely accessible at: https://www.mdpi.com/journal/land/special_issues/0D2H422FY0.

For citation purposes, cite each article independently as indicated on the article page online and as indicated below:

| Lastname, A.A.; Lastname, B.B. Article Title. *Journal Name* **Year**, *Volume Number*, Page Range. |
|---|

**ISBN 978-3-7258-2721-3 (Hbk)**
**ISBN 978-3-7258-2722-0 (PDF)**
**https://doi.org/10.3390/books978-3-7258-2722-0**

# Contents

**About the Editors** . . . . . . . . . . . . . . . . . . . . . . . . . . . . . . . . . . . . . . . . . . . . . **vii**

**Mohammad Farzamian, Maria Conceição Gonçalves and Tiago B. Ramos**
Salinity Monitoring and Modelling at Different Scales: An Editorial Overview
Reprinted from: *Land* **2024**, *13*, 1890, https://doi.org/10.3390/land13111890 . . . . . . . . . . . . **1**

**Juan Herrero and Carmen Castañeda**
Comparing Two Saline-Gypseous Wetland Soils in NE Spain
Reprinted from: *Land* **2023**, *12*, 1990, https://doi.org/10.3390/land12111990 . . . . . . . . . . . . **6**

**Raimundo Jiménez-Ballesta, Santos Cirujano-Bracamonte, Eduardo Palencia-Mayordomo and Mario Álvarez-Soto**
An Ecological Overview of Halophytes and Salt-Affected Soils at El Hito Saline Pond (Central Spain): Baseline Study for Future Conservation–Rehabilitation Measures
Reprinted from: *Land* **2024**, *13*, 449, https://doi.org/10.3390/land13040449 . . . . . . . . . . . . **23**

**Yin Zhang, Qingfeng Miao, Ruiping Li, Minghai Sun, Xinmin Yang, Wei Wang, et al.**
Distribution and Variation of Soil Water and Salt before and after Autumn Irrigation
Reprinted from: *Land* **2024**, *13*, 773, https://doi.org/10.3390/land13060773 . . . . . . . . . . . . **41**

**Panagiota Antonia Petsetidi and George Kargas**
Assessment and Mapping of Soil Salinity Using the EM38 and EM38MK2 Sensors: A Focus on the Modeling Approaches
Reprinted from: *Land* **2023**, *12*, 1932, https://doi.org/10.3390/land12101932 . . . . . . . . . . . . **59**

**Maria Catarina Paz, Nádia Luísa Castanheira, Ana Marta Paz, Maria Conceição Gonçalves, Fernando Monteiro Santos and Mohammad Farzamian**
Comparison of Electromagnetic Induction and Electrical Resistivity Tomography in Assessing Soil Salinity: Insights from Four Plots with Distinct Soil Salinity Levels
Reprinted from: *Land* **2024**, *13*, 295, https://doi.org/10.3390/land13030295 . . . . . . . . . . . . **86**

**Lorenzo De Carlo and Mohammad Farzamian**
Assessing the Impact of Brackish Water on Soil Salinization with Time-Lapse Inversion of Electromagnetic Induction Data
Reprinted from: *Land* **2024**, *13*, 961, https://doi.org/10.3390/land13070961 . . . . . . . . . . . . **103**

**László Pásztor, Katalin Takács, János Mészáros, Gábor Szatmári, Mátyás Árvai, Tibor Tóth, et al.**
Indirect Prediction of Salt Affected Soil Indicator Properties through Habitat Types of a Natural Saline Grassland Using Unmanned Aerial Vehicle Imagery
Reprinted from: *Land* **2023**, *12*, 1516, https://doi.org/10.3390/land12081516 . . . . . . . . . . . . **119**

**Magboul M. Sulieman, Fuat Kaya, Mohammed A. Elsheikh, Levent Başayiğit and Rosa Francaviglia**
Application of Machine Learning Algorithms for Digital Mapping of Soil Salinity Levels and Assessing Their Spatial Transferability in Arid Regions
Reprinted from: *Land* **2023**, *12*, 1680, https://doi.org/10.3390/land12091680 . . . . . . . . . . . . **142**

**Chengshen Yin, Quanming Liu, Teng Ma, Yanru Shi and Fuqiang Wang**
Bibliometric and Visualization Analysis of the Literature on the Remote Sensing Inversion of Soil Salinization from 2000 to 2023
Reprinted from: *Land* **2024**, *13*, 659, https://doi.org/10.3390/land13050659 . . . . . . . . . . . . **164**

**Vinod Phogat, Tim Pitt, Paul Petrie, Jirka Šimůnek and Michael Cutting**
Optimization of Irrigation of Wine Grapes with Brackish Water for Managing Soil Salinization
Reprinted from: *Land* **2023**, *12*, 1947, https://doi.org/10.3390/land12101947 . . . . . . . . . . . . . **188**

**Sabri Kanzari, Jiří Šimůnek, Issam Daghari, Anis Younes, Khouloud Ben Ali, Sana Ben Mariem and Samir Ghannem**
Modeling Irrigation of Tomatoes with Saline Water in Semi-Arid Conditions Using Hydrus-1D
Reprinted from: *Land* **2024**, *13*, 739, https://doi.org/10.3390/land13060739 . . . . . . . . . . . . . **217**

# About the Editors

**Maria da Conceição Gonçalves**

Maria da Conceição Gonçalves is a senior researcher at Instituto Nacional de Investigação Agrária e Veterinária (INIAV) in Portugal. She has a degree in Scientific Coordination, obtained with the presentation of the research program "Soil properties and interactions in the soil–water system". She holds a PhD in Agricultural Engineering from the Instituto Superior de Agronomia, University of Lisbon, Portugal. She is the national representative in the consortium of the H2020 European Joint Programme SOIL. Her main research interests are soil physics, soil hydraulic properties, soil solute transport parameters, the development of pedotransfer functions, soil health indicators, and modelling soil water and solute dynamics. She has authored 47 articles in ISI journals.

**Mohammad Farzamian**

Mohammad Farzamian is a researcher at Instituto Nacional de Investigação Agrária e Veterinária (INIAV) and University of Lisbon. He holds a PhD in Geophysics from the University of Lisbon. His research activities involve the development and application of geophysical methods along with remote and proximal sensing tools for vadose zone investigations; precision agriculture; saline intrusion; permafrost; and hydrogeophysics approaches. The main goal of his research is to contribute to more efficient field assessment techniques by coupling multi-scale investigations, from point-scale monitoring to geophysical surveys and remote sensing analyses.

**Tiago Brito Ramos**

Tiago Brito Ramos is a researcher at MARETEC , Instituto Superior Técnico, University of Lisbon. He holds a PhD in Biosystems Engineering from the Instituto Superior de Agronomia, University of Lisbon His main research interests are soil physics, irrigation water management, modeling soil water dynamics and solute transport at different scales, and the development of soil information systems. He has authored 72 articles in ISI journals.

land

MDPI

Editorial

# Salinity Monitoring and Modelling at Different Scales: An Editorial Overview

**Mohammad Farzamian [1], Maria Conceição Gonçalves [1] and Tiago B. Ramos [2,*]**

1 Instituto Nacional de Investigação Agrária e Veterinária, Avenida da República, Quinta do Marquês, 2780-157 Oeiras, Portugal; mohammad.farzamian@iniav.pt (M.F.); maria.goncalves@iniav.pt (M.C.G.)

2 Centro de Ciência e Tecnologia do Ambiente e do Mar (MARETEC-LARSyS), Instituto Superior Técnico, Universidade de Lisboa, Av. Rovisco Pais, 1, 1049-001 Lisboa, Portugal

* Correspondence: tiagobramos@tecnico.ulisboa.pt

## 1. Introduction

Soil salinization is a significant abiotic process affecting arid, semi-arid, and sub-humid regions worldwide. Estimates suggest that approximately 412 million ha of land are impacted by salinity, while sodicity affects about 618 million ha [1]. Most of these areas have a natural origin and play a crucial ecological role, significantly contributing to global biodiversity. However, estimates also indicate that around one-third of the world's irrigated land—approximately 70 million ha—is affected by soil salinization [1], with an annual expansion rate of about 1.0 to 2.0 million ha. This issue arises not only from human-induced processes but also poses a serious threat to global food production, potentially leading to hunger and exacerbating societal problems in salt-affected regions and beyond.

Due to the transient nature of the soil salinization process, which is influenced by various factors including meteorological conditions, soil properties, crop tolerance to salinity levels, and irrigation water management, effective monitoring is crucial for mitigating the harmful effects of soil salinity in agricultural areas, often unnoticed until it is too late. Conversely, in natural salt-affected regions, monitoring supports the characterization and conservation of ecosystems. Therefore, establishing monitoring frameworks for rapid, non-invasive, and cost-effective assessment of the spatial and temporal distribution of salt-affected areas at various scales has emerged as a top priority for research [2].

In this Special Issue titled "Salinity Monitoring and Modeling at Different Scales", we aimed to compile articles that showcase a diverse range of approaches and methods for monitoring soil salinity across various environments and scales. The 11 published articles illustrate a variety of applications for characterizing both natural and human-induced salt-affected areas. The articles focused on the characterization of natural salt-affected areas provide valuable insights into the importance of wetland conservation. Meanwhile, the articles addressing agricultural water management emphasize improving practices for soil conservation and the sustainability of soil and water resources. Overall, this Special Issue offers a wide range of perspectives on the theme of soil salinization.

**Citation:** Farzamian, M.; Gonçalves, M.C.; Ramos, T.B. Salinity Monitoring and Modelling at Different Scales: An Editorial Overview. *Land* **2024**, *13*, 1890. https://doi.org/10.3390/land13111890

Received: 26 October 2024
Accepted: 4 November 2024
Published: 12 November 2024



## 2. Contributions to the Special Issue

The articles in this Special Issue come from researchers worldwide, presenting cutting-edge research on salinity monitoring and management in various salt-affected regions. The studies span work conducted in countries including Portugal, Spain, Italy, France, Hungary, Türkiye, Tunisia, China, the USA, Australia, and Sudan. Collectively, these articles offer a comprehensive overview of various techniques for soil salinity monitoring and management, ranging from traditional sampling campaigns to advanced methods like proximal and remote sensing for soil salinity mapping, as well as the application of process-based modeling tools to predict salt transport in soil profiles and its impact on crops and yields.

Three contributions utilize traditional sampling campaigns to characterize landscape soil salinity and its relationship with ecological status. Herrero and Castañeda (List of Contributions) focus on small, ecologically significant wetlands in the Ebro Basin of north-eastern Spain. They conducted soil monitoring to collect samples and assess key parameters, including percent water saturation, equivalent calcium carbonate, gypsum content, soil cations and anions, and soil salinity. Salinity was measured as electrical conductivity using both a 1:5 soil-to-water ratio and saturation extracts in the laboratory. The article examines critical factors limiting life in two inland saline marshes and provides baseline data for guiding biodiversity protection. The findings emphasize the importance of comparing these conditions with other wetlands to better understand species-specific environmental requirements. Additionally, the article highlights the necessity of using appropriate analytical techniques for gypsum-rich soils to ensure accurate monitoring and effective conservation.

Similarly, Jiménez-Ballesta et al. (List of Contributions) investigate the relationship between soil properties and the presence and abundance of sub-communities within a saline pond in Castilla La Mancha, Spain. The authors characterized various top and subsoil layers from several soil profiles in the region, focusing on properties such as soil texture, electrical conductivity at a 1:5 soil-to-water ratio, and the concentrations of soil cations and anions, among other characteristics. They present baseline data essential for the conservation of local habitats, which include halophilous and gypsophilous vegetation.

In another study, Zhang et al. (List of Contributions) examine the impacts of agricultural water management on soil conservation by analyzing the spatial variation in soil salinity following autumn irrigation in the Hetao irrigation district of northern China. This region experiences widespread soil salinization due to excessive irrigation and challenging hydrogeological conditions. Autumn irrigation plays a crucial role in managing soil salinity levels; however, its effectiveness is significantly influenced by the shallow depth of the groundwater table in some areas. The study explored the distribution and variation in soil moisture and salinity in locations that received autumn irrigation compared to those that were not irrigated. Results indicated that autumn irrigation improved the uniform distribution of soil water and salt profiles compared to non-irrigated fields, resulting in higher leaching. However, unreasonable autumn irrigation raised groundwater levels, increasing the risk of salt movement to the surface during freeze–thaw cycles, which could negatively impact spring crop growth. These findings emphasize the need for effective water resource management to mitigate soil salinization in cold, dry areas.

Three other contributions to this Special Issue underscore the growing importance of electromagnetic induction (EMI) in soil salinity mapping and monitoring, highlighting advancements in both technology and its applications. EMI is a non-invasive method that measures apparent electrical conductivity without physical contact, making it ideal for mapping and monitoring soil salinity at depth on a field scale. Petsetidi and Kargas (List of Contributions) provide a comprehensive review of EM38 and EM38MK2 sensors, emphasizing their effectiveness in assessing soil salinity across various scales. By integrating these sensors with machine learning and remote sensing, the authors propose a method to more accurately predict spatiotemporal salinity variations, showcasing the potential of these sensors for large-scale agricultural management and environmental sustainability.

Paz et al. (List of Contributions) offer a comparative analysis between EMI and Electrical Resistivity Tomography (ERT) for predicting soil salinity in diverse environments. While ERT provides high-resolution images of the subsurface, its requirement for electrode installation and extensive cabling limits its application to field-scale investigations. Through field surveys conducted at four sites in Portugal with varying salinity levels, the study demonstrates that while ERT is more reliable in extreme salinity conditions, EMI offers a faster and sufficiently accurate alternative in most scenarios. This finding is very relevant for soil salinity management as it supports the practical implementation of EMI in large-scale soil salinity assessments.

De Carlo and Farzamian (List of Contributions) utilize time-lapse EMI measurements to monitor soil salinization under different irrigation strategies in tomato crops in Italy. By capturing changes in soil electrical conductivity over time, this study validates EMI as a robust tool for tracking salinization trends, particularly in response to brackish irrigation water. The ability to non-invasively collect data across extensive areas and in real-time further establishes EMI as an effective method for monitoring salinity dynamics and informing sustainable water management practices.

Three contributions to this Special Issue focused on remote sensing and the use of manned aerial vehicle (UAV) technology and machine learning. Pásztor et al. (List of Contributions) introduce an approach to indirectly assess salt-affected soil properties in protected salt meadows using UAV imagery. Through a machine learning model based on spectral indices and a digital elevation model, the researchers mapped soil indicators (total salt content, Na, and pH) in five habitat types across Hungarian salt meadows. This approach facilitates non-invasive environmental monitoring, allowing for the preservation of sensitive habitats while accurately estimating soil conditions.

Sulieman et al. (List of Contributions) apply machine learning techniques to map soil salinity in Sudan's arid regions, utilizing both optical and radar satellite data. By combining these data with digital elevation models as well as electrical conductivity measurements from the saturation extracts of soil samples, the study transfers soil salinity models across geographies with similar environmental conditions. The findings highlight model transferability for spatial salinity classification and demonstrate a framework for salinity assessment and land management planning in water-limited regions.

Offering a bibliometric overview, Yin et al. (List of Contributions) provide a comprehensive analysis of research trends in remote sensing applications for soil salinization from 2000 to 2023. The study tracks the field's evolution, highlighting the contributions of key authors, institutions, and nations. By detailing advancements in machine learning, UAV technology, and hyperspectral imaging, this review identifies trends and future directions of remote sensing applications for salinity management.

Lastly, two contributions to this Special Issue focused on using physical process-based modeling to enhance irrigation water management in salt-affected areas. Several modeling tools are now available to assess site-specific soil, water, and crop parameters while accounting for time-varying field conditions, including the timing and amount of irrigation as well as variable soil salinity. Among the available vadose zone models, the HYDRUS software package offers a range of state-of-the-art approaches for studying salinity dynamics in agricultural fields [3]. In this context, Phogat et al. (List of Contributions) use the HYDRUS (2D/3D) model to optimize irrigation management using brackish water in a vineyard in Southern Australia. After calibration and validation, the model evaluated the impact of irrigating with water of varying quality. The experiment involved blending and alternating the use of brackish waters with fresh waters, as well as testing spring irrigation as a viable method for salt leaching. The results indicate that leaching irrigation with high-quality river water at the beginning of the growing season enhances salt leaching efficiency and helps control salinity. Additionally, blending or alternating saline brackish water with non-saline surface water reduces salt deposition. The study advocates for early-season leaching irrigation as a strategic management option during droughts and emphasizes the need for ongoing monitoring to ensure long-term resilience in irrigated viticulture.

Similarly, Kanzari et al. (List of Contributions) use the HYDRUS-1D model to explore the combined effects of soil matric and osmotic potential stresses on water uptake by tomato roots grown in pots and in a field in Tunisia. The experiment involved irrigating with water of varying qualities, and the model played a crucial role in assessing how different salinity levels impacted tomato growth to varying degrees. However, incorporating crop growth models is necessary to expand the research to yield analysis.

## 3. Conclusions

The papers in this Special Issue offer a comprehensive overview of diverse methods and approaches for monitoring soil salinity, ranging from traditional sampling techniques to advanced methods such as proximal and remote sensing, which are used to characterize and map salt-affected areas across various global regions. Additionally, the articles showcase applications of process-based vadose zone models aimed at enhancing irrigation wate management, particularly when using brackish waters.

By bringing together a wide array of studies, methodologies, and approaches related to soil salinity monitoring and management, this Special Issue highlights the critical importance of these practices in addressing one of the most pressing abiotic stresses worldwide. It is hoped that this collection will inspire more interdisciplinary research on soil salinity and motivate researchers to pursue effective solutions to manage this growing challenge in an increasingly water-scarce world.

**Author Contributions:** Conceptualization, T.B.R. and M.F.; writing—original draft preparation, T.B.R. and M.F.; writing—review and editing, T.B.R., M.F. and M.C.G. All authors have read and agreed to the published version of the manuscript.

**Funding:** The support of FCT—Fundação para a Ciência e a Tecnologia, I.P., through grant attributed to T.B. Ramos (CEECIND/01152/2017) is acknowledged.

**Acknowledgments:** We are grateful to the academic and managing editors and reviewers for the collaboration and support necessary to ensure the academic excellence of this Special Issue.

**Conflicts of Interest:** The authors declare no conflicts of interest.

**List of Contributions**

1. Herrero, J.; Castañeda, C. Comparing Two Saline-Gypseous Wetland Soils in NE Spain. *Land* **2023**, *12*, 1990. https://doi.org/10.3390/land12111990
2. Jiménez-Ballesta, R.; Cirujano-Bracamonte, S.; Palencia-Mayordomo, E.; Álvarez-Soto, M. An Ecological Overview of Halophytes and Salt-Affected Soils at El Hito Saline Pond (Central Spain): Baseline Study for Future Conservation–Rehabilitation Measures. *Land* **2024**, *13*, 449. https://doi.org/10.3390/land13040449
3. Zhang, Y.; Miao, Q.; Li, R.; Sun, M.; Yang, X.; Wang, W.; Huang, Y.; Feng, W. Distribution and Variation of Soil Water and Salt before and after Autumn Irrigation. *Land* **2024**, *13*, 773. https://doi.org/10.3390/land13060773
4. Petsetidi, P.A.; Kargas, G. Assessment and Mapping of Soil Salinity Using the EM38 and EM38MK2 Sensors: A Focus on the Modeling Approaches. *Land* **2023**, *12*, 1932. https://doi.org/10.3390/land12101932
5. Paz, M.C.; Castanheira, N.L.; Paz, A.M.; Gonçalves, M.C.; Monteiro Santos, F.; Farzamian, M. Comparison of Electromagnetic Induction and Electrical Resistivity Tomography in Assessing Soil Salinity: Insights from Four Plots with Distinct Soil Salinity Levels. *Land* **2024**, *13*, 295. https://doi.org/10.3390/land13030295
6. De Carlo, L.; Farzamian, M. Assessing the Impact of Brackish Water on Soil Salinization with Time-Lapse Inversion of Electromagnetic Induction Data. *Land* **2024**, *13*, 961. https://doi.org/10.3390/land13070961
7. Pásztor, L.; Takács, K.; Mészáros, J.; Szatmári, G.; Árvai, M.; Tóth, T.; Barna, G.; Koós, S.; Kovács, Z.A.; László, P.; et al. Indirect Prediction of Salt Affected Soil Indicator Properties through Habitat Types of a Natural Saline Grassland Using Unmanned Aerial Vehicle Imagery. *Land* **2023**, *12*, 1516. https://doi.org/10.3390/land12081516
8. Sulieman, M.M.; Kaya, F.; Elsheikh, M.A.; Başayiğit, L.; Francaviglia, R. Application of Machine Learning Algorithms for Digital Mapping of Soil Salinity Levels and Assessing Their Spatial Transferability in Arid Regions. *Land* **2023**, *12*, 1680. https://doi.org/10.3390/land12091680
9. Yin, C.; Liu, Q.; Ma, T.; Shi, Y.; Wang, F. Bibliometric and Visualization Analysis of the Literature on the Remote Sensing Inversion of Soil Salinization from 2000 to 2023. *Land* **2024**, *13*, 659. https://doi.org/10.3390/land13050659
10. Phogat, V.; Pitt, T.; Petrie, P.; Šimůnek, J.; Cutting, M. Optimization of Irrigation of Wine Grapes with Brackish Water for Managing Soil Salinization. *Land* **2023**, *12*, 1947. https://doi.org/10.3390/land12101947
11. Kanzari, S.; Šimůnek, J.; Daghari, I.; Younes, A.; Ali, K.B.; Mariem, S.B.; Ghannem, S. Modeling Irrigation of Tomatoes with Saline Water in Semi-Arid Conditions Using Hydrus-1D. *Land* **2024**, *13*, 739. https://doi.org/10.3390/land13060739

## References

1. FAO. *Status of the World's Soil Resources. Main Report*; Food and Agriculture Organization of the United Nations and Intergovernmental Technical Panel on Soils: Rome, Italy, 2015.
2. Hopmans, J.W.; Qureshi, A.S.; Kisekka, I.; Munns, R.; Grattan, S.R.; Rengasamy, P.; Ben-Gal, A.; Assouline, S.; Javaux, M.; Minhas, P.S.; et al. Critical knowledge gaps and research priorities in global soil salinity. *Adv. Agron.* **2021**, *169*, 1–191.
3. Šimůnek, J.; van Genuchten, M.T.; Šejna, M. Recent developments and applications of the HYDRUS computer software packages. *Vadose Zone J.* **2016**, *15*, 1–25. [CrossRef]

land

MDPI

Article

# Comparing Two Saline-Gypseous Wetland Soils in NE Spain

**Juan Herrero * and Carmen Castañeda**

Estación Experimental de Aula Dei, Consejo Superior de Investigaciones Científicas, Ave. Montañana 1005, 50059 Zaragoza, Spain; ccastaneda@eead.csic.es
* Correspondence: jhi@eead.csic.es

**Abstract:** Small (<1 $km^2$) saline wetlands scattered across the landscape often go unnoticed or are threatened by urbanization or other interventions, despite their role as biodiversity shelters. This study is needed to show methods for monitoring this specific kind of wetland, and to guide the selection of analytical techniques. We provide data and comparisons for salient soil traits of two quasi-pristine gypsiferous and saline wetlands named Farrachuela (FA) and Agustín (AG). The soil characteristics presented in this article are a more sensitive indicator of their ecological status than some of the most used indicators, such as birds and plants. We found significant differences between the two saladas in percent water saturation, equivalent calcium carbonate, gypsum content, and soil salinity expressed as electrical conductivity both of 1:5 soil-to-water ratio and of saturation extracts. The differences were also significant in the concentrations of $Mg^{2+}$, $Na^{+}$, and $Cl^{-}$, while they were non-significant for $Ca^{2+}$, $HCO_3{}^{2-}$, and $SO_4{}^{2-}$. The mean contents of the six ions were lower in FA than in AG. Both pH and sodium adsorption ratios were significantly different between the two wetlands. The data are mainly examined and plotted by displaying their non-parametric statistics, a synoptic approach that will allow us to monitor the evolution of the wetlands against both traditional agricultural pressures and emerging green energy infrastructures. Last but not least, we discuss the shortcomings of some standard laboratory methods when applied to gypsum-rich soils.

**Keywords:** arid land; athalassohaline; gypsum; Natura 2000 network; soil salinity

**Citation:** Herrero, J.; Castañeda, C. Comparing Two Saline-Gypseous Wetland Soils in NE Spain. *Land* **2023**, *12*, 1990. https://doi.org/10.3390/land12111990

Academic Editors: Maria da Conceição Gonçalves, Mohammad Farzamian and Tiago Brito Ramos

Received: 25 September 2023
Revised: 25 October 2023
Accepted: 27 October 2023
Published: 30 October 2023



## 1. Introduction

European inland saline habitats have been quasi-eliminated [1], with many of the surviving ones reduced in size or badly degraded. This is also the case in the arid Central Ebro Basin, Spain [2]. The scarcity of regulations or incentives for protecting soil diversity occurs in several European countries [3] and is more pronounced for inland saline wetlands in general.

This scarcity also happens for the lands whose composition, color, vegetation or other characteristics are due to an abundance of gypsum ($CaSO_4 \bullet 2H_2O$). These whitish gypsum soil landscapes, known by local people in NE Spain as chesas, occur mainly in warm countries around the world [4], but also in cold regions with gyprock outcrops, e.g., [5,6]. Chesas have also been traditionally disdained due to their low agricultural production and geotechnical characteristics. Irrigation of these lands, needed in dry areas for agricultural production, is technically unfeasible by inundation due to the solubility of gypsum, and only the pressurized application of water has alleviated the drawbacks. The disdain is counteracted by the agricultural pressure and by the increasing attention of scientists to the gypsophilous plants protected by environmental rules [7]. Thus, saline wetlands located in gypseous soils pose a double challenge to life, and are therefore of twofold interest to science. Moreover, the study and description of these saline wetlands face methodological issues in their lab analyses—often overlooked in the literature—derived from the abundance of gypsum [4,8].

The old judgment on both the wetlands and the gypseous soils based on their little or no agricultural value is evolving towards legal protection due to the biodiversity they harbor and their role in ecological balances. In addition, insights into biodiversity and ecosystem services raise the need to move beyond single-lake thinking [9] and to include the small wetlands [10], a line of reasoning that also applies to gypseous soils. Together with the coexistence with agriculture, including irrigation and intensive animal husbandry, there now arise threats from the "green energy" sources with several kinds of impacts, such as those recently reported in [11,12].

The Ebro Basin, located in northeastern Spain (Figure 1), has a dry climate and a high presence of gypsum rocks 1.9 $10^6$ hectares of gypsiferous lithofacies [13] together with many landscape evidences of salinity, such as saline ravines associated with springs or oozing sources, and especially many endorheic saline wetlands and saline depressions. Some of these wetlands had artisanal salterns until one century ago. Nowadays, these athalassohaline wetlands —locally named saladas— are hot spots for nature conservation due to the rare or singular habitats they host. Scientific interest in these wetlands can be traced back to the 19th century. The data set [14] lists many works describing the physical conditions, geology, soils, vegetation, and other features of the saladas of Sástago-Bujaraloz in the Central Ebro Basin, an endorheic complex included in the RAMSAR list. Salada Agustín is included in this complex. By contrast, salada Farrachuela located in the chesas of the Barbastro Gypsum Formation, about 70 km NE of salada Agustín (Figure 1), is isolated and went unnoticed by scientists until recently. This kind of wetland is found in areas with arid or Mediterranean climates, but is also comparable to wetlands developed in gypsum karst depressions in cold climates [6].

**Figure 1.** The shaded area is the Ebro Basin, where the two saline wetlands compared, Farrachuela (FA) and Agustín (AG), are located on gypsiferous substrates.

Protection of the European saladas should be implemented throughout the Natura 2000 network of Special Conservation Areas (SCA). Farrachuela is located within SCA "ES2410074 Yesos de Barbastro", but the salada went unnoticed in the data form. Agustín, located within SCA "ES2430082 Monegros", has been protected by excluding it and its surrounding lands from irrigation.

Several investigations, e.g., [15], ratify the traditional knowledge that salinity together with intermittent flooding are the main conditioners of life in the saladas. On the other hand, many published reports and investigations on saline wetlands overlook the shortcomings of some routine concepts and methods unsuited for analyzing gypsum-rich materials, like the soils or sediments of the saladas studied here.

This paper aims to: (i) compare certain compositional features, especially the contents of the major ions relevant for life, at two gypseous and hypersaline wetlands located in the Ebro Basin, NE Spain (Figure 1); and (ii) discuss the adequacy of some common basic operations in the analytical determinations of the gypseous soils.

Beyond the scientific interest of the above objectives, we emphasize their usefulness for defining the living conditions of the protected organisms harbored by these wetlands, and monitoring for undesirable alterations.

## 2. Materials and Methods

### 2.1. The Context of the Wetlands Studied

We compare the isolated wetland Farrachuela (FA) to Agustín (AG) which belongs to the Sástago-Bujaraloz complex. The two hypersaline wetlands lie about 70 km apart (Figure 1), and both of them occupy Quaternary depressions formed by dissolution of rocks rich in gypsum and more soluble salts, with the remaining residuum in their bottoms. A conspicuous network of surficial cracks appearing after each rain episode demonstrates the inundation, while the fresh plant residues visible to the naked eye evidence the recent biological activity in the soil. Contrariwise, the microcrystalline gypsum, i.e., the flour-like gypsum described elsewhere [4], is not observed within the two saladas studied, but does occur in the areas surrounding them. This fact mirrors the contrasting soil hydric regime of the almost permanently dry soils in the Central Ebro Basin against the persistently moist or wet soils in the enclaved saladas.

The two saladas occupy flat-bottomed valleys, which allows for preferential water circulation and accumulation. FA stands isolated in the landscape at the outcrops of the Barbastro Gypsum Formation from the Eocene–Oligocene age [16], whose dipping strata facilitate deep percolation of rainfall. AG is part of the largest group of saline wetlands in the Ebro Basin, with a total of 149 after the inventory of [17], a number that is in stark contrast with the absence of wetlands around FA. The near-horizontal Miocene strata of gypsum and limestone interbedded with lutites of the Bujaraloz Formation in Monegros [18] allow the occurrence of karstic depressions aligned along flat-bottomed valleys and low-lying areas hosting wetlands. Other differentiating factors are the presence of shallow saline aquifers (aquitards) in Monegros [19], responsible for groundwater discharge into the intermittently flooded depressions, and the wind regime that influences the excavation and shaping of the playa lakes in this area.

Winter cereals were cultivated in the watersheds of both saladas (Figures A1 and A2 in the Appendix A) at the time of the samplings, i.e., July 2013 for Farrachuela; and February, March, April, July, and August of the years 1979–1980 and 1999–2000 for Agustín. The plants collected (Table 1) are well known as salinity and waterlogging tolerant or as gypsophilous, and they illustrate the ecology of these wetlands. The floristic composition has remained stable up to the present.

**Table 1.** Plants collected in Farrachuela [20] and in Agustín [21].

| Farrachuela, 1981 | Agustín, 2004 to 2007 |
| --- | --- |
| *Hornungia procumbens* (L.) Hayek | *Aeluropus littoralis* (Gouan) Parl. |
| *Helianthemum salicifolium* (L.) Mill. | *Artemisia herba-alba* Asso |
| *Frankenia pulverulenta* L. | *Arthrocnemum macrostachyum* (Moric.) Moris in Moris and Delponte |
| *Lepidium subulatum* L. | *Atriplex halimus* L. |
| *Phragmites australis* (Cav.) Trin. ex Steud. | *Frankenia pulverulenta* L. |
| | *Gypsophila struthium* subsp. *hispánica* (Willk.) G. López |
| Farrachuela, 2013 | *Hordeum marinum* Huds. |
| *Suaeda spicata* (Willd.) Moq. | *Kochia prostrata* = *Bassia prostrata* (L.) Beck in Rchb. |
| *Salsola soda* L. | *Limonium* sp. pl. |
| *Puccinellia festuciformis* (Host) Parl. | *Puccinellia fasciculata* (Torrey) E.P. Bicknell |
| *Frankenia pulverulenta* L. | *Salsola vermiculata* L. |
| | *Suaeda vera* subsp. *braun-blanquetii* Castrov. and Pedrol |

The aerial photographs from 1956 of both saladas attest that irrelevant anthropic interventions occurred prior to the years of the soil samplings presented in this study. We deem AG representative of those saladas of Sástago-Bujaraloz that remained in an acceptable state of conservation, even threatened by flooding with fresh water from forthcoming irrigation. The land around FA, cleared more than two hundred years ago, bears barley and sparse olive trees, with no irrigation planned.

### 2.2. Climate

Climate data were obtained from the nearest automatic weather stations, in Tamarite, located 11.5 km south of Farrachuela, and Valfarta located 8.5 km NNW of Agustín. Both stations belong to the Spanish SIAR network. For the period 2004–2021, the mean annual rainfall in the Farrachuela area was 350 mm, the mean temperature was 14 °C, and the mean evapotranspiration rate was 1041 mm. The same mean temperature and precipitation of 359 mm were obtained in Agustín area for the same period, with a higher mean evapotranspiration rate (1258 mm).

Using the terminology of the Soil Survey Staff [22] and the data from the two weather observatories, the soil moisture regime at both saladas is aquic and the most likely soil temperature regimes are mesic for Farrachuela and thermic for Agustín.

### 2.3. Sampling

We collected soil samples by hand augering down to a depth of 2 m in Farrachuela, and until reaching an impenetrable layer in Agustín. All the samples were examined visually, including with a magnifying glass, and by touch both in the field and in the laboratory. Then, the samples were air-dried and passed through a 2 mm sieving mill to obtain the fine earth fraction, i.e., <2 mm Ø, used for further analyses. Table 2 shows details of the augerings, and the number of analyses.

**Table 2.** Location and size of the two wetlands and data of their sampling.

| Name of the Salada | Farrachuela | Agustín |
|---|---|---|
| Sheet name of the National Topographic Map of Spain at 1:25 000 scale | Tamarite | Bujaraloz |
| Geographical coordinates (N, E) | 41.8845, 0.3742 | 41.4311, −0.1091 |
| Surface area, ha | 3.4 | 68.1 |
| Elevation, meters above sea level | 400 | 329 |
| Number of augerings | 4 | 24 |
| Augering mean deep, cm | 200 | 63 |
| Number of soil samples for chemical analyses | 32 | 133 |

### 2.4. Analyses

The official methods of the Agricultural Ministry of Spain [23] provided the methodology for the chemical determinations. Calcium carbonate ($CaCO_3$) equivalent (CCE) was measured by gasometry, and gypsum was titrated as per [24].

All the determinations of electrical conductivity are reported as dS $m^{-1}$ at 25 °C. We measured the electrical conductivity (EC1:5) of the aqueous extracts at an earth-to-water ratio weight:weight of 1:5. Moreover, we prepared water-saturated pastes of the fine earth [25], recording the water saturation percentage (SP, %) and measuring the electrical conductivity of the extract (ECe), and its pH (pHe). The major ions in the saturation extracts were determined using the classical methods specified in [20].

We report the ionic contents (ionic-C, mmolc $L^{-1}$) as the sum of cations + anions divided by two in order to counterbalance the deviations to electroneutrality, frequent when analyzing highly concentrated solutions. The sodium adsorption ratio SAR (mmolc $L^{-1})^{0.5}$ is calculated from the concentrations in mmolc $L^{-1}$ of $Na^+$, $Ca^{2+}$, and $Mg^{2+}$ by the expression SAR = $Na^+/[(Ca^{2+} + Mg^{2+})/2]^{0.5}$.

### 2.5. Statistical Procedures

We display and analyze most of our data through graphical methods, representing the measures of exploratory data analysis by means of the intuitive boxplots proposed by [26]. Our boxplots, drawn following [27], include the 95% confidence intervals for the medians estimated as per [28]. Figure A3 in Appendix A sketches the meanings of all the parts of the boxplot diagrams presented in this paper.

We calculate the regression lines by the Ordinary Least Squares (OLS) method with $p = 0.05$, and we check their equations with a non-parametric simple regression using Theil's method [29,30] in order to cope with the habitual non-Gaussian distributions of our data, the frequent outliers, and the limited effectiveness of some data batches.

## 3. Results and Discussion

### 3.1. Saturation Percentages

The mean SP in the 16 samples from FA with this determination is 59.0%, far from the 39.3% in the 133 samples from AG. The other statistics and the distributions of SP also are very distinct between the two saladas (Figure 2). The use of SP to compare soil characteristics is sounder than the classical particle size distribution (PSD) determinations, which, in some materials, produce artifactual results, as argued in Section 3.9.

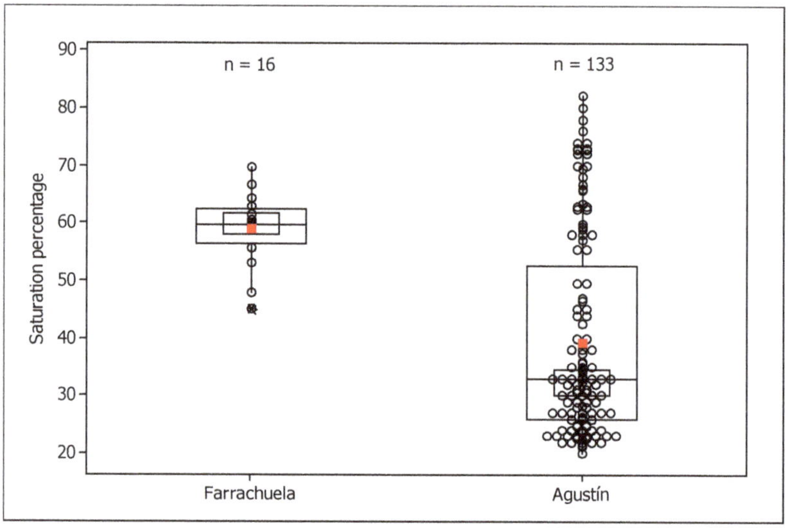


**Figure 2.** Distribution and main statistics of SP in Farrachuela and Agustín. The red squares mark the means; for the meaning of all parts, see Figure A3.

The saturated paste extraction approximates the field capacity of the soil much better than other, more diluted, soil-to-water ratios, while making water extraction feasible with simple equipment [31,32]. Thus, the extract of the saturated paste is broadly used [22,23,25] to assess the salt tolerance of plants by measuring the electrical conductivity of the extract (ECe). Also, eventually, its ionic composition is determined to appraise the effects of individual ions or their co-occurrence, as is the case of the structural stability of the soil. Many studies of soil salinity report SP because of its interest in understanding the hydric behavior of the soil. The simplicity of the preparation of the saturated paste, based on the skill of the operator, and the saturation point determined by feel and experience, together with the use of few—if any—sophisticated laboratory tools, probably contribute to SP being overlooked in some studies.

However, even if the composition of saturated extracts was not to be analyzed, the SP is worth recording because SP can be a surrogate for textural composition, or "a quantitative expression of soil texture" as noted by [33], and represents key functional soil properties, like the available water capacity [34]. This is an old issue in soil science when "dealing with single-valued expressions instead of trying to interpret the complex series of numbers represented by the mechanical analysis" [35], as in the class–moisture equivalent diagram of [36]. When dealing with soil salinity, SP has been used in recent attempts [37] at estimating ECe from extracts at fixed dilution ratios. Also, SP is a relevant soil characteristic for precision agriculture, as noted in [38], with SP performing better than organic matter, $CaCO_3$, ECe, and pH for yield prediction. Furthermore, SP can help to check consistency over time in sample preparation [32], which we also did successfully to compare two operators [39].

The surrogate role of SP for texture is especially useful in gypsum-rich soils where the standard methods for PSD determination are unsuitable [4,40] and many substitute methods are unfeasible in routine labs. In [8], hand textural class estimates were used to check the proposed method for PSD determination. After the above considerations, we did not determine PSD, and we claim SP as a strong indicator of hydric behavior and a reliable surrogate for textural composition [40–42].

### *3.2. Major Mineral Components*

The boxplots of calcium carbonate equivalent (CCE) and gypsum for each salada (Figure 3) show their distribution along the samples, and the prominence of both components. The mean percent rounded to the nearest integer for FA and AG are 24% and 10% for CCE while for gypsum they are 30% and 62%, respectively. Then, the remaining material, i.e., silica and silicates—including the mineralogical clays—plus organic components is <50% on average. These compositions agree with the contrasting average SP of 59% in FA, and 39% in AG (Figure 2), since gypsum and calcium carbonate have a much lower water-holding capacity than the other minority components, like clay and organic matter.

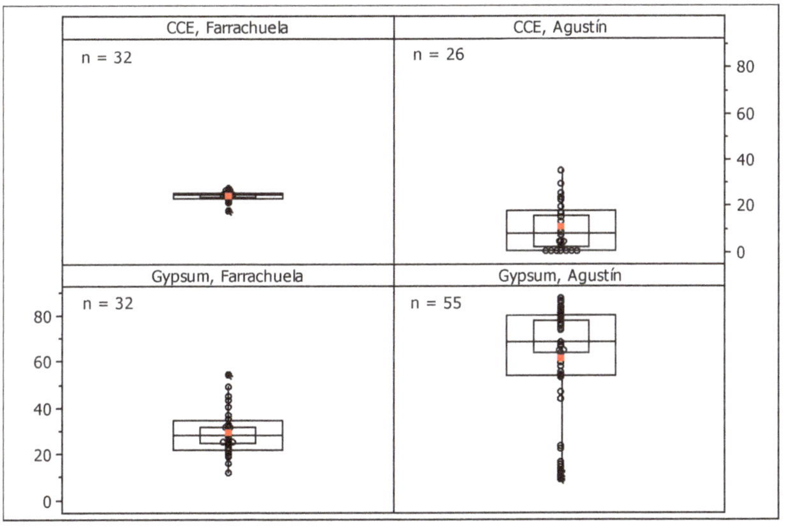


**Figure 3.** Boxplots of CCE and gypsum in the analyzed samples from Farrachuela and Agustín, with the number (n) of samples computed. The red squares mark the means; for the meaning of all parts, see Figure A3.

The abundance of gypsum in the soils and geological materials governs the chemistry of the soil solution and the exchange complex due to their saturation in $Ca^{2+}$. The contents of gypsum (Figure 3) in the samples analyzed—with a minimum of 12.2% in FA and 9.5% in AG—largely guarantee $Ca^{2+}$ saturation, preventing clay dispersion.

### *3.3. Soil Salinity*

In the present study, a common visual symptom of soil salinity is the white bright efflorescences mostly produced by evapoconcentration during dry periods at the bottoms and margins of both FA and AG. These efflorescences are often composed mainly of crystals of gypsum and more soluble salts, as evidenced in the field by naked-eye and hand-lens observations, as well as by taste. The halophilous vegetation also attests the soil salinity.

The visual symptoms do not quantify the salt stock in the soils. Such stock is a key indicator of the ecological status of the soil, which can be tracked by recording the changes in the salt content through time. To establish this stock, the soil salt must be extracted at fixed-ratio dilution, higher than the saturated paste [25,31] as is evident for hypersaline soils [43]. To appraise the salt stock of the soil, we use the easy, unsophisticated, and reproducible determinations of EC1:5 [25]. Such attributes make EC1:5 well-suited for future comparative assessments.

The mean of EC1:5 in 32 extracts of FA was 7.18 dS $m^{-1}$, versus 8.78 dS $m^{-1}$ in the 72 extracts from AG for which EC1:5 was analyzed (Figure 4). We consider these conductivities a proxy for the total content of highly soluble salts that can allow tracking the evolution of the soil's salt stock. The 1.68 dS $m^{-1}$ of the difference between the means of EC1:5 seems moderate. However, AG is a more salt-stressing environment than FA, as shown by the vegetation (Table 1) and by the mean ECe of 37.04 dS $m^{-1}$ for FA against 68.58 dS $m^{-1}$ for AG. The difference is more pronounced than in EC1:5, and the same is true for the other statistics graphed in Figure 4. The longer range of the distributions in AG seems sound if considering the 20-fold surface area of AG versus FA (Table 2), which allows for more variable conditions in AG.

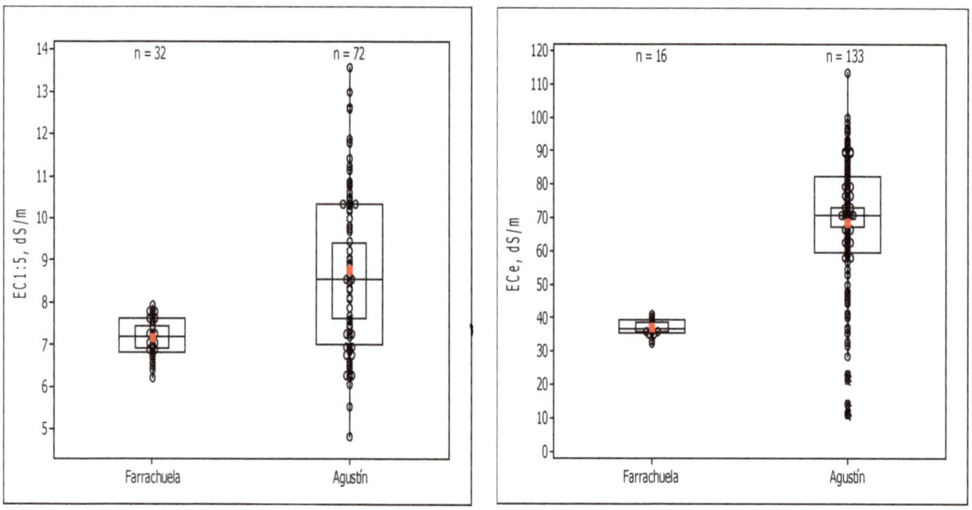


**Figure 4.** Boxplots of EC1:5 and ECe for the two saladas compared. The red squares mark the means; for the meaning of all parts, see Figure A3.

The much higher values of EC1:5 and ECe in AG than in FA (Figure 4) agree with the species of halophytes recorded at both saladas (Table 1). The absence of *Arthrocnemum macrostachyum* in FA is relevant because this plant lives in the more often inundated areas of AG where ECe is >80 dS $m^{-1}$, a soil salinity not reached in FA. The difference is also pronounced if only the upper soil layer is considered, with about 37 dS $m^{-1}$ in FA versus 85 dS $m^{-1}$ in AG. Accordingly, *A. macrostachyum*, the most salinity perennial tolerant plant, lives in AG and not in FA.

The scatterplot of ECe on EC1:5 (Figure 5) does not show the inflection associated with gypsum [43] because all samples have EC > 2.25 dS $m^{-1}$ in both EC1:5 and ECe. The wide scattering of the values—seen in the plot—hampers the estimation of ECe from the plain EC1:5. If desired, estimates of ECe can be made from EC1:5 by the method in [44]. In this article, we apply the easy transformation of EC1:5 by SP as per [45], i.e., by calculating regressions of the shape $ECe = a + b \times EC1{:}5 \times (500/SP)^q$ with empirical powers of $q$. The best correlation coefficients (Table A1, in Appendix A) were 0.863 for FA, and 0.928 for AG.

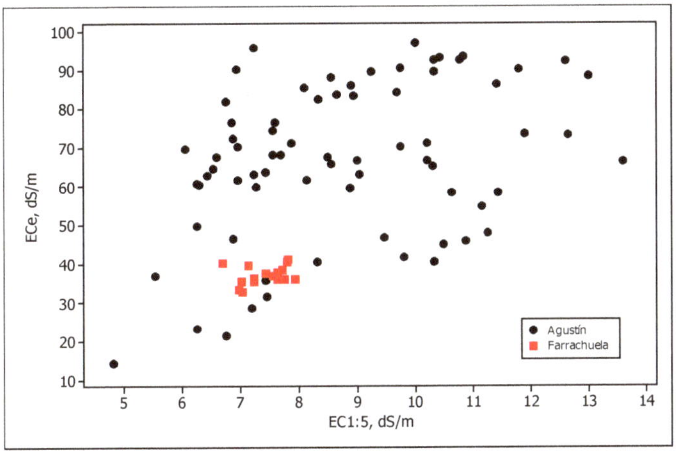


**Figure 5.** Scatterplot of ECe on EC1:5 for the two saladas.

For the present article, the equations allowed us to check the coherence between ECe and EC1:5. The future studies of these wetlands will have to decide whether or not to estimate ECe from EC1:5 with this kind of equation. The reason is that ECe is the standard indicator of saline stress on plants, used to express a plant's tolerance to soil salinity, a key feature for crop feasibility [32]. The equations in Table A1 allowed us to check the consistency between ECe and EC1:5 and will help to decide in future studies of these wetlands whether or not to estimate ECe from EC1:5 with these kinds of equations.

### *3.4. Soil Classification*

The inspection of the auger samples and their analyses for ECe and gypsum content enabled us to identify the diagnostic horizons defined by the Soil Survey Staff [22] without the need to open a trench, a much more soil disturbing procedure. According to the data of the ECe and gypsum content, and following step-by-step the statements in [22], the soils of FA and AG have the Salic and Gypsic horizon in conjunction. Thus, the soils of both wetlands are Gypsic Aquisalids.

### *3.5. Ionic Contents of the Saturation Extracts*

Bicarbonate, titrated in all the extracts, was always below the limits of detection. Potassium contents are irrelevant. The pairs of boxplot diagrams in Figure 6 compare the content of each of the ions $Ca^{2+}$, $Mg^{2+}$, $Na^{+}$, $HCO_3^{-}$, $SO_4^{2-}$, and $Cl^{-}$ in the saturation extracts of FA to AG. The range of all the distributions is broader for AG, which also happens for the interquartile ranges.

Of note are: (i) the greater surface area of AG and the greater number of samples than in FA, and (ii) the greater surface area of AG allowing for more diverse hydric conditions including the redistribution of water by the wind. These conditions are mirrored by the occurrence of the highly halophilic *A. macrostachyum* in AG. Both saladas are similar with regard to the predominance of $SO_4^{2-}$ and $Mg^{2+}$, but the concentrations of $Cl^{-}$ and $Na^{+}$ are significantly lower in Farrachuela, hence its lower salinity.

Figure 6 shows that the median concentrations are significantly lower in FA than in AG for the six ions in question, but the difference is insignificant for sulfate. The mean concentrations are also lower in FA than in AG for the ions analyzed. This happens when computing all the analyses from AG extracts (Figure 6), and also if computing only the extracts with the six ions analyzed (Table 3). This Table also shows numerically the differences between the mean contents of each individual ion in the two saladas. The greatest difference occurs in $Cl^{-}$, followed by $Na^{+}$ and $SO_4^{2-}$, and then by $Mg^{2+}$, with 594, 410, 317, and 224 mmolc $L^{-1}$, respectively. We deem irrelevant the differences in $Ca^{2+}$ and

$HCO_3^-$. The differences between the two wetlands for $Mg^{2+}$, $Na^+$, $SO_4^{2-}$, and $Cl^-$ agree with their distinct physical characteristics outlined in the second paragraph of Section 2.1. Of note is the detection of magnesium by taste in the efflorescences of AG, agreeing with the boxplot in Figure 6. The abundance of magnesium should be due to evapoconcentration from the parental rocks of gypsum that contain this element.

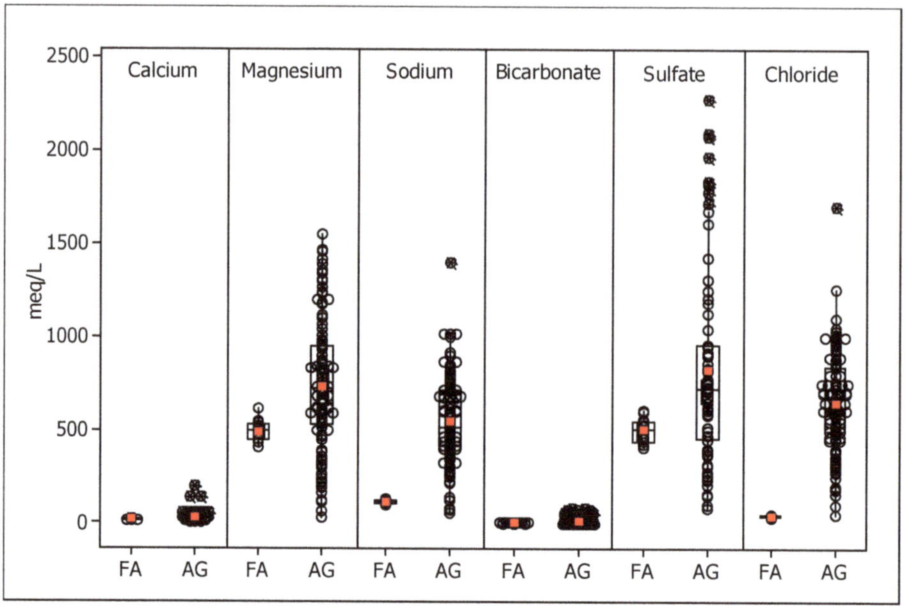


**Figure 6.** Pairs of boxplots for the six main ions in all the analyzed saturation extracts from Farrachuela (FA) and Agustín (AG). The red squares mark the means, for the meaning of all parts, see Figure A3.

**Table 3.** Mean contents (mmolc $L^{-1}$) of six major ions for: (a) the 16 saturation extracts from FA, (b) all the analytical determinations from AG as shown in Figure 6, (c) differences between these means, (d) means of the 76 extracts from AG with the six ions titrated, and (e) differences between the means of all determinations from AG and the means of FA.

| | $Ca^{2+}$ | $Mg^{2+}$ | $Na^+$ | $Cl^-$ | $SO_4^{2-}$ | $HCO_3^-$ |
|---|---|---|---|---|---|---|
| | | | | mmolc $L^{-1}$ | | |
| (a) Farrachuela | 18.9 | 492.3 | 113.0 | 40.7 | 503.1 | 2.2 |
| (b) all determinations from Agustín * | 29.0 | 734.0 | 546.4 | 646.4 | 820.6 | 16.7 |
| (c) differences between saladas, (b) minus (a) | 10.1 | 241.7 | 433.4 | 605.7 | 317.5 | 14.5 |
| (d) Agustín, only the 76 samples with the six ions titrated | 27.5 | 716.7 | 523.4 | 634.5 | 820.6 | 3.0 |
| (e) differences, b minus d | 1.5 | 17.3 | 23.0 | 11.9 | --- | 13.7 |

* The number of determinations in AG were 132 for $Ca^{2+}$, $Na^+$, and $Cl^-$; 124 for $Mg^{2+}$; and 76 for $SO_4^{2-}$.

Table 3 also shows the consistency of the results obtained by computing all the available titrations or only the ones from the samples with the six ions analyzed. This feature is interesting for long-term tracking, when the homogeneity of the number of ions analyzed along successive campaigns cannot be guaranteed.

### 3.6. SAR and pH

Due to their salinity and frequent inundation, the saladas are either bald or seasonally populated by halophytes, and are unusable for the cash crops feasible in the Ebro Basin. Notwithstanding, we present (Table 4) the statistics of SAR and pH of the saturation extracts in both wetlands, showing their significant difference. The values of SAR in Farrachuela are well below the classical threshold of SAR = 13 for clay dispersion in soils. By contrast, the values of SAR in Agustín are largely in excess of 13, denoting the unfeasibility of cash crops. The pH values are significantly different between the two wetlands (Table 4); however, pH by itself should not be a problem for most plants, as it ranges from 7.28 to 8.44 in the two saladas. These basic pH values are allowable for most plants. For the infiltration behavior, SAR must be considered jointly with pH and salinity [45].

**Table 4.** Statistics of SAR and pH in the saturation extracts of samples from Farrachuela (FA) and Agustín (AG).

| | Number of Analyzed Samples | Minimum | Maximum | Median | 95% Confidence Interval for The Median | |
|---|---|---|---|---|---|---|
| | | | | SAR $(\mathrm{mmolc/L})^{0.5}$ | | |
| FA | 16 | 6.70 | 7.80 | 7.00 | 6.80 | 7.30 |
| AG | 124 | 10.41 | 45.08 | 27.96 | 26.80 | 28.94 |
| | | | | pH | | |
| FA | 16 | 7.92 | 8.44 | 8.26 | 8.17 | 8.36 |
| AG | 52 | 7.28 | 8.33 | 7.92 | 7.71 | 8.00 |

For SAR, it does not make sense to apply the commonly used threshold SAR < 13 $(\mathrm{mmol_c\ L^{-1}})^{0.5}$ to judge the sustainability of irrigation, since for most soil samples the ECe is >>20 dS $\mathrm{m^{-1}}$ (Table 4). Then, the SAR/EC ratios do not affect the clay dispersion and no associated imperviousness would happen [45,46]. Moreover, the contents of mineralogical clay are low, as the sum of the means of gypsum and calcium carbonate equivalent are 54% for Farrachuela, and 72% for Agustín. The mineralogical clays are thus in the minority, reducing the value of SAR to predict the hydrological behavior of these gypsum-rich soils often saturated in water. More relevant to the hydraulic behavior is the horizontal microstructure in FA and AG, similar to the layering described in other saladas in Monegros.

### 3.7. Relationship between Ionic Concentration and ECe

Figure 7A shows the concordance between both saladas regarding the relation of ionic contents (ionic-C, mmolc $\mathrm{L^{-1}}$) in the extracts, as defined in Section 2.4., versus ECe. The tusk-shaped distribution in this figure illustrates the increase in the ratio of ions to ECe for the high ionic concentrations, due to the occurrence of neutral and other ionic pairs [47]. The dilution of the more concentrated extracts, required for the analytical methods, also contributes to the scattering of values in the high concentrations.

For the two saladas, the distributions approach a straight line after transforming both variables using decimal logarithms (Figure 7B), as proposed in [48]. This reinforces the consistency of the data. Table A2 in the Appendix A shows the regression equations obtained from these transformations.

### 3.8. Relationship between the Concentration of Individual Ions and ECe

The saturation extracts of AG showed a good correlation coefficient (R) of the ions between them as well as between the ECe and the ions (Table 5). These coefficients in AG are higher than their counterparts in FA, except the correlation of ECe with $SO_4^{2-}$, which is slightly higher in FA than in AG.

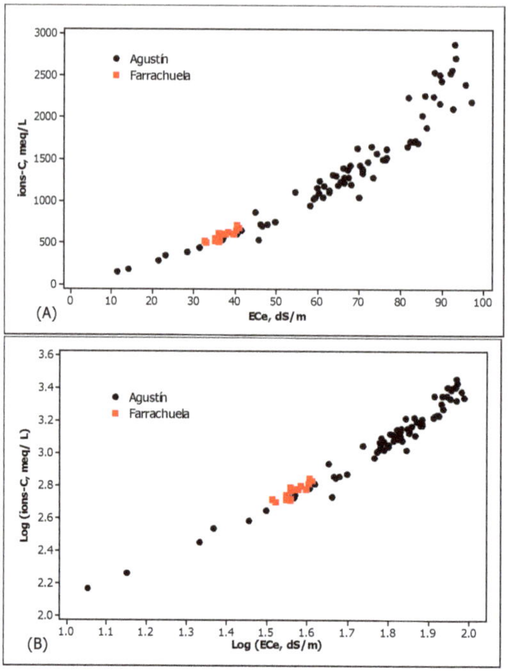


**Figure 7.** (**A**) Scatterplot of the ionic contents over ECe in the saturation extracts; (**B**) Log–log scatter plot of the same parameters.

**Table 5.** Correlation coefficients between ECe and the main ions, and between them. The *p*-value is < 0.001, except if non-significance (n.s.) is indicated.

| | ECe | $Mg^{2+}$ | $Na^+$ | $SO_4{}^{2-}$ |
|---|---|---|---|---|
| | | Farrachuela | | |
| $Mg^{2+}$ | 0.822 | | | |
| $Na^+$ | 0.792 | 0.815 | | |
| $SO_4{}^{2-}$ | 0.815 | 0.840 | n.s. | |
| $Cl^-$ | n.s. | n.s. | n.s. | n.s. |
| | | Agustín | | |
| $Mg^{2+}$ | 0.913 | | | |
| $Na^+$ | 0.936 | 0.901 | | |
| $SO_4{}^{2-}$ | 0.794 | 0.907 | 0.792 | |
| $Cl^-$ | 0.945 | 0.888 | 0.957 | 0.723 |

For the saturation extracts of FA, the correlation of ECe was significant ($p < 0.001$) with $Mg^{2+}$, $Na^+$, and $SO_4{}^{2-}$, but not with $Cl^-$. By contrast, in AG the correlation ECe/$Cl^-$ was the highest, as shown in Table 5.

The coefficients of determination ($R^2$) attained by the OLS regression of the average ionic content versus ECe (Table 5) are the same as when using Theil calculations, while the standard errors using the OLS regression are smaller than with Theil regressions. The OLS method is thus recommended for estimating the ionic content from the ECe in the saturation extracts of both saladas within the ranges indicated in Table 6. These equations could help to streamline the environmental tracking of the saladas.

**Table 6.** Regression equations of the shape "ionic-C = $a + b \times$ ECe" with the coefficient of determination ($R^2$, %) and standard error (S) calculated for 16 and 76 soil samples from Farrachuela and Agustín, respectively. The methods of calculation are ordinary least squares (OLS), and Theil using the median of the interceptors.

| | Method | *a* | *b* | $R^2$ | S | Range of Ionic-C |
|---|---|---|---|---|---|---|
| | | | | % | mmolc $L^{-1}$ | |
| Farrachuela | OLS | −213.1 | 21.59 | 76.2 | 29.60 | 502.4 to 705.2 |
| | Theil | −200.4 | 21.33 | 76.2 | 32.71 | |
| Agustín | OLS | −620.6 | 30.30 | 88.8 | 218.8 | 147.8 to 2876.0 |
| | Theil | −629.6 | 29.50 | 88.8 | 281.1 | |

### *3.9. The Specificity in the Analytics of Gypseous Materials*

In fresh soil samples of the gypseous horizons of the soils studied, the abundant gypsum crystals are bound together by fine material, the strength of which increases as the moisture content decreases. This is a well-known characteristic of many soils [49]. The coarse gypsum crystals are easily separated, intact, using the fingers when the sample is wet, but when air dried, these crystals break easily —mostly along cleavage plains— before becoming detached from the mass. The breaking happens even if a dry soil sample is gently crushed between a wooden board and a wooden rolling pin, and, of course, in mechanical shakers and sieving mills, until they eventually pass through the sieve. Milling may result in the entire sample passing through the 2mm sieve, making the time at which the operation is interrupted arbitrary. For this reason, the content of coarse gypsum crystals and other gypsum fragments must be appraised before drying and sieving in the lab.

In gypsum-rich soils, the low hardness and cohesion of the gypsum crystals lessens the value and significance of the particle size separates obtained after milling. Moreover, the abundance of gypsum and its solubility ($\approx$2.4 g $L^{-1}$) immediately cause the sedimentation of clays when being dispersed for particle size separations [4,8]. The judgments about other methods for PSD determinations based on X–ray fluorescence, e.g., [50] or its combination with FT-NIR spectroscopy, e.g., [51], or with Vis-NIR and pXRF [52] are unsound if these methods are calibrated against either pipette or wet sieving and hydrometer, which do not work in gypseous soils. The method proposed by [8], based on sonication in a 7:3 ethanol to water solution—a procedure available in few labs—allows dispersion of fine earth, but the significance of the coarse and fine earth remains arguable in the soils studied. In the common circumstances of gypsum-rich soils, the lab determination of PSD should not be taken as representing the behavior of the undisturbed soil. These shortcomings led to coining two terms in lieu of texture: "Coarse Gypsum Material" and "Fine Gypsum Material" [49].

The issues discussed in the above two paragraphs indicate the non-sense of using some routine analytical methods with gypseous sediments and soils. Then, [4] proposed the elimination of PSD for assessing the hydric behavior of gypseous soils. Specifically, we support considering saturation percentage as a sound and useful surrogate for textural composition. Accordingly, we did not separate the clay fraction, whose determination would also be impractical, cumbersome and probably irrelevant for future monitoring.

## 4. Conclusions

The article provides a snapshot of the main limiters to life in two inland saline marshes. We hope it provides a baseline for protecting of these shelters of biodiversity. The intuitive graphical procedures used in this article allow for a quick and detailed statistical comparison of the soil characteristics that are decisive for life in the saline wetlands studied. The features compared are: (i) contents of gypsum and calcium carbonate; (ii) soil salinity as EC1:5; (iii) saturation percentage; and (iv) pH, electrical conductivity and ionic contents in the saturation extracts. The data will constitute a baseline —the only one at present— for tracking possible compositional changes in the two saladas studied. The calculation methods and graphical presentation are easily transposable to other saline wetlands, and may allow for multiple comparisons in the future.

The characteristics studied herein ought to be contrasted with other saline wetlands in order to delimit the environmental conditions for the life of specific plants, microbes, or other organisms. Hopefully, the correlations presented in this article will support the management and conservation measures to be implemented by environmental authorities in charge of the wetlands.

In situ recording of variable soil parameters is advisable. For some of them, the technology is available and is continuously improving. However, the implementation of recording seems unlikely in the near future due to the limited capacity of the conservation agencies responsible.

The misuse of analytical techniques that are wrong or unsound for gypsum-rich soils, like the ones discussed in the present article, is a key point to be: (i) avoided when conducting future analyses, and (ii) checked when scrutinizing heritage data.

**Author Contributions:** Conceptualization, J.H.; Methodology, C.C.; Writing—original draft, J.H.; Writing—review & editing, C.C.; Funding acquisition, C.C. All authors have read and agreed to the published version of the manuscript.

**Funding:** This research was made possible by the grant PID2021-127170OB-I00 funded by MCIN/AEI/10.13039/501100011033 and by "ERDF A way of making Europe", and the grant TED2021-130303B-I00 funded by MCIN/AEI/10.13039/501100011033 and by the "European Union NextGeneration EU/PRTR".

**Data Availability Statement:** The data used in this study are available upon request from the authors.

**Acknowledgments:** Thanks to the botanists of the University of Lleida J. Pedrol and J.A. Conesa for determining the plants collected in July 2013.

**Conflicts of Interest:** The authors declare no conflict of interest.

## Appendix A

**Figure A1.** Farrachuela, 21 July 2013. In the foreground the gypsophilous vegetation, then a plowed field and the salada bottom, partially inundated. In the background, the gyprock outcrops.

**Figure A2.** Salada Agustín, 12 October 1979 at 4p.m. In the foreground, a plowed field, followed by vegetation of *Suaeda vera*. Behind it the area covered by *Arthrocnemum macrostachyum*, often flooded, around the bottom occupied by water.

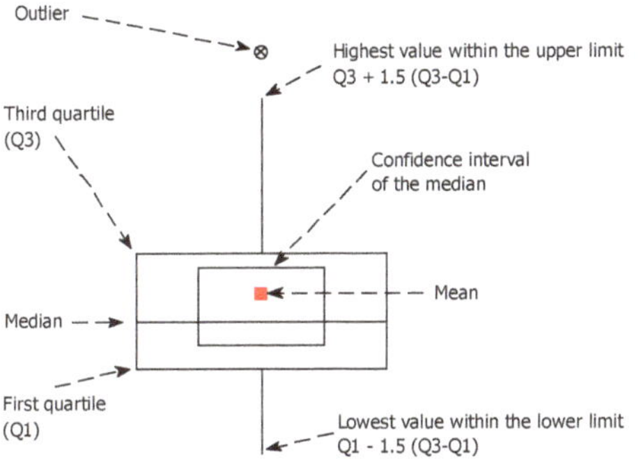


**Figure A3.** Sketch of a boxplot.

**Table A1.** Pearson's correlation coefficients (R) obtained with OLS regression equations of the shape $ECe = a + b \times EC1{:}5 \times (500/SP)^q$ for several arbitrary values of $q$.

| | **Farrachuela** | **Agustín** |
|---|---|---|
| $q$ | **R** | |
| 0 | 0.291 | 0.360 |
| 0.50 | 0.862 | 0.887 |
| 0.55 | 0.863 | 0.902 |
| 0.60 | 0.857 | 0.913 |
| 0.65 | 0.848 | 0.920 |
| 0.70 | 0.837 | 0.925 |
| 0.80 | 0.812 | 0.928 |
| 0.90 | 0.787 | 0.925 |
| 1 | 0.763 | 0.919 |

**Table A2.** Regression equations of the shape *log ionic-C* = *a* + *b* × *log ECe* calculated by OLS and Theil methods for 16 and 76 soil samples from Farrachuela and Agustín, respectively.

| Salada | Method | *a* | *b* | $R^2$, % |
|---|---|---|---|---|
| Farrachuela | OLS | 0.659 | 1.34 | 75.4 |
| | Theil | 0.701 | 1.32 | 75.4 |
| Agustín | OLS | 0.541 | 1.42 | 95.5 |
| | Theil | 0.302 | 1.54 | 95.5 |

## References


1. Cížková, E.; Navrátilová, J.; Martinát, S.; Navrátil, J.; Frazier, R.J. Impact of water level on species quantity and composition grown from the soil seed bank of the inland salt marsh: An ex-situ experiment. *Land* **2020**, *9*, 533. [CrossRef]
2. Domínguez-Beisiegel, M.; Herrero, J.; Castañeda, C. Saline wetlands' fate in inland deserts: An example of 80 years' decline in Monegros, Spain. *Land Degrad. Dev.* **2013**, *24*, 250–265. [CrossRef]
3. Köninger, J.; Panagos, P.; Jones, A.; Briones, M.J.I.; Orgiazzi, A. In defence of soil biodiversity: Towards an inclusive protection in the European Union. *Biol. Conserv.* **2022**, *268*, 109475. [CrossRef]
4. Casby-Horton, S.; Herrero, J.; Rolong, N.A. Gypsum soils—Their morphology, classification, function, and landscapes. *Adv. Agron.* **2015**, *130*, 231–290. [CrossRef]
5. Goryachkin, S.V.; Spiridonova, I.A.; Sedov, S.N.; Targulian, V.O. Boreal soils on hard gypsum rocks: Morphology, properties, and genesis. *Eurasian Soil Sci.* **2003**, *36*, 691–703.
6. Tuyukina, T.Y. Geochemical studies of northern taiga (gypsum) karst ecosystems and their high vulnerability to natural and anthropogenic hazards. *Environ. Geol.* **2009**, *58*, 269–274. [CrossRef]
7. Palacio, S.; Cera, A.; Escudero, A.; Luzuriaga, A.L.; Sánchez, A.M.; Mota, J.F.; Pérez-Serrano, M.; Merlo, M.E.; Martínez-Hernández, F.; Salmerón-Sánchez, E.; et al. Recent and ancient evolutionary events shaped plant elemental composition of edaphic endemics: A phylogeny-wide analysis of Iberian gypsum plants. *New Phytol.* **2022**, *235*, 2406–2423. [CrossRef]
8. Pearson, M.J.; Monteith, S.E.; Ferguson, R.R.; Hallmark, C.T.; Hudnall, W.H.; Monger, H.C.; Reinsch, T.G.; West, L.T. A method to determine particle size distribution in soils with gypsum. *Geoderma* **2015**, *237–238*, 318–324. [CrossRef]
9. Heino, J.; Alahuhta, J.; Bini, L.M.; Cai, Y.; Heiskanen, A.S.; Hellsten, S.; Kortelainen, P.; Kotamaki, N.; Tolonen, K.T.; Vihervaara, P.; et al. Lakes in the era of global change: Moving beyond single-lake thinking in maintaining biodiversity and ecosystem services. *Biol. Rev.* **2021**, *96*, 89–106. [CrossRef]
10. Lehner, B.; Doll, P. Development and validation of a global database of lakes, reservoirs and wetlands. *J. Hydrol.* **2004**, *296*, 1–22. [CrossRef]
11. Prăvălie, R.; Sîrodoev, I.; Ruiz-Arias, J.; Dumitraşcu, M. Using renewable (solar) energy as a sustainable management pathway of lands highly sensitive to degradation in Romania. A countrywide analysis based on ex-ploring the geographical and technical solar potentials. *Renew. Energy* **2022**, *193*, 976–990. [CrossRef]
12. Wang, G.; Li, G.; Liu, Z. Wind farms dry surface soil in temporal and spatial variation. *Sci. Total Environ.* **2023**, *857*, 159293. [CrossRef]
13. Alberto, F.; Navas, A. La participación de los yesos en la salinización de las aguas superficiales de la cuenca del Ebro. I. Cartografía de síntesis de las formaciones con yesos. *An. Aula Dei* **1986**, *18*, 7–18. Available online: http://hdl.handle.net/10261/13516 (accessed on 3 September 2023).
14. Herrero, J.; Castañeda, C. A Dataset of Scientific References on Saline Wetlands of Monegros, NE Spain. Mendeley Data. 2023. Available online: https://data.mendeley.com/datasets/kxbgpc8y5r/3 (accessed on 20 September 2023).
15. Reid, R.P.; Oehlert, A.M.; Suosaari, E.P.; Demergasso, C.; Chong, G.; Escudero, L.V.; Piggot, A.M.; Lascu, I.; Palma, A.T. Electrical conductivity as a driver of biological and geological spatial heterogeneity in the Pu-quios, Salar de Llamara, Atacama Desert, Chile. *Sci. Rep.* **2021**, *11*, 12769. [CrossRef]
16. Pardo, G.; Villena, J. Aportación a la geología de la región de Barbastro. *Acta Geol. Hisp.* **1979**, *14*, 289–292.
17. Castañeda, C.; Herrero, J.; Conesa, J.A. Distribution, morphology and habitats of saline wetlands: A case study from Monegros. *Geol. Acta* **2013**, *11*, 371–388. [CrossRef]
18. Salvany, J.M.; García-Vera, M.A.; Samper, J. Geología e hidrogeología de la zona endorreica de Bujaraloz-Sástago (Los Monegros, provincias de Zaragoza y Huesca). *Acta Geol. Hisp.* **1995**, *30*, 31–51.
19. Samper-Calvete, F.J.; García-Vera, M.A. Inverse modeling of groundwater flow in the semiarid evaporitic closed basin of Los Monegros, Spain. *Hydrogeol. J.* **1998**, *6*, 33–49. [CrossRef]
20. Herrero, J.; Castañeda, C.; Velayos, M. Salada Farrachuela, a saline wetland in Tamarite de Litera, Spain. *Bol. Real Soc. Esp. Hist. Nat.* **2020**, *114*, 67–80. [CrossRef]
21. Conesa, J.A.; Castañeda, C.; Pedrol, J. Las saladas de Monegros y su entorno. In *Hábitats y Paisaje Vegetal*; Consejo de Protección de la Naturaleza de Aragón: Zaragoza, Spain, 2011; 540p, ISBN 978-84-89862-76-0.

22. Soil Survey Staff. *Keys to Soil Taxonomy*, 13th ed.; USDA-Natural Resources Conservation Service: Washington, DC, USA, 2022.
23. M.A.P.A. *Métodos Oficiales de Análisis*; Tomo III; Ministerio de Agricultura, Pesca y Alimentación, Secretaría General Técnica: Madrid, Spain, 1994; 662p.
24. Artieda, O.; Herrero, J.; Drohan, P.J. A refinement of the differential water loss method for gypsum deter-mination in soils. *Soil Sci. Soc. Am. J.* **2006**, *70*, 1932–1935. [CrossRef]
25. USDA; United States Salinity Laboratory Staff. Diagnosis and Improvement of Saline and Alkali Soils. Agriculture Handbook no. 60. Reprint 1969. 1954. Available online: https://www.ars.usda.gov/ARSUserFiles/20360500/hb60_pdf/hb60complete.pdf (accessed on 24 September 2023).
26. Tukey, J.W. *Exploratory Data Analysis*; Addison-Wesley: Reading, MA, USA, 1977.
27. Chambers, J.M.; Cleveland, W.S.; Kleiner, B.; Tukey, P.A. *Graphical Methods for Data Analysis*; Reissued in 2018 by CRC Press; Chapman and Hall: New York, NY, USA, 1983.
28. Hettmansperger, T.P.; Sheather, S.J. Confidence intervals based on interpolated order statistics. *Stat. Probab. Lett.* **1986**, *4*, 75–79. [CrossRef]
29. Daniel, W.W. *Applied Nonparametric Statistics*, 2nd ed.; Pws-Kent Publishing Company: Boston, MA, USA, 1990.
30. Glaister, P. A comparison of best fit lines for data with outliers. *Int. J. Math. Educ. Sci. Technol.* **2005**, *36*, 110–117. [CrossRef]
31. Bower, C.A.; Wilcox, L.V. Soluble salts. In *Methods of Soil Analysis, Part 2 Chemical and Microbiological Properties*; Black, C.A., Ed.; American Society of Agronomy: Madison, WI, USA, 1965; pp. 933–951.
32. Hanson, B.R.; Grattan, S.R.; Fulton, A. *Agricultural Salinity and Drainage*; Water Management Series, Publication 113375; University of California Davis: Davis, CA, USA, 2006.
33. Stiven, G.A.; Khan, M.A. Saturation percentage as a measure of soil texture in the Lower Indus Basin. *J. Soil Sci.* **1966**, *17*, 255–263. [CrossRef]
34. Minasny, B.; McBratney, A.B.; Field, D.J.; Tranter, T.; McKenzie, N.J.; Brough, D.M. Relationships between field texture and particle-size distribution in Australia and their implications. *Aust. J. Soil Res.* **2007**, *45*, 428–437. [CrossRef]
35. Briggs, L.J.; McLane, J.W. Moisture equivalent determinations and their application. *Agron. J.* **1910**, *2*, 138–147. [CrossRef]
36. Bodman, G.B.; Mahmud, A.J. The use of the moisture equivalent in the textural classification of soils. *Soil Sci.* **1932**, *33*, 363–374. [CrossRef]
37. Gharaibeh, M.A.; Albalasmeh, A.A.; Hanandeh, A.E. Estimation of saturated paste electrical conductivity using three modelling approaches: Traditional dilution extracts; saturation percentage and artificial neural networks. *Catena* **2021**, *200*, 105141. [CrossRef]
38. Elsharkawy, M.M.; Sheta, A.S.; D'Antonio, P.; Abdelwahed, M.S.; Scopa, A. Tool for the establishment of agro-management zones using GIS techniques for precision farming in Egypt. *Sustainability* **2020**, *14*, 5437. [CrossRef]
39. Herrero, J.; Castañeda, C. Changes in soil salinity in the habitats of five halophytes after 20 years. *Catena* **2013**, *109*, 58–71. [CrossRef]
40. Herrero, J.; Mandado, J. Clarifications on statements in Badía et al. (2013), Geoderma 193–194, 13–21. *Geoderma* **2016**, *275*, 82–83. [CrossRef]
41. Al-Kayssi, A.W. Quantifying soil physical quality by using indicators and pore volume-function characteris-tics of the gypsiferous soils in Iraq. *Geoderma Reg.* **2022**, *30*, e00556. [CrossRef]
42. Moret-Fernández, D.; Herrero, J. Effect of gypsum content on soil water retention. *J. Hydrol.* **2015**, *528*, 122–126. [CrossRef]
43. Herrero, J.; Weindorf, D.C.; Castañeda, C. Two fixed ratio dilutions for soil salinity monitoring in hypersaline wetlands. *PLoS ONE* **2015**, *10*, e0126493. [CrossRef]
44. Carpena, O.; Guillén, M.G.; Fernández, F.G.; Caro, M.G. Saline soil classification using the 5:1 aqueous extract. In *Transactions of the 9th International Congress of Soil Science*; American Elsevier Publishing Company: Adelaide, Australia, 1968; Volume I, pp. 483–490.
45. Sumner, M.E.; Rengasamy, P.; Naidu, R. Sodic Soils: A Reappraisal. In *Sodic Soils: Distribution, Properties, Management, and Environmental Consequences*; Sumner, M.E., Naidu, R., Eds.; Oxford University Press: New York, NY, USA, 1998; pp. 3–17.
46. Quirk, J.P.; Schofield, R.K. Landmark Papers: No. 2. The effect of electrolyte concentration on soil permeability. *Eur. J. Soil Sci.* **2013**, *64*, 8–15. [CrossRef]
47. Darab, K.; Csillag, J.; Painter, I. Studies on the ion composition of salt solutions and of saturation extracts of salt-affected soils. *Geoderma* **1980**, *23*, 95–111. [CrossRef]
48. Marion, G.M.; Babcock, K.L. Predicting specific conductance and salt concentration in dilute aqueous solutions. *Soil Sci.* **1976**, *122*, 181–187. [CrossRef]
49. Schoeneberger, P.J.; Wysocki, D.A.; Benham, E.C.; Soil Survey Staff. *Field Book for Describing and Sampling Soils*; Version 3.0., Reprint 2021; National Soil Survey Center: Lincoln, NE, USA, 2012. Available online: https://www.nrcs.usda.gov/sites/default/files/2022-09/field-book.pdf (accessed on 29 August 2023).
50. Zhu, Y.; Weindorf, D.C.; Zhang, W. Characterizing soils using a portable X-ray fluorescence spectrometer: 1. Soil texture. *Geoderma* **2011**, *167–168*, 167–177. [CrossRef]

51. Wang, S.Q.; Li, W.D.; Li, J.; Liu, X.S. Prediction of soil texture using FT-NIR spectroscopy and PXRF spectrometry with data fusion. *Soil Sci.* **2013**, *178*, 626–638. [CrossRef]
52. Gozukara, G.; Akça, E.; Dengiz, O.; Kapur, S.; Alper, A. Soil particle size prediction using Vis-NIR and pXRF spectra in a semiarid agricultural ecosystem in Central Anatolia of Türkiye. *Catena* **2022**, *217*, 106514. [CrossRef]

land

MDPI

Article

# An Ecological Overview of Halophytes and Salt-Affected Soils at El Hito Saline Pond (Central Spain): Baseline Study for Future Conservation–Rehabilitation Measures

Raimundo Jiménez-Ballesta [1,*], Santos Cirujano-Bracamonte [2], Eduardo Palencia-Mayordomo [3] and Mario Álvarez-Soto [2]

[1] Department of Geology and Geochemistry, Autónoma University of Madrid, 28049 Madrid, Spain
[2] Consultores en Biología de la Conservación S.L., 28004 Madrid, Spain; santoscirujano@gmail.com (S.C.-B.); mario.alv.soto@gmail.com (M.Á.-S.)
[3] Fundación Global Nature, 28231 Madrid, Spain; epalencia@fundacionglobalnature.org
* Correspondence: profe.raimundojimenez@gmail.com

**Abstract:** In an attempt to boost the potential ecological viability of wetlands, this study aimed to discover the relationship between soil salinity and vegetation composition in a quasi-pristine saline pond, "El Hito Lagoon". This wetland is situated in the largest continuous natural semi-arid steppe land of western Europe (specifically in Castilla La Mancha, Central Spain). Several soil profiles and a series of surface samples (0–10 cm) extracted from a systematic network throughout the saline pond were described, sampled, and analyzed. The most significant results included the detection of elevated levels of soil salinity, with distinctive sub-areas of extreme elevated surface salinity where the pH reading peaked at 9.89 and the electrical conductivity was higher than 40 (dS/m). The very high content of total available P displayed quite an irregular scatter within the soil profile. Specifically, the range oscillated between 8.57 mg/kg and 388.1 mg/kg, several samples having values greater than 100 mg/kg. An aspect that the abundant presence of *Salsola soda*, a plant frequently found growing in nutrient-rich wetlands, was able to confirm.

**Keywords:** semi-arid environment; Natura 2000 network; halophytic flora; salt stress; soil salinity; ecological environment; basin filling

**Citation:** Jiménez-Ballesta, R.; Cirujano-Bracamonte, S.; Palencia-Mayordomo, E.; Álvarez-Soto, M. An Ecological Overview of Halophytes and Salt-Affected Soils at El Hito Saline Pond (Central Spain): Baseline Study for Future Conservation–Rehabilitation Measures. *Land* **2024**, *13*, 449. https://doi.org/10.3390/land13040449

Academic Editors: Maria da Conceição Gonçalves, Mohammad Farzamian, Tiago Brito Ramos and Shiliang Liu

Received: 28 February 2024
Revised: 22 March 2024
Accepted: 29 March 2024
Published: 31 March 2024



## 1. Introduction

In semi-arid and arid regions, soil salinization and environmental problems go hand in hand, due to the severe effects of salinity on agricultural productivity and long-term sustainable development. Indeed, soil salinization normally produces severe effects on soil quality through changes in soil structure and chemistry, soil biology, and crop productivity and yield. Under such conditions, plants such as halophytes can survive and complete their life cycle in the presence of significant concentrations of soluble salts. So, it is generally accepted that identifying saline soils is of great importance for protecting land resources and ecological niches and for sustainable agricultural development.

The issue of soil salinity is a result of both natural and anthropogenic causes. Salts are naturally present in soil due to the weathering of parent minerals.

Although the soil-science community has generated abundant information on the current state of soil salinity [1–5], the idea still persists that salt-affected soils of high salinity are of low quality. Nevertheless, wetlands are among the most productive ecosystems on Earth and provide various key ecosystem services for humans and other flora and fauna. Hence, there exists a need for appropriate conservation policies and a more comprehensive understanding of both wetland ecosystem services (especially saline ecosystems) and the drivers behind their change.

Globally speaking, about 900 million ha of soil in over 100 countries has been affected by salinization or sodification [6–8]. With that global perspective in mind, soil salinization management is vital to the achievement of "Zero Hunger" (SDG2) and "Life on Land" (SDG15) among other sustainable development goals.

Generally, saline wetlands are important for the conservation of many endemic and rare plant species and the migration of water birds, its salinity being a result of both natural and anthropogenic causes [9–11] (Nachshon, 2018; Corwin, 2021; Stavi et al., 2021). Halophytes are plants that can survive in saline wetlands where the salt concentration exceeds 200 mM of NaCl (~20 dS $m^{-1}$) [12] (Flowers and T. D. Colmer). When the soil has a shallow water table, then the salinity of the soil solution will be strongly affected by the salinity of the groundwater. In this context, it is worth asking what are the main soils' properties as well as their edaphic and biogeochemical processes in geographic and geological areas such as the El Hito saline wetland. The aim is to assess whether the use of autochthonous species from saline environments may have advantages due to their better tolerance to edaphoclimatic conditions, besides having several uses.

Salinity is a continuous and a complex global problem that cannot be solved simply. And as is known, the salinity problem is increasing rapidly throughout the world. Hence, in this study (necessary for monitoring this specific kind of wetland), the aim is to analyze the compositional features, especially the contents of the main elements and ions relevant to understanding the source of salinization and its spatial variability, and to design a successful plant community. The specific questions concerned the following three points: (1) whether different soil compositions will emerge as the primary variables for explaining the presence and abundance of sub-communities within the saline pond; (2) whether there are consistent patterns across surface horizon soil salinity levels; (3) which factors are more than any other exerting a substantial influence on the structure of abundant and rare sub-communities. With that in mind, the following specific objectives were addressed: (1) to report relevant features such as pH, electrical conductivity, and other soil properties of a hypersaline wetland located in the municipalities of El Hito-Montalbo; (2) to identify and to evaluate the soil salinity of the study site with special emphasis on the relation between soil salinity and plants; (3) to relate those parameters to the different types of natural plants, in order to detect the main problems concerning salinity versus halophytes, while investigating new proposals for conservation–rehabilitation measures.

## 2. Materials and Methods

### 2.1. Description of the Study Area

The study area is El Hito Pond (Cuenca, Spain), located between 39°52′28″ to 2°42′06″ North latitude and 39°51′08″ to 2°40′56″ South latitude (Figure 1). It is an endorheic, palustral, ephemeral depression, with little topographic contrast, and a floodable lagoon basin with an estimated volume of around 250–573 ha [13–15]. Dry and hot summer and cold winter seasons are characterized by low levels of irregular precipitation, in which potential evapotranspiration exceeds precipitation, with such high rates of evaporation causing salts to accumulate on the soil surface. Mean annual precipitation stands at 587 mm and the average temperature at 11.7 °C. The relief of the surrounding area is largely flat and sometimes undulating, where the main soil types are Cambisols, Luvisols, Calcisols, Regosols, and Leptsols [16].

The dominant species of the vegetation are all salt-tolerant. Fortunately, there is detailed information on saline continental Mediterranean wetlands such as El Hito and their plant characteristics [15,17–21]. Figures 2 and 3 show photographs of various halophilic plant communities colonizing El Hito Pond.

**Figure 1.** Orthophoto of the study area in Castilla La Mancha at (**A**) summer time; (**B**) winter time.

**Figure 2.** *Puccinellia festuciformis* formations now colonize extensive areas of the basin, as the pond waters recede over increasingly shorter periods.

Among other species found in this wetland and of great biogeographical interest is the endemic presence of Phanerogam *Limonium soboliferum* around the lagoon. Likewise, numerous insects, including the orthopteroid *Roseliana oporina*, known as the "Cuenca Meadow Bush-Cricket", and the small crustacean *Branchinecta orientalis*, now threatened with extinction, are also prevalent in the same location.

Regarding the conservation figures, we can mention that the natural area around El Hito lagoon is included in the Natura 2000 Network (ZEPA-LIC-ZEC ES0000161) and in the Ramsar List of Wetlands of International Importance.

**Figure 3.** *Salsola soda* and *Salicornia ramosissima* communities at the edges of the flooded area. In orange, *Puccinellia festuciformis* formations.

### 2.2. Soil Samples Collected

Soil selection for sampling was dependent on certain soil-related features, such as drainage, presence or absence of salt crust at the soil surface, and especially vegetation type and density (Figure 4). For example, during sampling, the epsomite efflorescences appeared in the form of fibrous or hairlike acicular crust, while the structure of the hexahydrite efflorescences tended to be columnar. The soil profiles were described and sampled, having previously been opened up with a backhoe machine (Figure 5). Both the soil profiles and the surface sampling points were geo-referenced using Global Positioning System (GPS) coordinates. Probably, the most striking visual expressions of salinity at El Hito are, on one hand, the efflorescence on the soil surface (in dry periods) and, on the other, the halophilous vegetation (Figure 4). The samples were bagged, properly labelled, and transported to the laboratory for analysis. In total, 49 final soil samples were collected, consisting of 18 from several soil profiles and 31 from surface soil horizons.

**Figure 4.** *Cont.*

**Figure 4.** Criteria for sampling: (**A**) grassland interspersed with formations of *Lygeum spartum*, *Puccinellia festuciformis*, and *Elymus curvifolius*; (**B**) bare soil with dry remnants of formations of *Puccinellia festuciformis*; (**C**) *Puccinellia festuciformis*; (**D**) *Salsola soda*, *Hordeum marinum*, and *Puccinellia festuciformis*; (**E**) shrinkage cracks with a few examples of *Salsola soda*; (**F**) grassland of *Puccinellia festuciformis* and annual plants of nitrified soils; (**G**) *Lygeum spartum* formations; (**H**) saline crust in bare soil; (**I**) sparse grassland of *Salicornia ramossisima*; (**J**) grassland with *Salsola soda* and *Salicornia ramosissima*.

**Figure 5.** All soil profiles for description and sampling purposes were dug with a backhoe machine.

### 2.3. Soil Sample Analysis

Soil samples were dried and sieved at 2 mm (separating out thick fragments and roots from the remainder of the material). The same samples were then divided into two parts: one to determine the general physico-chemical properties of the soil; the other to determine the elemental spectrum via X-ray fluorescence (XRF) spectroscopy. Likewise, the physico-chemical properties were determined with the techniques listed in Table 1.

**Table 1.** Methods used for the soil sample analysis.

| Parameter | Method | References |
|---|---|---|
| Texture | Touch | [22] |
| pH | pH meter measurements of 1:2.5 soil/water suspension | [23] |
| Electric conductivity (EC) | Conductivity meter measurements of 1:5 soil/water suspension | [24] |
| $CaCO_3$ | Bernard method with a calcimeter | |
| Organic matter (OM) | Dichromate digestion (Walkley and Black) | [25] |
| CEC | Percolation with ammonium acetate solution at pH = 7 | [26] |
| P | The Olsen method | [27] |
| N | The Kjeldahl distillation method | [28] |
| $Ca^{2+}$, $Mg^{2+}$, $Na^+$, $K^+$ | Inductively coupled plasma optical emission spectrometry (ICP-OES) | |
| $SO_4{}^{2-}$, $Cl^-$ | Ion chromatography | |
| Mineralogy | X-ray diffraction | |
| Elements | X-ray fluorescence spectrometers | |

### 2.4. Statistical Analysis

Statistical analysis was performed with the Statistical Product and Service Solutions (SPSS) 19.0 software package for Windows, SPSS Inc., IL, USA, with the institutional license of the Fundación Global Nature (Spain). The following analyses were performed: mean, minimum and maximum, median, standard deviation (SD), coefficient of variation (CV), and kurtosis of pH and EC results.

## 3. Results

All the pedons under study were deep and had no shallow water table. The soil profiling and field sampling results were in morphological terms monotonous; the morphology is of the Az-C type, with neither cambic (Bw) horizons nor, of course, argillic (Bt) horizons detected, typical of the nearby environment that borders the lagoon [29]. Indeed, the pedons showed similar morphological characteristics emanating from similar ecological conditions and modes of formation. No redoximorphic features (iron mottle due to reductive/waterlogged soil conditions) were detected.

In general, the structure was weak to moderate, in subangular blocks alternating with massive clods and with a single grain (with no structure); the weak structure or its complete absence could be attributed to the presence of $Na^+$ that acted as a dispersant, slowing down hydraulic conductivity [26,27,30,31]. What was really happening, as previously explained by He et al. [32], was that the suspended particles had the potential to clog the soil pores and consequently to diminish the capillary effects within the soil. Indeed, little porosity was detected (except in some sandier horizons) and no stoniness in each horizon of the soil profiles. Dispersal of wet, clay particles within the soil was observed, covering the macropores of the soil surface. The consequence was that root growth through the soil and surface crust was blocked, restricting plant emergence and thereby limiting root growth in salt-affected soils.

All the above contributed to the definition of "pedogenon" characteristics. Indeed, following the concept of "genon", defined by Boulaine [33] as "a soil volume comprising all the pedons that have the same structure, the same characteristics and result from the same pedogenesis", the concept of "soil genoforms" has more recently appeared in the work of Rossiter and Bouma [34], understood as "soil classes", identified in the soil classification system used for detailed soil mapping of local areas. Following the classification of soil classes, the term "pedogenon" emerged, proposed as a conceptual soil taxon derived from a set of quantitative state variables representing the soil-forming factors [35,36]. If each pedogenon is characterized by a soil type formed under a dominant parent material occupying a unique position in the landscape, then at El Hito there is a characteristic pedogenon that can, finally, be used as a benchmark soil.

Table 2 presents the physico-chemical parameters of El Hito pedogenon. The general morphology is the Az-$C_1$-$C_1$ type (and therefore poorly developed), where the A horizon is resting upon C material. Referring to the Keys to Soil Taxonomy [37] (Soil Survey Staff, 2014), the pedogenon was classified as Aridisols, with Great group Aquisalids, correlating to Solonchaks of the IUSS Working Group WRB [16].

**Table 2.** Physico-chemical properties of the pedogenon in the study area.

| Pedogenon | Horizon | pH ($H_2O$) | EC (dS/m) | OC (%) | N (%) | C/N | P (mg/kg) | $CaCO_3$ (%) |
|---|---|---|---|---|---|---|---|---|
| El Hito | Az | 9.06 | 10.38 | 1.10 | 0.16 | 1.9 | 155.31 | 10.5 |
| | C1 | 8.75 | 6.19 | 0.25 | 0.04 | 6.3 | 95.62 | 11.6 |
| | C2 | 8.69 | 8.12 | 0.23 | 0.03 | 7.6 | 41.31 | 4.8 |

With the above in mind, the pedogenon of El Hito lagoon can be characterized by its dominant parent materials, silt, gypsum, and margogypsum, nurturing its natural vegetation of Mediterranean salt meadows and grasslands growing on flat terrain. The soil texture is heavy (clay silty), and the predominant soil color is whitish to greyish. Cracks, crust, and efflorescence appear in the dry season. The soil class is Typic Aquilasids [33,37]. Soil pH peaked at a value of 9.06, which is highly alkaline in nature. Electrical conductivity (EC1:5) was recorded at a value of 10.38 dS/m.

The dominant cations within salt-affected soils are sodium ($Na^+$), calcium ($Ca^{2+}$), magnesium ($Mg^{2+}$), and potassium ($K^+$), and the dominant anions are chloride ($Cl^-$), sulphate ($SO_4{}^{2-}$), carbonate ($CO_3{}^{2-}$), bicarbonate ($HCO^-$), and nitrates ($NO_3{}^-$) [24]. Its contents are high in the case study (Table 3). Based on the above values, it can be assumed that $Mg^{2+}$ and sometimes $Ca^{2+}$ dominates, followed by $Na^+$ and finally $K^+$. Regarding the anions, sulphate and chloride mainly dominate and, to a lesser extent, carbonate and bicarbonate; in that way, although carbonate concentration may not be negligible, $Na^+$ is probably responsible for the high pH values. In other words, sodium bicarbonate remains within the water, producing trace amounts of soda that raises the pH.

**Table 3.** Results of the analyses of the paste saturation extract.

| Pedogenon | Horizon | $Ca^{2+}$ | $Mg^{2+}$ | $Na^+$ | $K^+$ | $SO_4^{2-}$ | $Cl^-$ | $CO_3H^-$ | $CO_3^{2-}$ |
|---|---|---|---|---|---|---|---|---|---|
| | | | | | mg/kg | | | | |
| El Hito | Az | 5621 | 6800 | 930 | 754 | 32,652 | 7110 | 138 | 147 |
| | C1 | 3627 | 5250 | 1150 | 503 | 18,017 | 7503 | 96 | 101 |
| | C2 | 3559 | 6350 | 1275 | 450 | 25,125 | 9750 | 106 | 113 |

Biomass accumulation is limited in many saline wetlands and soil organic matter is unsurprisingly also limited at El Hito Pond. The high temperatures within the zone cause fresh residual vegetation to decompose, which can therefore explain the low organic matter (OM) value. The calcium content probably decreased, because of a competitive effect with magnesium. Generally, soils are considered to be salinized when a soil saturation extract has an ECe value of 4 dS/m or greater. As can be seen in the majority of the surface sample results, the ECe values ranged between 1.39 and 41 dS/m (Table 4), clearly confirming that El Hito site samples were highly salinized in 31 surface soil samples.

**Table 4.** Results of pH and electrical conductivity (ECe) analyses on 31 surface soil samples. The characteristics of the type of vegetation/use are also shown.

| Soil Surface | Vegetation Type | Coordinates | pH ($H_2O$) | ECe (dS/m) |
|---|---|---|---|---|
| C1 | *Puccinellia festuciformis* meadows | 39.877087–2.701919 | 9.42 | 41.00 |
| C2 | *Salsola soda* meadows | 39.876511–2.701869 | 9.48 | 30.70 |
| C3 | Bare soil with *Salsola soda* | 39.872515–2.702211 | 9.61 | 32.70 |
| C4 | *Puccinellia festuciformis* meadows | 39.870211–2.700909 | 9.71 | 16.26 |
| C5 | *Aeluropus litoralis* meadows | 39.867876–2.698168 | 8.62 | 2.18 |
| C6 | *Puccinellia festuciformis* meadows | 39.866100–2.695747 | 8.74 | 2.89 |
| C7 | Bare soil with *Salsola soda* | 39.866407–2.692898 | 9.89 | 39.70 |
| C8 | *Salsola soda* and *Salicornia ramosissima* | 39.869053–2.687970 | 9.71 | 4.99 |
| C9 | *Lygeum spartum* formations | 39.871532–2.684968 | 8.78 | 2.58 |
| C10 | Grassland among *Lygeum spartum* | 39.871252–2.684295 | 9.25 | 5.83 |
| C11 | Grassland among *Lygeum spartum* | 39.875732–2.678837 | 8.54 | 1.39 |
| C12 | Fallow with Bassia scoparia | 39.865365–2.673279 | 8.79 | 2.10 |
| C13 | Elymus repens meadows | 39.866324–2.674148 | 8.68 | 4.00 |
| C14 | *Aeluropus litoralis* meadows | 39.866599–2.674488 | 8.37 | 1.96 |
| C15 | Fallow | 39.860569–2.678548 | 8.79 | 1.99 |
| C16 | *Frankenia laevis* and *Plantago coronopus* | 39.860592–2.679418 | 9.65 | 27.40 |
| C17 | *Salicornia ramosisima* and *Puccinellia caespitosa* | 39.862437–2.679445 | 9.54 | 12.97 |
| C18 | Bare soil | 39.863784–2.682094 | 9.51 | 20.00 |
| C19 | *Salsola soda* meadows | 39.863516–2.684056 | 9.46 | 8.00 |
| C20 | Grassland among *Lygeum spartum* | 39.858481–2.686284 | 8.96 | 2.13 |
| C21 | Bare soil-fallow | 39.858233–2.694175 | 8.93 | 3.94 |
| C22 | Grassland among *Lygeum spartum* | 39.858594–2.699989 | 8.29 | 2.13 |

Table 4. Cont.

| Soil Surface | Vegetation Type | Coordinates | pH ($H_2O$) | ECe (dS/m) |
|---|---|---|---|---|
| C23 | Fallow with *Salsola kali* | 39.860900–2.709894 | 8.42 | 2.33 |
| C24 | Fallow | 39.867409–2.711967 | 8.64 | 2.06 |
| C25 | Fallow | 39.872950–2.710535 | 8.92 | 0.74 |
| C26 | Bare soil-fallow | 39.878496–2.709827 | 8.99 | 0.49 |
| C27 | *Puccinellia* caespitosa meadows | 39.876401–2.695190 | 9.59 | 16.89 |
| C28 | *Salsola soda* meadows | 39.874595–2.701872 | 9.60 | 9.50 |
| C29 | Bare soil with *Salsola soda* | 39.872515–2.702211 | 8.53 | 8.06 |
| C30 | *Puccinellia festuciformis* meadows | 39.866100–2.695747 | 8.44 | 2.69 |
| C31 | Bare soil with *Salsola soda* | 39.866407–2.692898 | 9.67 | 21.30 |

Careful observation of the above data highlights the lateral anisotropy of soil salinity, which was attributed to the temporal fluctuations of salt content following the appearance and/or displacement of soluble salts within the soil. Those salinity variations constitute an important driving factor of vegetation zonation.

The calculated statistics included mean, standard deviation (SD), coefficient of variation (CV), minimum, maximum, and kurtosis, as seen in Table 5. The saline soils of El Hito Pond display high spatial variability in soil salinity at a field scale, caused by temporal and spatial variations of external factors. It is worth noting that the SD of ECe (12.15) was larger than that of the pH ($H_2O$) (0.49), with the coefficient of variation (CV) of ECe (113%) larger than that of the pH ($H_2O$) (5%); a CV value lower than 10% indicated low variability. Both data sets had large values of kurtosis with the ECah0 (6.64) larger than ECav0 (4.53), with both being leptokurtic (value is greater than + 1.0) (Table 5).

Table 5. Basic summary statistics of the measured pH ($H_2O$) and ECe (dS/m).

| Variable | Mean | Median | Min | Max | SD | CV | Kurtosis |
|---|---|---|---|---|---|---|---|
| pH ($H_2O$) | 9.08 | 8.96 | 8.29 | 9.89 | 0.49 | 0.05 | −1.48 |
| ECe (dS/m) | 10.67 | 4.00 | 0.49 | 41.00 | 12.15 | 1.13 | 0.38 |

The graph in Figure 6 below represents the linear relationship between soil pH and soil EC for the depth of 0–10 cm. Visual observation and statistical analyses indicated that the soil pH had a significant positive relationship with the soil EC (pH = 5.403–0.152 EC). The correlation coefficient was −0.88.

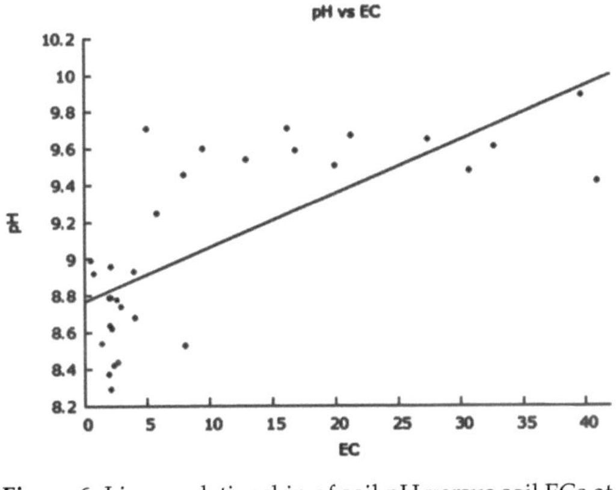


Figure 6. Linear relationship of soil pH versus soil ECe at 0–10 cm.

The elemental spectrum of the pedogenon is presented in Table 6. The most abundant element was S, whose presence ranged between 36.88 (%) and 27.94 (%), expressed as $SO_3$. Ca came second in quantitative terms, ranging between 28.21 and 31.12 of CaO (%) and then, in much smaller proportions, Mg, Si, Al, Na, K, Fe, and Cl, in descending order.

**Table 6.** Results of the chemical elemental analysis.

| Pedogenon | Horizon | $SiO_2$ | $Al_2O_3$ | $Fe_2O_3$ | CaO | MgO | $Na_2O$ | $K_2O$ | $SO_3$ | Cl | PPC |
|---|---|---|---|---|---|---|---|---|---|---|---|
| | | (%) | | | | | | | | | |
| El Hito | Az | 4.15 | 1.02 | 0.27 | 30.63 | 4.02 | 0.33 | 0.24 | 36.88 | 0.19 | 21.74 |
| | C1 | 4.01 | 0.83 | 0.27 | 31.12 | 2.30 | 0.35 | 0.23 | 33.59 | 0.21 | 23.09 |
| | C2 | 6.45 | 1.28 | 0.48 | 28.21 | 9.34 | 0.40 | 0.31 | 27.94 | 0.25 | 24.56 |

Mineralogically, the three fundamental mineral components were gypsum, carbonates, and more labile salts, i.e., epsomite ($SO_4Ca{\bullet}7\ H_2O$) and hexahydrite ($SO_4Ca{\bullet}6\ H_2O$).

Finally, if the five salinity classes established in 1983 by [34,38] (Table 7) are applied to the samples, almost half may be considered extremely saline soils; around one third, very saline soils; approximately 11%, moderately saline soil; only 4% can be considered slightly saline soils; finally, only 4% can be termed non-saline soils.

**Table 7.** Classification of El Hito lagoon soils based on electrical conductivity (EC) following the criteria of Duran [38]. Number of saline samples and percentage within the lagoon.

| EC (dS/m) | Salinity Classes | Number of Soil Samples | % Soil Samples |
|---|---|---|---|
| $\leq 0.6$ | Non-saline soil | 1 | 3.7 |
| $0.6 < EC \leq 1$ | Slightly saline soil | 1 | 3.7 |
| $1 < EC \leq 2$ | Moderately saline soil | 3 | 11.1 |
| $2 < EC \leq 4$ | Very saline soil | 9 | 33.3 |
| >4 | Extremely saline soil | 13 | 48.2 |

The electrical conductivity values for the surface soils, which must be borne in mind for ecological purposes, therefore indicated that a large portion of the total wetland area was classified as saline soil.

## 4. Discussion

### *4.1. On the Origin of Soil Salinity in El Hito Saline Pond*

El Hito saline pond occupies a flat-depressed zone, which favors preferential water circulation and accumulation. As within other European salt steppes, the natural process of soil formation shows specific patterns. Salts evidently accumulate, leading to salinization, although temporal dynamics were observed within specific areas, which explains the substantial variability of the data throughout the lagoon. In the opinion of Boettinger and Richardson [39], wet soils and wetlands are in general not prevalent in dry and seasonally dry climates, but their morphology and characteristics are considerably different from those in more humid climates. Indeed, the major pedogenic process involved in the formation of saline wetland soils in semi-arid climates, such as El Hito, is salinization.

The combined effects of high levels of evaporation together with insufficient leaching leads to the accumulation of salts within the soil and consequently to the development of salt-affected soils [40]. The only prerequisite is water with sufficient salts for further accumulation. The levels of silt and clay favor drastic impermeability. Shao and Zhang [41] suggested that an increase in silt content results in the thickening of the bound-water film and a reduction in soil porosity.

El Hito saline pond is located within areas where water accumulates, closely related to an endorheic basin (i.e., with little or no external drainage). There are briefly three factors or processes shaping the distribution and genesis of El Hito: geomorphology of water concentration, lithological sources of sediments, and evapo-concentration of salts, or in an environmental setting, pedogenesis. But globally speaking, a distinction is made between the primary (natural) and secondary (human-induced) processes. In the primary formation process, the nature of the parent geological material plays an important role, alongside the natural climatic conditions (arid or semi-arid), i.e., a lack of precipitation and an excess of evapotranspiration. On the contrary, the secondary process is caused by human activities [4,42–44].

Indeed, El Hito Pond is filled with a shallow body of water (maximum depth 15–20 cm) [45] for several months, approximately between November and May. Fundamentally, surface runoff replenishes the water layer, favored by the depressed topography, so that it is only during the rainy seasons when the lagoon retains water. Therefore, the primary drivers of soil salinization are closely related to both the mineralogical and the chemical characteristics of the parent materials and to both the topography and the type of climate. The process is well known: salty water rising through the soil profile by capillary action contributes to the salts, which remain in the soil when the water evaporates [46]. Given the topographic situation of the wetland, water accumulation occurs. The summer evaporation rates are higher than the precipitation rates within the region, so the normal pattern over many years is more or less rapid water evaporation. The salinity levels therefore rise as evaporation continues, and the surrounding soil salinization is continuous as water salinization gradually adds to soil salinity levels. Anthropic sources have been cited in other areas of the world [4,10,43,47–49], as are also glimpses of anthropic effects in the case of El Hito Pond. In any case, the evaporation process accelerates salt encrustation and formation, and the salts rise by capillarity action to the surface of the soil where crusts may even form.

### 4.2. Discussion on Halophyte Adaptation to Salt-Affected Soils

There is, to the best of our knowledge, limited research on soil salinity levels and the specialized vegetation that can prosper under such conditions. Halophyte formations generally reflect the chemical composition and the levels of soil salinity under natural conditions quite accurately. Soil pH and EC, in particular, emerged as primary environmental factors shaping the abundance of some plant species.

One of the identifying features of El Hito saline pond soils is their salt accumulation levels, especially within the surface horizons. The origin of the salt must be linked to the materials (gypsum and silt and gypsum marl) and their composition, as well as to the semi-arid climatic conditions that facilitate the water evaporation and saline precipitation. The following salt types were identified: epsomite ($SO_4Mg{\bullet}7H_2O$), mirabilite ($SO_4Na_2{\bullet}10H_2O$), bloedite [$(SO_4)_2Na_2Mg{\bullet}4H_2O$], halites (ClNa), etc. [14]. Sometimes, together with those saline accumulations, the same salts form crust-shaped accumulations (Figures 2–4), a type of crusting layer that forms at different stages (between approximately 0–10 cm) where salts accumulate at the onset of the dry summer season.

Soil salinity has a twofold influence on plant growth: high toxic ion concentrations and negative water potential [50,51]. Simply mentioning the word salinity in the context of soil can refer to problematic issues, since it negatively affects the growth of most plants. However, the so-called halophytic plants are salt-tolerant plants, that can normally prosper at levels of approximately 20 dS $m^{-1}$ of EC [52,53]. Authors such as Rozema and Schats [54] pointed out that plant growth could be stimulated within salinity ranges between 15 and 25 dS/m.

In general, plants growing under saline conditions have to adapt to three types of stress: (a) water stress (caused by osmotic pressure); (b) mineral toxicity stress caused by salt; and (c) alterations in mineral balance. Excess salts can inhibit water extraction through plant roots, i.e., less water is available to the plant, with a consequent reduction in soil productivity. Additionally, the salts provoke swelling and dispersion of the colloidal particles (caused by an excess of exchangeable Na), which generates both water infiltration and root penetration problems, as well as complicating the emergence of the seedlings.

Finally, in the summer, coinciding with maximum evapotranspiration, capillary rise occurs, generating saline crust formations on the surface.

The plants that colonize El Hito Pond are salt tolerant. The distribution of different species is dependent upon the duration of flooding. In this regard, plant tolerance to waterlogging can be summarized as follows: *Lygeum* < *Frankenia* < *Puccinellia* < *Salicornia* = *Salsola*. Principio del formulario

Crops such as potatoes, wheat, *Vitis vinifera*, *Citrus*, *Prunus*, *Lactuca*, and *Cucumis* and plants such as *Lavandula* and *Calendula* are damaged by soil salinization. In contrast, the growth and the development of plants such as *Salsola* is, on one hand, common at El Hito saline pond, which is nothing strange in itself given that its growth is not negatively affected by salt stress; on the other hand, plants and crops commonly found within the environment of El Hito Pond are negatively affected, because of the disruption of certain physiological processes such as an imbalance in the ratios of macronutrients, nutritional deficiency, and even Na phytotoxicity [55,56]. In fact, saline stress may somehow affect processes such as photosynthesis and respiration, causing inhibition, wilting, drying, and the demise of entire plant organs of most plants and vegetable crops [57,58]. Indeed, from the biological point of view, soil salinization processes affect several ecological soil functions. Thus, respiration and microbial activity are both worth mentioning, insofar as they interact with processes related to organic matter, decomposition, and the N cycle, eventually affecting microbial activity [53,59,60].

Globally speaking, soil salinity is an important variable that plays a relevant role in reducing fertility levels. Soil salinization is therefore a very severe problem that is mainly due to irrigation strategies and other intensive agricultural activities [61,62]. It becomes an issue for agricultural productivity, but not within ecological niches where saline soils are produced by natural processes.

### 4.3. Ecosystem Services of El Hito Saline Pond: Proposal for Conservation–Rehabilitation Measures

Ecosystem services have become a prominent concept in international policy and research agendas [63]. Wetlands are among the most diverse and productive and therefore the most valuable ecosystems [64–68] in the world. Various ecosystem services can co-exist within one ecosystem [69]. El Hito Pond constitutes a small iconic environment representative of the most diverse and productive ecosystems within the Mediterranean region. A fascinating result of interactions between aquatic and terrestrial ecosystems, it provides a wide range of services and benefits to local inhabitants and visitors, including climate regulation, water purification, and flood hazard reduction; although its recreational and cultural value may be highlighted above all, they are in addition to its role as a natural wildlife habitat.

There were until practically a few decades ago hardly any environmental concerns voiced in the society over the importance of wetlands, much less so whether these are saline in nature. In the latter case, they have rather been disdained, based on the fact that they were practically useless for agricultural purposes. With that in mind, given that our findings have demonstrated clearly varied levels of soil salinity and ionic compositions across the lagoon, soil management may consider the spatial heterogeneity of saline soil types. Our results therefore provide scientific guidance for soil management and restoration of the lagoon. Future studies should address the quantification of temporal changes to soil salt-affection patterns in and around El Hito Pond.

Halophytic plants, which are salt-tolerant species, can survive under values of EC of 20 dS/m [52], while [54] pointed out that plant growth can be stimulated within a salinity range of 15 and 25 dS/m. This series of species is equipped with adaptive mechanisms to survive in saline environments; they include biochemical, physiological, anatomical, and morphological characteristics [52,70,71].

Some scholars [72–76] have maintained that halophytes were viable alternatives for food, fodder, bioenergy, ornamental, and pharmaceutical uses. According to Duarte and

Caçador [72], the benefits of halophyte ecosystems are acknowledged, insofar as they improve soil health and ecosystem biodiversity and can store large amounts of carbon, thereby increasing the ecosystem resilience to climate change and offering green solutions to slow climate change. Both [77,78] pointed out that halophytes have long been used for pharmaceutical purposes, in view of their bioactive compounds with medical properties.

Halophytes are considered precious natural resources with potential economic value as grain, vegetable, fruit, medicine, animal feed, and biofuel feedstocks and in greening and coastal protection [79]. In this way, plants from the genus *Salsola* are known to be used in traditional medicine; for example, *Halothamnus somalensis* (N.E.BR.) Botsch (*Salsola somalensis* N.E.Br.) has been used as hypotensive, antibacterial, and anticancer agents. *Salsola* is a halophyte with succulent vegetative structures that can be defined as a successful salt-absorbing species. The genus can also help with the restoration and reclamation of degraded salty areas and saline soils [80]. A halophyte native to the Mediterranean basin, *Salsola soda* L., is considered to be a "biodesalinating companion plant" to tomatoes and peppers in the saline soils of central Italy [81]. The potential use of *Salsola soda* for the phytostabilization of polluted areas has been shown by [79,82,83]. Recently, Ref. [84] stated that planting halophytes in salt-affected areas can improve soil quality and restore biodiversity. It also produces valuable products, such as animal feeds and renewable energy sources. Therefore, the presence of *Salsola soda* at El Hito should be understood as a paradigmatic example of a species to conserve, since it is perfectly adapted to a dominant saline soil environment.

It is expected that as climate change brings drier and warmer conditions to the region, sparser water resources will lower or completely exhaust El Hito lagoon water levels. It is therefore necessary to propose conservation–remediation measures.

*Lygeum spartum* L. has recently been introduced into areas with high soil salinity levels [85]. Popularly known as Albardín, *Lygeum spartum* is an herbaceous plant with an extensive root system that shows spontaneous growth in saline soils and reduces soil erosion, while enhancing soil phytostabilization. It is therefore used in landscape restoration and for erosion control in the southeastern Iberian Peninsula according to [86], especially in semi-arid areas.

However, Ref. [87] pointed out that the effect of salinity on germination varies considerably with temperature regimes, while [88] through a field experiment concluded that that *L. spartum* can be used as a bio-indicator of soil salt type. Ref. [89] demonstrated that an increase in salinity induces both a reduction in the percentage of germinating seeds and a delay in the initiation of the germination process. More recently, Refs. [90,91] stated that salinity, temperature, and their interaction affected the germination percentage of *L. spartum* seeds. According to our findings, late autumn and early winter sowings are recommended when salinity and temperature stresses are reduced, when plant canopies do not cover the soil, roots are insufficiently developed, and the soil has no protection against erosion (Figure 7).

**Figure 7.** The formations of esparto grass (*Lygeum spartum*) known as *"albardinares"* are typical of the Iberian "steppes", located in the highest area around El Hito saline pond that the waters leave untouched. These formations protect the soil from erosion and delay siltation within the pond.

## 5. Conclusions

The different habitats of El Hito saline pond whose conservation is a priority interest in the European Union are of high environmental value. It is a hypersaline wetland, where salinity, hydric conditions, and halophilous vegetation contrast distinctly with the surrounding non-saline gypseous land and its gypsophilous vegetation. It is nevertheless evident that this wetland is fragile in nature, and a better understanding is therefore required of how to promote and to adapt appropriate management strategies. In this paper, useful baseline data have been presented for its conservation.

**Author Contributions:** R.J.-B. and S.C.-B. conceived and designed this study; R.J.-B., S.C.-B. and E.P.-M. participated in the collection of soil samples; software, E.P.-M. and M.Á.-S.; formal analysis S.C.-B.; writing— original draft preparation, R.J.-B. and S.C.-B. All authors have read and agreed to the published version of the manuscript.

**Funding:** The present work has been financed with funds from the LIFE EL HITO project (LIFE 20NAT/ES/000035).

**Data Availability Statement:** The data that support the findings of this study are available upon request from the corresponding author. The data are not publicly available due to privacy or ethics restrictions.

**Acknowledgments:** We thank the members of the management team for their support and assistance in its completion.

**Conflicts of Interest:** Santos Cirujano Bracamonte, currently serves as President of the Global Nature Foundation (FGN), a non-profit organization based in Spain that works on nature conservation and biodiversity both nationally and internationally. Founded in 1993, its main mission is to promote sustainable development and environmental conservation through collaboration with various sectors, including the business sector, public administrations, and civil society. It develops projects in areas of high ecological value and at risk of degradation, particularly focusing on the restoration and management of wetlands. Mario Álvarez-Soto was employed by Consultores en Biología de la Conservación S.L. He participated in the study by contributing to the generation of GIS maps. The role of the company was to provide assistance and support in data generation and manuscript preparation. Eduardo Palencia-Mayordomo was employed in Global Nature Foundation, working primarily in field work. The remaining authors declare that the research was conducted in the absence of any commercial or financial relationships that could be construed as a potential conflict of interest.

## References


1. Rengasamy, P. World salinization with emphasis on Australia. *Comp. Biochem. Phys. A Mol. Integr. Physiol.* **2005**, *141*, 337–348. [CrossRef] [PubMed]
2. Castañeda, C.; Herrero, J. Assessing the degradation of saline wetlands in an arid agricultural region in Spain. *Catena* **2008**, *72*, 205–213. [CrossRef]
3. Kargas, G.; Chatzigiakoumis, I.; Kollias, A.; Spiliotis, D.; Kerkides, P. An Investigation of the Relationship between the Electrical Conductivity of the Soil Saturated Paste Extract ECe with the Respective Values of the Mass Soil/Water Ratios 1:1 and 1:5 (EC1:1 and EC1:5). In Proceedings of the 3rd EWaS International Conference on "Insights on the Water-Energy-Food Nexus", Lefkada Island, Greece, 27–30 June 2018; Volume 2, p. 661.
4. Omuto, C.T.; Vargas, R.R.; El Mobarak, A.M.; Mohamed, N.; Viatkin, K.; Yigini, Y. *Mapping of Salt-Affected Soils: Technical Manual*; FAO: Rome, Italy, 2020. [CrossRef]
5. Herrero, S.; Castañeda, C. Comparing Two Saline-Gypseous Wetland Soils in NE Spain. *Land* **2023**, *12*, 1990. [CrossRef]
6. Tanji, K.K. (Ed.) *Agricultural Salinity Assessment and Management*; Manual Reports on Engineering Practices, 71; ASCE: New York, NY, USA, 1990.
7. Rath, K.M.; Fierer, N.; Murphy, D.V.; Rousk, J. Linking Bacterial Community Composition to Soil Salinity along Environmental Gradients. *ISME J.* **2019**, *13*, 836–846. [CrossRef] [PubMed]
8. Li, X.; Kang, Y. Agricultural utilization and vegetation establishment on saline-sodic soils using a water–salt regulation method for scheduled drip irrigation. *Agric. Water Manag.* **2020**, *231*, 105995. [CrossRef]
9. Nachshon, U. Cropland soil salinization and associated hydrology: Trends, processes and examples. *Water* **2018**, *10*, 1030. [CrossRef]
10. Stavi, I.; Thevs, N.; Priori, S. Soil salinity and sodicity in drylands: A review of causes, effects, monitoring, and restoration measures. *Front. Environ. Sci.* **2021**, *9*, 330. [CrossRef]
11. Corwin, D.L. Climate change impacts on soil salinity in agricultural areas. *Eur. J. Soil Sci.* **2021**, *72*, 842–862. [CrossRef]
12. Flowers, T.J.; Colmer, T.D. Salinity tolerance in halophytes. *New Phytol.* **2008**, *179*, 945–963. [CrossRef]
13. Soriano-Hernando, O.; Álvarez-Cobelas, M. (Eds.) *Limnología de las Lagunas de la Cuenca del Guadiana*; Grupo de Investigación del Agua, Serie Limnoiberia: Madrid, Spain, 2016.
14. Manchado, E.M.; Sánchez-Palencia, Y.; García-Romero, E.; Torres, T.; Ortiz, J.E.; Suárez, M. Estudio mineralógico de los sedimentos de la Laguna de El Hito (Cuenca). *Macla* **2018**, *23*, 53–54.
15. Cirujano, S.; Alvarez-Soto, M. Caracterización y cambios de la vegetación en un humedal salino continental. Laguna de El Hito (Cuenca). *Acta Bot. Malacit.* **2023**, *48*, 74–92.
16. IUSS Working Group WRB. *World Reference Base for Soil Resources 2014 Update 2015. International Soil Classification System for Naming Soils and Creating Legends for Soil Maps*; World Soil Resource Report No. 106; FAO: Rome, Italy, 2015.
17. Castroviejo, S.; Cirujano, S. Sarcocornietea en La Mancha (España). *Anal. Jard. Bot. Madr.* **1980**, *37*, 143–154.
18. Cirujano, S. Las lagunas manchegas y su vegetación I. *Anal. Jard. Bot. Madr.* **1980**, *37*, 155–192.
19. Cirujano, S. Las lagunas manchegas y su vegetación II. *Anal. Jard. Bot. Madr.* **1981**, *38*, 187–232.
20. Cirujano, S. *Flora y Vegetación de las Lagunas y Humedales de la Provincia de Cuenca*; Real Jardín Botánico: Madrid, Spain, 1995.
21. Cirujano, S.; Medina-Domingo, L.; Chirino-Argenta, M. *Plantas Acuáticas de las Lagunas y Humedales de Castilla-La Mancha*; Junta Comunidades Castilla-La Mancha: Toledo, Spain, 2002.
22. Ritchey, E.L.; McGrath, J.M.; Gehring, D. Determining Soil Texture by Feel. Agriculture and Natural Resources Publications 139. 2015. Available online: https://uknowledge.uky.edu/anr_reports/139 (accessed on 23 January 2024).
23. Peech, M.; Alexander, L.T.; Dean, L.A.; Reed, J.F. *Methods of Soil Analysis for Soil Fertility Investigations*, 1st ed.; United States Department of Agriculture: Washington, DC, USA, 1947; p. 25.
24. Richards, L.A. *Diagnosis and Improvements of Saline and Alkali Soils*; Agriculture Handbook N°. 60; USDA: Washington, DC, USA, 1954.
25. Nelson, D.W.; Sommers, L.E. Total carbon, organic carbon and organic matter. In *Methods of Soil Analysis, Part 2, Chemical and Mineralogical Properties*, 2nd ed.; Page, A.L., Miller, R.H., Keeney, D.R., Eds.; American Society of Agronomy: Madison, WI, USA, 1982; pp. 539–557.
26. Thomas, G.W. Exchangeable cations. In *Methods of Soil Analysis, Part 2*; Page, A.L., Miller, R.H., Keeney, D.R., Eds.; American Society of Agronomy, Soil Science Society of America: Madison, WI, USA, 1982; pp. 159–165.
27. Olsen, S.R.; Cole, C.V.; Watanabe, F.S. *Estimation of Available Phosphorus in Soils by Extraction with Sodium Bicarbonate*; USDA Circular No. 939; US Government Printing Office: Washington, DC, USA, 1954.
28. Bremner, J.M.; Mulvaney, C.S. Total nitrogen. In *Methods of Soil Analysis. Part 2*; Page, A.L., Miller, R.H., Keeney, D.R., Eds.; Agronomy Monograph 9, American Society of Agronomy: Madison, WI, USA, 1982; pp. 1149–1170.
29. Jiménez-Ballesta, R.; Pérez-De-Los-Reyes, C.; Amorós, A.; Bravo, S.; Navarro, F.J.G. Pedodiversity in Vineyards of Castilla-La Mancha, Spain. In Proceedings of the XIII Congres International Terroir, Zaragoza, Spain, 18–22 June 2018; pp. 324–329, ISBN 978-84-09-03040-8.

30. Qadir, M.; Noble, A.D.; Schubert, S.; Thomas, R.J.; Arslan, A. Sodicity induced land degradation and its sustainable management: Problems and prospects. *Land Degrad. Dev.* **2006**, *17*, 661–676. [CrossRef]
31. Rengasamy, P. Soil chemistry factors confounding crop salinity tolerance-a review. *Agronomy* **2016**, *6*, 53. [CrossRef]
32. He, B.; Cai, Y.; Ran, W.; Zhao, X.; Jiang, H. Spatiotemporal heterogeneity of soil salinity after the establishment of vegetation on a coastal saline field. *Catena* **2015**, *127*, 129–134. [CrossRef]
33. Boulaine, J. Sol, Pedon et Genon. Concepts et definitions. *Bull. AFES* **1969**, *5*, 7–10.
34. Rossiter, D.G.; Bouma, J. A new look at soil phenoforms—Definition, identification, mapping. *Geoderma* **2018**, *314*, 113–121. [CrossRef]
35. Dobarco, M.R.; McBratney, A.; Minasny, B.; Malone, B. A modelling framework for pedogenon mapping. *Geoderma* **2021**, *393*, 115012. [CrossRef]
36. Jang, H.J.; Dobarco, M.R.; Minasny, B.; McBratney, A.; Jones, E. Developing and testing of pedogenons in the lower Namoi valley, NSW, Australia. *Geoderma* **2022**, *428*, 116182. [CrossRef]
37. *Soil Survey Staff, Keys to Soil Taxonomy*, 12th ed.; United States Department of Agriculture, Natural Resources Conservation Service: Washington, DC, USA, 2014.
38. Durand, J.H. *Les Sols Irrigables; Etude Pedologique*; Presses Universitaires de France: Paris, France, 1983.
39. Boettinger, J.L.; Richardson, J.L. Saline and Wet Soils of Wetlands in Dry Climates. In *Wetland Soils: Genesis, Hydrology, Landscapes, and Classification*; Richardson, J.L., Vepraskas, M.J., Eds.; Lewis Publishers CRC Press LLC.: Boca Raton, FL, USA, 2001; pp. 383–390.
40. Schaetzl, R.J.; Anderson, S. *Soils, Genesis and Geomorphology*; Cambridge University Press: Cambridge, UK, 2005.
41. Shao, M.A.; Zhang, B.W. Experimental study on the infiltration of crude oil into disturbed soil. *J. Soil Sci.* **2009**, *5*, 781–787. [CrossRef]
42. Rangesamy, P.; Marchuk, A. Cation ratio of soil structural stability (CROSS). *Soil Res.* **2011**, *49*, 280–285. [CrossRef]
43. Omuto, C.T.; Vargas, R.R.; El Mobarak, A.A.; Mapeshoane, B.E.; Koetlisi, K.A.; Ahmadzai, H.; Abdalla Mohamed, N. Digital Soil Assessment in Support of a Soil Information System for Monitoring Salinization and Sodification 693 in Agricultural Areas. *Land Degrad. Dev.* **2022**, 33, 1204–1218. [CrossRef]
44. Hopmans, W.; Qureshi, A.S.; Kisekka, I.; Munns, R.; Grattan, S.R.; Rengasamy, P.; Ben-Gal, A.; Assouline, S.; Javaux, M.; Minhas, P.S.; et al. Critical knowledge gaps and research priorities in global soil salinity. *Adv. Agron.* **2021**, *169*, 1–191. [CrossRef]
45. Sastre, A.; Martínez, S.; Jurado, A.; Acaso, E. *Geohidrología del Humedal de El Hito (prov. de Cuenca)*; Fernández Uría, A., Ed.; IX Simposio de Hidrogeología: Elche, Spain, 2008; pp. 305–314.
46. Rengasamy, P. World salinisation with emphasis on Australia. *J. Exp. Bot.* **2006**, *57*, 1017–1023. [CrossRef] [PubMed]
47. Datta, K.K.; de Jong, C. Adverse effect of waterlogging and soil salinity on crop and land productivity in northwest region of Haryana, India. *Agric. Water Manag.* **2002**, *57*, 223–238. [CrossRef]
48. Daliakopoulos, I.N.; Tsanis, I.K.; Koutroulis, A.; Kourgialas, N.N.; Varouchakis, A.E.; Karatzas, G.P.; Ritsema, C.J. The threat of soil salinity: A European scale review. *Sci. Total Environ.* **2016**, *573*, 727–739. [CrossRef] [PubMed]
49. Ondrasek, G.; Rengel, Z. Environmental salinization processes: Detection, implications & solutions. *Sci. Total Environ.* **2021**, *754*, 142432. [CrossRef]
50. Dikilitas, M.; Karakas, S. Crop Production for Agricultural Improvement. In *Behavior of Plant Pathogens for Crops under Stress during the Determination of Physiological, Biochemical and Molecular Approaches for Salt Stress Tolerance*; Ashraf, M., Ed.; Springer: Berlin/Heidelberg, Germany, 2012; pp. 417–441.
51. Otlewska, A.; Migliore, M.; Dybka-Stępień, K.; Manfredini, A.; Struszczyk-Świta, K.; Napoli, R.; Białkowska, A.; Canfora, L.; Pinzari, F. When Salt Meddles Between Plant, Soil, and Microorganisms. *Front. Plant Sci.* **2020**, *11*, 553087. [CrossRef]
52. Flowers, T.J.; Colmer, T.D. Plant salt tolerance: Adaptations in halophytes. *Ann. Bot.* **2015**, *115*, 327–331. [CrossRef]
53. Meng, X.; Zhou, J.; Sui, N. Mechanisms of salt tolerance in halophytes: Current understanding and recent advances. *Open Life Sci.* **2018**, *13*, 149–154. [CrossRef] [PubMed]
54. Rozema, J.; Schats, H. Salt tolerance of halophytes, research questions reviewed in the perspective of saline agriculture. *Environ. Exp. Bot.* **2013**, *92*, 83–95. [CrossRef]
55. Acosta-Motos, J.R.; Ortuño, M.F.; Bernal-Vicente, A.; Diaz-Vivancos, P.; Sanchez-Blanco, M.J.; Hernandez, J.A. Plant responses to salt stress: Adaptive mechanisms. *Agronomy* **2017**, *7*, 18. [CrossRef]
56. Morton, M.J.L.; Awlia, M.; Al-Tamimi, N.; Saade, S.; Pailles, Y.; Negrão, S.; Tester, M. Salt stress under the scalp–dissecting the genetics of salt tolerance. *Plant J.* **2019**, *97*, 148–163. [CrossRef] [PubMed]
57. Cheeseman, J.M. The evolution of halophytes, glycophytes and crops, and its implications for food security under saline conditions. *New Phytol.* **2015**, *206*, 557–570. [CrossRef]
58. Zörb, C.; Geilfus, C.-M.; Dietz, K.-J. Salinity and crop yield. *Plant Biol.* **2019**, *21*, 31–38. [CrossRef]
59. Singh, M.; Nara, U.; Kumar, A.; Choudhary, A.; Singh, H.; Thapa, S. Salinity tolerance mechanisms and their breeding implications. *J. Genet. Eng. Biotechnol.* **2021**, *19*, 173. [CrossRef]
60. Dong, Y.; Chen, R.; Petropoulos, E.; Yu, B.; Zhang, J.; Lin, X.; Feng, Y. Interactive effects of salinity and SOM on the ecoenzymatic activities across coastal soils subjected to a saline gradient. *Geoderma* **2022**, *406*, 115519. [CrossRef]
61. Akramkhanov, A.; Martius, C.; Jin Park, S.; Hendrickx, J.M.H. Environmental factors of spatial distribution of soil salinity on flat irrigated terrain. *Geoderma* **2011**, *163*, 55–62. [CrossRef]

62. Zheng, C.; Jiang, D.; Liu, F.; Dai, T.; Liu, W.; Jing, Q.; Cao, W. Exogenous nitric oxide improves seed germination in wheat against mitochondrial oxidative damage induced by high salinity. *Environ. Exp. Bot.* **2009**, *67*, 222227. [CrossRef]
63. Schroter, M.; Stumpf, K.H.; Loos, J.; van Oudenhoven, A.P.E.; Böhnke-Henrichs, A.; Abson, D.J. Refocusing ecosystem services towards sustainability. *Ecosyst. Serv.* **2017**, *25*, 35–43. [CrossRef]
64. Costanza, R.; De Groot, R.; Sutton, P.; Van der Ploeg, S.; Anderson, S.J.; Kubiszewski, I.; Farber, S.; Turner, R.K. Changes in the global value of ecosystem services. Global Environmental Change. *Hum. Policy Dimens.* **2014**, *26*, 152–158. [CrossRef]
65. Kingsford, R.T.; Basset, A.; Jackson, L. Wetlands: Conservation's poverty cousins. *Aquat. Conserv. Mar. Freshw. Echoes.* **2016**, *26*, 892–916. [CrossRef]
66. Chaplin-kramer, R.; Sharp, R.P.; Weil, C.; Bennett, E.M.; Pascual, U.; Arkema, K.K.; Brauman, K.A.; Bryant, B.P.; Guerry, A.D.; Haddad, N.M.; et al. Global modeling of nature's contributions to people. *Science* **2019**, *336*, 255–258. [CrossRef] [PubMed]
67. Mengist, W.; Soromessa, T.; Feyisa, G.L. A global view of regulatory ecosystem services: Existed knowledge, trends, and research gaps. *Ecol. Process.* **2020**, *9*, 461. [CrossRef]
68. Hong, Z.D.; Ding, S.Y.; Zhao, Q.H.; Qiu, P.W.; Chang, J.L.; Peng, L.; Wang, S.Q.; Hong, Y.Y.; Liu, G.J. Plant trait-environment trends and their conservation implications for riparian wetlands in the Yellow River. *Sci. Total Environ.* **2021**, *767*, 144867. [CrossRef]
69. Chang, L.; Zhao, Z.B.; Jiang, L.X.; Li, Y.F. Quantifying the Ecosystem Services of Soda Saline-Alkali Grasslands in Western Jilin Province, NE China. *Int. J. Environ. Res. Public Health* **2022**, *19*, 4760. [CrossRef]
70. Yuan, M.; Huang, L.L.; Chen, J.H.; Wu, J.; Xu, Q. The emerging treatment landscape of targeted therapy in non-small-cell lung cancer. *Signal Transduct. Target. Ther.* **2019**, *4*, 61. [CrossRef]
71. Rahman, M.M.; Shahrivar, A.A.; Hagare, D.; Maheshwari, B. Impact of Recycled Water Irrigation on Soil Salinity and Its Remediation. *Soil Syst.* **2022**, *6*, 13. [CrossRef]
72. Cassaniti, C.; Romano, D.; Flowers, T.J. The response of ornamental plants to saline irrigation water. *Water Manag. Pollut. Altern. Strateg.* **2012**, *131*, 158.
73. Alvarez, S.; Sanchez-Blanco, M.J. Comparison of individual and combined effects of salinity and deficit irrigation on physiological, nutritional and ornamental aspects of tolerance in Callistemon laevis plants. *J. Plant Physiol.* **2015**, *185*, 65–74. [CrossRef] [PubMed]
74. Koyro, H.-W.; Lieth, H.; Gul, B.; Ansari, R.; Huchzermeyer, B.; Abideen, Z.; Hussain, T.; Kahn, M. Importance of the Diversity within the Halophytes to Agriculture and Land Management in Arid and Semiarid Countries. In *Sabkha Ecosystems 4: Cash Crop Halophyte and Biodiversity Conservation*; Khan, M.A., Böer, B., Münir Öztürk, M., Abdessalaam, T.Z., Clüsener-Godt, M., Gul, B., Eds.; Springer: Berlin/Heidelberg, Germany, 2014; pp. 175–198.
75. Sharma, R.; Wungrampha, S.; Singh, V.; Pareek, A.; Sharma, M.K. Halophytes as bioenergy crops. *Front. Plant Sci.* **2016**, *7*, 1372. [CrossRef] [PubMed]
76. Duarte, B.; Caçador, I. Iberian Halophytes as Agroecological Solutions for Degraded Lands and Biosaline Agriculture. *Sustainability* **2021**, *13*, 1005. [CrossRef]
77. Giordano, R.; Saii, Z.; Fredsgaard, M.; Hulkko, L.S.S.; Poulsen, T.B.G.; Thomsen, M.E.; Henneberg, N.; Zucolotto, S.M.; Arendt-Nielsen, L.; Papenbrock, J.; et al. Pharmacological Insights into Halophyte Bioactive Extract Action on Anti-Inflammatory, Pain Relief and Antibiotics-Type Mechanisms. *Molecules* **2021**, *26*, 3140. [CrossRef] [PubMed]
78. Ferreira, M.J.; Pinto, D.C.; Cunha, A.; Silva, H. Halophytes as medicinal plants against human infectious diseases. *Appl. Sci.* **2022**, *12*, 7493. [CrossRef]
79. Ksouri, R.; Ksouri, W.M.; Jallali, I.; Debez, A.; Magne, C.; Hiroko, I.; Abdelly, C. Medicinal halophytes: Potent source of health promoting biomolecules with medical, nutraceutical and food applications. *Crit. Rev. Biotechnol.* **2012**, *32*, 289–326. [CrossRef] [PubMed]
80. Chauhan, M.; Gopal, B. Saline Wetlands of the Arid Zone of Western India. In *The Wetland Book II*; Finlayson, C.M., Milton, G.R., Prentice, R.G., Davidson, N.C., Eds.; Springer: Dordrecht, The Netherlands, 2018. [CrossRef]
81. Colla, G.; Rouphael, Y.; Fallovo, C.; Cardarelli, M.; Graifenberg, A. Use of Salsola soda as a companion plant to improve greenhouse pepper (*Capsicum annuum*) performance under saline conditions. *N. Z. J. Crop Hortic. Sci.* **2006**, *34*, 283–290. [CrossRef]
82. Milic, D.; Lukovic, J.; Ninkov, J.; Zeremski-Skoric, T.; Zoric, L.; Vasin, J.; Milic, S. Heavy metal content in halophytic plants from inland and maritime saline areas. *Cent. Eur. J. Biol.* **2012**, *7*, 307–317. [CrossRef]
83. Lorestani, B.; Cheraghi, M.; Yousefi, N. Accumulation of Pb, Fe, Mn, Cu and Zn in plants and choice of hyperaccumulator plant in the industrial town of Vian, Iran. *Arch. Biol. Sci.* **2011**, *63*, 739–745. [CrossRef]
84. Abdellaoui, R.; Elkelish, A.; El-Keblawy, A.; Mighri, H.; Boughalleb, F.; Bakhshandeh, E. Editorial: Halophytes: Salt stress tolerance mechanisms and potential use. *Front. Plant Sci.* **2023**, *14*, 1218184. [CrossRef] [PubMed]
85. Nedjimi, B.; Daoud, Y.; Carvajal, M.; Martínez-Ballesta, M.C. Improvement of the adaptation of *Lygeum spartum* L. to salinity under the presence of calcium. *Commun. Soil Sci. Plant Anal.* **2010**, *41*, 2301–2317. [CrossRef]
86. García-Fuentes, A.; Salazar, C.; Torres, J.A.; Cano, E.; Valle, F. Review of communities of *Lygeum spartum* L. in the southeastern Iberian Peninsula (western Mediterranean). *J. Arid Environ.* **2001**, *48*, 323–339. [CrossRef]
87. Khan, M.A.; Ungar, I.A. Influence of salinity and temperature on the germination of Haloxylon recurvum Bunge ex. Boiss. *Ann. Bot.* **1996**, *78*, 547–551. [CrossRef]
88. Rogel, J.A.; Silla, R.O.; Ariza, F.A. Edaphic characterization and soil ionic composition influencing plant zonation in a semiarid Mediterranean salt marsh. *Geoderma* **2001**, *99*, 81–98. [CrossRef]

89. Meot-Duros, L.; Magné, C. Effect of salinity and chemical factors on seed germination in the halophyte *Crithmum maritimum* L. *Plant Soil* **2008**, *313*, 83–87. [CrossRef]
90. Nedjimi, B. Effect of salinity and temperature on germination of *Lygeum spartum* L. *Agric. Res.* **2013**, *2*, 340–345. [CrossRef]
91. Nedjimi, B. *Lygeum spartum* L.: A review of a candidate for West Mediterranean arid rangeland rehabilitation. *Rangel. J.* **2016**, *38*, 493–499. [CrossRef]

land

Article

# Distribution and Variation of Soil Water and Salt before and after Autumn Irrigation

Yin Zhang [1,2], Qingfeng Miao [1], Ruiping Li [1,*], Minghai Sun [2], Xinmin Yang [3], Wei Wang [3], Yongping Huang [4] and Weiying Feng [5,*]

1 College of Water Conservancy and Civil Engineering, Inner Mongolia Agricultural University, Hohhot 010018, China; zhangyin1987@emails.imau.edu.cn (Y.Z.); imaumqf@imau.edu.cn (Q.M.)
2 Department of Water Conservancy and Civil Engineering, Hetao College, Bayannur 015000, China; jxc@emails.imau.edu.cn
3 Urat Sub-Center, Inner Mongolia Hetao Irrigation District Water Conservancy Development Centre, Bayannur 014400, China; nmgwfq@emails.imau.edu.cn (X.Y.); sl@emails.imau.edu.cn (W.W.)
4 Water Conservancy Research Institute of Bayannur City, Bayannur 015000, China; nmgwys@emails.imau.edu.cn
5 School of Materials Science and Engineering, Beihang University, Beijing 100191, China
* Correspondence: nmglrp@imau.edu.cn (R.L.); fengweiying@buaa.edu.cn (W.F.)

**Abstract:** Autumn irrigation is a key measure for alleviating soil salinity and promoting sustainable agricultural development in the Hetao Irrigation district; however, only a part of farmland is irrigated in autumn during the non-growth period of crops, which leads to the redistribution of soil water and salt between autumn-irrigated land (AIL) and adjacent non-autumn-irrigated land (NAIL) after autumn irrigation. To explore the distribution and variation of soil water and salt in different positions of AIL and NAIL after local autumn irrigation and reveal the interaction range between AIL and NAIL, field experiments were carried out for two years in typical test areas. The results showed that compared with non-autumn irrigation, autumn irrigation improved the distribution uniformity of soil water and salt profiles in both horizontal and vertical directions; after autumn irrigation, the water content of the soil at the nearest sampling point to the boundary in the AIL increased the least, but the desalination rate was the greatest, while the water and salt contents of the soil within 45 m from the sampling points to the boundary in the NAIL both increased significantly. NAIL received the drainage of AIL and made the groundwater level after the rise in AIL fell quickly back, but unreasonable autumn irrigation caused the groundwater level of AIL to remain at a high level before freezing, exacerbating the risk of groundwater carrying salts to the surface soil during the freezing and thawing period, detrimental to the growth of crops in the next spring. The research results are of great significance to the rational use of farmland water resources and the improvement of soil salinization in cold and dry areas.

**Keywords:** Hetao Irrigation District; autumn irrigation; dry drainage; water and salt movement

**Citation:** Zhang, Y.; Miao, Q.; Li, R.; Sun, M.; Yang, X.; Wang, W.; Huang, Y.; Feng, W. Distribution and Variation of Soil Water and Salt before and after Autumn Irrigation. *Land* **2024**, *13*, 773. https://doi.org/10.3390/land13060773

Academic Editors: Claude Hammecker, Tiago Brito Ramos, Maria da Conceição Gonçalves and Mohammad Farzamian

Received: 23 March 2024
Revised: 28 April 2024
Accepted: 27 May 2024
Published: 30 May 2024



## 1. Introduction

Irrigation is a major use of water in agriculture. Thus far, about 70% of freshwater on Earth is used for agriculture, of which about 90% is used for irrigation [1,2]. In arid and semi-arid regions with insufficient precipitation, irrigation is essential to agricultural productivity. It can not only meet the water demand of crops, increase food production, and provide possibilities for regional and even global food security but also increase farmers' income, improve agricultural profitability, and promote economic prosperity [3–5]. However, the use of saline irrigation water and chemical fertilizers increases soil salinity, and, combined with improper irrigation water and drainage practices, this usually results in increased soil salinization, threatening crop growth and reducing agricultural productivity [6,7]. According to statistics, more than 20% of irrigated land worldwide is impacted by soil salinization, and the total loss caused by irrigation-related salinity is USD 27.3 billion per year. If left unattended, by 2025, the global irrigation area threatened by soil salinization may expand to more than 50% [8–10]. Therefore, a correct and profound understanding of the problem of soil salinization brought on by irrigation is essential for achieving sustainable agricultural development.

The Hetao Irrigation District (HID) is one of the largest irrigation districts in China, with an irrigated area of 570,000 ha, and it is an important grain and oil production base in the country. Surface water flood irrigation is the common irrigation method for the region, but, due to its location in arid and semi-arid regions, extremely high evaporation precipitation ratios (about 10:1), shallow water table depths (the average annual groundwater level is 1.5–2 m), long-term irrigation from the Yellow River (the average amount of annual total dissolved salts in irrigation water is 0.5 g/L), and imperfect drainage systems result in the severe salinization of soil in the root zone [11]. In order to reduce the harm of soil salinization in the root zone to the growth of crops in the following year and adjust the soil moisture in the field, every year, after harvesting the crops in the autumn and before the soil freezes, the irrigation district uses water flood irrigation via the Yellow River to leach out the soil salinity in the root zone. After a long period of production practice, autumn irrigation has become an important local irrigation method [12,13]. However, due to the wide range of autumn irrigation methods, irrigation period concentration (October–November each year), and the large irrigation quota (approximately one-third of annual water consumption), there has been a sharp increase in groundwater levels after irrigation. In the case of poor drainage conditions, groundwater levels easily remain high for a long time before freezing, which, in turn, leads to the secondary salinization of soil and affects the spring sowing of crops in the following year [14–16].

Dry drainage (also known as internal drainage) is a method that can alleviate secondary soil salinization in irrigated farmland. This method mainly discharges excess water and salt from irrigated land to nearby fallow land (fallow land is land that has been permanently or seasonally fallowed) through groundwater flow and then consumes water in the soil under evaporation, while salt is stored in fallow land [17]. Compared with traditional artificial drainage, this method has more advantages in cost and environmental protection [18]. In the past 30 years, dry drainage has been widely used in Pakistan, Iran, and China [19–21]. Wu et al. [22] studied the effectiveness of dry drainage using a combination of remote sensing, a conceptual model, and field experiments in HID and observed that excess water and salt in the irrigated land migrating to the fallow land through groundwater and dry drainage was effective in controlling salt levels in the irrigated land. However, the effect of dry drainage is easily affected by climatic conditions, the ratio of irrigated land to fallow land, and the evaporation capacity of the land. To further prove the effectiveness of dry drainage [23,24], Wang et al. [25] conducted field observations for five years in a 2900 ha experimental field at Yonglian Experimental Station in HID and found that fallow land received excess water and salt from the surrounding irrigated cropland. Moreover, the salt content of the soil profile increased significantly, and the salt accumulation of irrigated cropland exhibited an accelerating trend with the weakening of the evaporative capacity of fallow land. Liu et al. [26] investigated the water–salt migration between cropland and adjacent wasteland during the growing season in HID, and the results showed that during the irrigation period, irrigation (and precipitation) promoted the flow of water and salt from cropland to wasteland. However, during the intervals of irrigation, when the evapotranspiration of farmland was greater than that of wasteland, the lateral water and salt flux was reversed. Although the above scholars revealed water and salt movement between irrigated and fallow lands at different scales, these studies were conducted under fully irrigated conditions (full irrigation means that all cropped lands are irrigated).

In recent years, due to the national directive on water conservation, the average annual water diversion of HID has been reduced from 5.2 billion $m^3$ to 4.0 billion $m^3$, based on which, the irrigation district has adopted the measure of "partial autumn irrigation", which mainly means that farmlands planted with late-season crops, such as sunflower, are not irrigated in autumn seasons, but are irrigated in spring after the soil melts the following year [27,28]. Simultaneously, influenced by the planting structure and farmers' willingness, the area of non-autumn-irrigated land (NAIL) is increasing, while the area of autumn-irrigated land (AIL) is relatively decreasing. AIL and NAIL are adjacent to each other and distributed in an interleaved manner [29]. To a certain extent, the working principle of NAIL is similar to that of fallow land under full irrigation. After irrigation, NAIL can receive drainage from surrounding irrigated land, causing the water table in the nearby irrigated land to fall back quickly, but, because the NAIL will be irrigated in the spring during the following year, drainage water durations are much shorter compared to fallow land. Peng et al. [30] studied the characteristics of water–salt movement in farmland after local autumn irrigation conditions in HID and found that the salts that were washed out of the AIL entered the NAIL through the flow of groundwater, increasing soil salt content in the NAIL. Although the study revealed that water–salt movement between AIL and NAIL was influenced by the proportion of AIL in the whole region, the scope of the interaction of water–salt movement between AIL and NAIL was not clear. Therefore, the objective of this study was as follows: to monitor changes in soil moisture, salinity, and groundwater in different positions of AIL and NAIL before and after autumn irrigation; quantitatively analyze the redistribution of water and salinity in different locations of AIL and NAIL and the distance of interaction; and reveal the scope of mutual influence between AIL and NAIL to provide scientific information for the better management of irrigation water and the control of soil salinization in the future.

In response to the current situation of autumn irrigation in HID, we conducted field monitoring of water and salt distributions at different distances between AIL and nearby ANIL. Our objectives were to (1) reveal the distribution of and changes in soil water and salt at different distances between AIL and NAIL before and after autumn irrigation, (2) find the distances at which AIL and NAIL interacted with each other under dry drainage conditions, and (3) provide a theoretical basis for rational autumn irrigation and soil salinization prevention.

## 2. Materials and Methods

### 2.1. Study Area Description

The study area (40°55′04″ N, 108°30′29″ E, 985.9 m) is located in Xixiaozhao Town, UradQianqi, Bayannur City, Inner Mongolia Autonomous Region, China, and belongs to the Wulate Irrigation District of HID (Figure 1a,b). Its climate is classified as follows: mid-temperate continental climate, dry and windy weather, sufficient sunshine, less precipitation, substantial evaporation, and a short frost-free period. The annual mean temperature of the research region is 6~8 °C; the average wind speed is 2.8 m/s; the average sunshine hours is 3202 h; the annual precipitation is 200~250 mm, with the majority falling between June and September (accounting for 79% of the annual precipitation); the annual pan evaporation (E20) is 2173 mm [31]; the frost-free period is about 130 d; the annual average groundwater depth is 1.8 m; and the maximum freezing depth is 1 m [32]. The soil texture of the 0~200 cm soil stratum is mainly sandy loam, silty loam, and loamy sand. The specific physical properties are shown in Table 1. The soil bulk density was determined via the ring knife method. Soil texture was determined using a HELOS & RODOS fully automated dry particle size analyzer (Sympatec GmbH, Dresden, Germany) to measure soil particle gradation, in accordance with the soil texture triangle map of the United States Department of Agriculture [33].

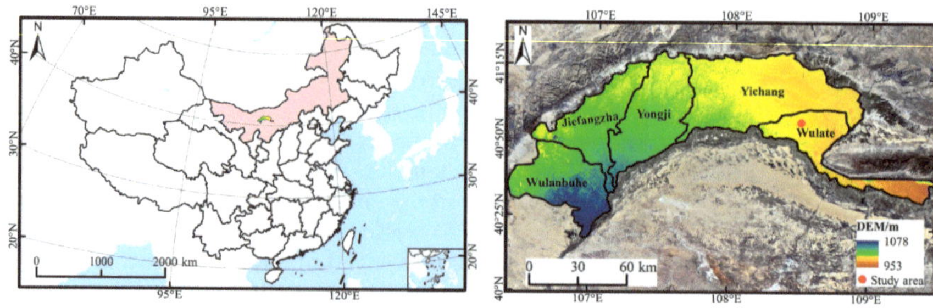

(a) (b)

**Figure 1.** The location of Hetao Irrigation District in Inner Mongolia, China, (**a**) and the location of the study area (**b**).

**Table 1.** Soil physical properties of the study area from 0 to 200 cm.

| Land Types | Soil Depth (cm) | Particle Distribution (%) | | | Soil Bulk Density (g cm$^{-3}$) | Soil Texture |
|---|---|---|---|---|---|---|
| | | Clay (<0.002 mm) | Silt (0.002–0.05 mm) | Sand (0.05–2 mm) | | |
| AIL | 0–5 | 2.29 | 43.18 | 54.53 | 1.51 | Sandy loam |
| | 5–20 | 2.45 | 43.30 | 54.25 | 1.55 | Sandy loam |
| | 20–40 | 2.25 | 55.62 | 42.13 | 1.57 | Silty loam |
| | 40–60 | 1.11 | 55.66 | 43.23 | 1.52 | Silty loam |
| | 60–80 | 0.79 | 54.67 | 44.54 | 1.53 | Silty loam |
| | 80–100 | 0.21 | 27.22 | 72.57 | 1.61 | Loamy sand |
| | 100–150 | 0.21 | 21.34 | 78.45 | 1.62 | Loamy sand |
| | 150–200 | 0.36 | 10.38 | 89.26 | 1.62 | Loamy sand |
| NAIL | 0–5 | 2.68 | 40.11 | 57.21 | 1.50 | Sandy loam |
| | 5–20 | 2.26 | 30.22 | 67.52 | 1.55 | Sandy loam |
| | 20–40 | 3.75 | 60.60 | 35.64 | 1.60 | Silty loam |
| | 40–60 | 4.90 | 54.73 | 40.37 | 1.54 | Silty loam |
| | 60–80 | 1.89 | 63.22 | 34.90 | 1.51 | Silty loam |
| | 80–100 | 0.85 | 52.67 | 46.47 | 1.55 | Silty loam |
| | 100–150 | 0.25 | 21.31 | 78.44 | 1.62 | Loamy sand |
| | 150–200 | 0.27 | 20.55 | 79.18 | 1.63 | Loamy sand |

Notes: AIL is autumn-irrigated land and NAIL is non-autumn-irrigated land.

## 2.2. Experimental Design

The fields adjacent to the AIL and NAIL were selected as the test area and tested from October to December 2021 and October to December 2022. The east side of the test area belongs to Huaimu Village, which comprises AIL, and the west side belongs to Beigedu Village, which comprises NAIL. AIL and NAIL are separated by the agricultural canal and the field road (Figure 2). The average ground elevation of the test area was 985.89 m. The ground elevation of the AIL was slightly higher than that of the NAIL, and the maximum elevation difference was 5 cm. The shape of the test area was rectangular, measuring 378.6 m long from the east to the west and 30 m wide from the north to the south. The test area covered about 11,360 m$^2$, with a 50% share of AIL and a 50% share of NAIL. With the exception of the interface between AIL and NAIL, the other three sides of the AIL were irrigated land, while the other three sides of the NAIL were non-irrigated land. Before the autumn harvest, the main crop planted in the experimental field was sunflower, and, after the autumn harvest, the experimental field was not turned over and leveled. Six soil water and salt sampling points (parallel to the boundary and 5, 20, 45, 92, 139, and 186 m from the boundary) were set up in AIL and NAIL, respectively, and they were named with respect to their distance from the boundary. The six sampling points in the AIL were named Q5, Q20, Q45, Q92, Q139, and Q186 and each sampling point in NAIL was named W5, W20,

W45, W92, W139, and W186. There were two replicates perpendicular to the boundary, with a total of 24 sampling points. A total of five groundwater level observation wells were laid out in the test area. Two observation wells, Q1 and Q2, were laid out from the west to the east in the AIL and three groundwater observation wells, W1, W2, and W3, were laid out from the east to the west in the NAIL. The canals on the east side of wells W1 and W3 are both diversion canals for NAIL. During autumn irrigation, the diversion openings of these canals are closed, and no water passes through the canals.

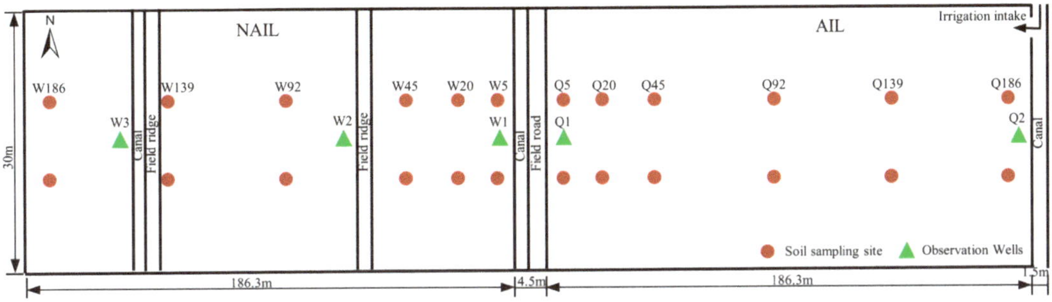


**Figure 2.** Schematic diagram of the test site's layout (AIL is autumn-irrigated land and NAIL is non-autumn-irrigated land).

### 2.3. Data Collection and Measurement

The test area was irrigated with water from the Yellow River on 18–19 October 2021 and 16–18 November 2022, and the irrigation method was flood irrigation. The irrigation water volume was determined using a trapezoidal water-measuring weir, which was 4600 $m^3$/ha and 4200 $m^3$/ha in 2021 and 2022, respectively, and the conductivity of the irrigation water was determined using a DDS-307A conductivity meter (Hangzhou Qiwei Instrument Co., Ltd., Hangzhou, China), which was 0.73 dS/m and 0.75 dS/m in the two years, respectively. Before and after autumn irrigation, soil samples were gathered at the set sampling points on 13 October and 25 November 2021 and 27 October and 15 December 2022, respectively. The soil sampling depth was 200 cm. There was a total of 8 layers: 2, 10, 30, 50, 70, 100, 150, and 200 cm. The soil samples were collected using soil drills, and soil water content (SWC) and salinity were measured. The SWC was obtained via the drying method, while soil salinity was described by means of soil electrical conductivity (SEC) [21]. The soil samples were dried naturally and then fully ground; then, they were passed through a 1 mm sieve to make a leaching solution with a soil–water ratio of 1:5, and conductivity (conductivity at 25 °C) was determined using a DDS-307A conductivity meter. Groundwater levels in all observation wells were automatically collected using a TD-Diver (Model DI801) groundwater level monitor manufactured (Chengdu Yaohua Technology Co., Ltd., Chengdu, China), and it was set to be collected at 1 h intervals. The meteorological data of the test period were gathered from the China Meteorological Data Network (http://data.cma.cn/ (accessed on 10 August 2023)), and these data were used to compute the daily reference evapotranspiration (ETo) using the FAO-56 Penman–Monteith equation [34]. The daily values of the temperature, reference evapotranspiration, and precipitation during the two-year test period are shown in Figure 3. In 2021 and 2022, the daily mean temperature continued to be lower than 0 °C from 29 November and 28 November, respectively, and the soil entered the freezing period. During the two-year sampling period, the total precipitation was 6.4 mm and 9.4 mm and the total reference evapotranspiration was 66 mm and 59 mm, respectively.

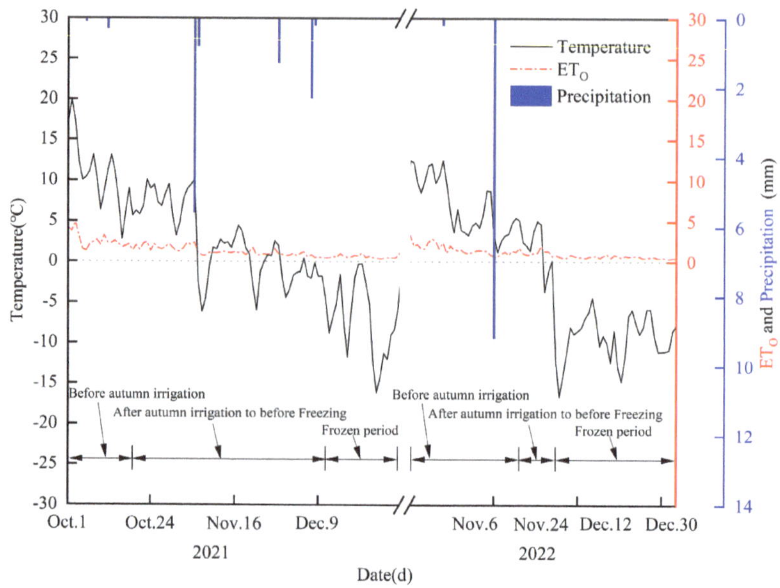


**Figure 3.** Daily values of temperature, reference evapotranspiration ($ET_O$), and precipitation during the 2021 and 2022 experimental periods.

### 2.4. Research Methods

#### 2.4.1. Changes in Soil Water Content

The variation in SWC was defined as the percentage of SWC variation after autumn irrigation compared to before autumn irrigation. The calculation formula is as follows:

$$\Delta\theta = (\theta_2 - \theta_1) \times 100\% / \theta_1 \quad (1)$$

where $\Delta\theta$ is the variation rate of SWC (%), $\theta_2$ is the SWC after autumn irrigation (%), and $\theta_1$ is the SWC before autumn irrigation (%). If $\Delta\theta > 0$, this indicates an increase (%). If $\Delta\theta < 0$, this indicates a decrease (%).

#### 2.4.2. Soil Desalination Rate

The soil desalination rate was defined as the percentage of soil salinity reduction after autumn irrigation compared to before autumn irrigation. The calculation formula is as follows:

$$\Delta SEC = (SEC_1 - SEC_2) \times 100\% / SEC_1 \quad (2)$$

where $\Delta SEC$ is the desalination rate (%), $SEC_1$ is the soil salinity before autumn irrigation (dS/m), and $SEC_2$ is the soil salinity after autumn irrigation (dS/m). If $\Delta SEC > 0$, desalination occurred (%). If $\Delta SEC < 0$, salt accumulation occurred (%).

#### 2.4.3. Estimation of Groundwater Table Depths at Soil Sampling Sites

Groundwater table depth was estimated at each sampling point in the NAIL according to Darcy's law:

$$J = \Delta H / L \quad (3)$$

$$Z_s = H_{sg} - H_{sw} = H_{sg} - (H_i - JL_{i-s}) \quad (4)$$

where $J$ is the hydraulic gradient between observation wells, $\Delta H$ is the groundwater level elevation difference between observation wells (m), and $L$ is the horizontal distance between observation wells (m). $Z_s$ is the groundwater table at each soil sampling point (s = 5, 20,

45, 92, 139, 186) in the NAIL (m), $H_{sg}$ is the ground elevation of each soil sampling point (m), $H_{sw}$ is the water table elevation of each soil sampling point (m), $H_i$ is the groundwater level elevation at the observation wells closer to the boundary ($i = w_1, w_2$) (m), and $L_{i-s}$ is the horizontal distance between the observation wells closer to the boundary and the soil sampling point (m).

### 2.5. Data Processing Methods

The Kriging interpolation method in Surfer15.0 software was used to plot the spatial and temporal distribution of soil water and salt and Origin2022 software was used to draw the change in soil water and salt at different depths. SPSS19.0 software was applied to test the differentiation of the data (one-way ANOVA).

## 3. Results

### 3.1. Distribution Characteristics of Water–Salt in the Soil Profiles

#### 3.1.1. Distribution Characteristics of Water Content in the Soil Profiles

The distribution of SWC was similar for two years in AIL and NAIL before and after autumn irrigation, as shown in Figure 4. Before autumn irrigation, the SWC of AIL and NAIL exhibited a Z-shaped (Figure 4a,c) and an inverted C-shaped distribution (Figure 4e,g) in the vertical direction, respectively. The maximum values appeared in the depth range of 50 ± 20 cm and 70 ± 30 cm and the minimum values appeared in the 100 cm soil layer and the surface layer (2 cm). The lowest value of SWC in AIL was not on the surface but in the 100 cm soil layer, which was a result of the relatively large amount of sand in the 100 cm soil layer (Table 1), resulting in the poor water retention capacity of the soil layer. The SWC of the 0~200 cm soil layer in AIL and NAIL changed between 9.7~20.3% and 14.3~22.6% in 2021 and between 16.3~25.7% and 16.3~28% in 2022, respectively. After autumn irrigation, the vertical distribution of SWC in AIL and NAIL was basically the same as that before autumn irrigation, and the SWC of all soil layers increased (except for the 2 cm soil layer with a horizontal distance of 92~186 m from the boundary in the 2021 NAIL). However, compared with that before autumn irrigation, for two years, the vertical maximum difference of SWC in AIL decreased by 13% and 37.4% (Figure 4b,d), while that of NAIL increased by 61.3% and 12.4%, respectively (Figure 4f,h). This shows that autumn irrigation improves the vertical distribution uniformity of SWC in AIL but reduces the vertical distribution uniformity of SWC in NAIL.

In the horizontal direction, before autumn irrigation, there was no significant difference in SWC at each distance sampling point in the AIL and NAIL ($p > 0.05$). The SWC varied between 15~18.3% (Figure 4a) and 20.7~23.2% (Figure 4c) for two years at each sampling site for the AIL and between 16.6~19.3% (Figure 4e) and 21.8~24.1% (Figure 4g) for NAIL. After autumn irrigation, the SWC for two years for AIL increased to 25.4 ± 0.2% (Figure 4b) and 27 ± 0.3% (Figure 4d), and the maximum difference was 84.2% and 67.6% lower than that before autumn irrigation. For NAIL, the SWC at 5, 20, and 45 m from the boundary was significantly higher than for the other three sampling points (there was no significant difference between the other three sampling points) ($p < 0.05$). The SWC at the three sampling sites within 45 m from the boundary in 2021 was higher than the mean values of the other three sampling sites by 14.6, 13.6, and 11.9% (Figure 4f), and it was 13.2%, 13.1%, and 11.7% higher than in 2022 (Figure 4h). The closer the sampling point to the boundary, the higher the SWC. Compared with before autumn irrigation, the maximum difference of SWC in the two years increased by 41.1% and 57.6%, respectively. This shows that autumn irrigation improved the horizontal distribution uniformity of SWC in AIL, but the horizontal distribution uniformity of SWC in NAIL was worse.

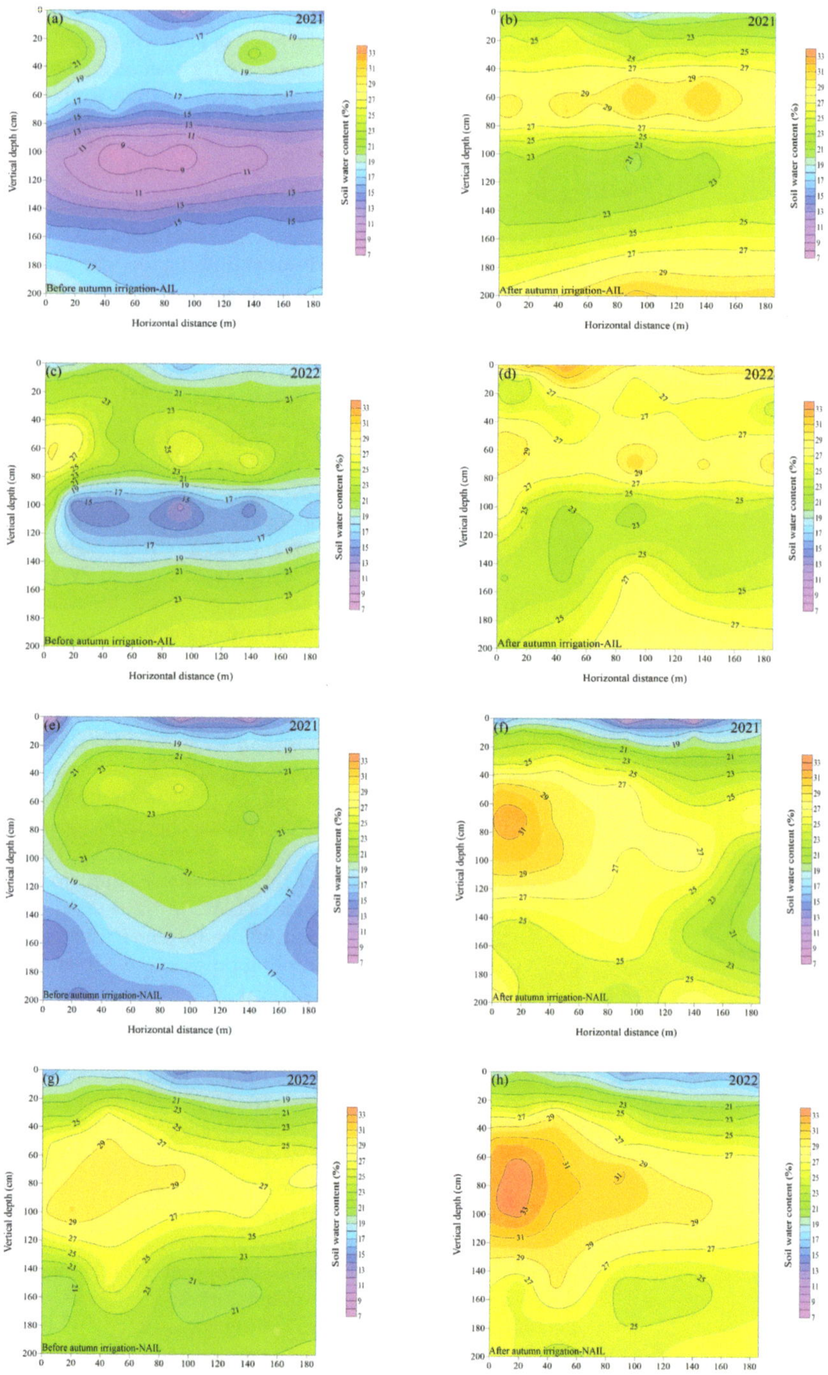

**Figure 4.** Distribution of soil water content in autumn-irrigated land (AIL, **a–d**) and non-autumn-irrigated land (NAIL, **e–h**) before and after autumn irrigation in 2021 and 2022.

3.1.2. Distribution Characterization of Salt in the Soil Profiles

The two-dimensional distribution of soil salinity in AIL and NAIL over two years was basically the same, as shown in Figure 5. Before autumn irrigation, due to evapotranspiration during the time of crop growth, soil salinity in the top layer (0~10 cm) of both AIL and NAIL was higher in the two years, ranging between 0.95~1.1 dS/m and 0.3~0.42 dS/m (Figure 5a,c) and between 0.68~0.86 dS/m and 0.41~0.54 dS/m (Figure 5e,g), respectively. With the increase in soil depth, soil salinity decreased significantly in the 10~50 cm soil layer and gradually decreased in the 50~200 cm soil layer, with minimum values of 0.25, 0.1 dS/m, and 0.29, 0.12 dS/m. The salts in the horizontal direction all accumulated near 0~5 m from the boundary, and the farther the distance, the fewer the salts. This may be a result of the accumulation of salt eluted from a distance away from the boundary with drainage water to the vicinity of the boundary after the last irrigation. After autumn irrigation, soil salts in the 0~50 cm soil layer exhibited a significant decrease (Figure 5b,d) and increase (Figure 5f,h) in AIL and NAIL, respectively, and the changes in soil salts in all other soil layers were small. For the entire profiles, the salinity of the AIL was close in the horizontal direction, varying between 0.24~0.33 dS/m and 0.11~0.16 dS/m in the two years, respectively, while the salinity of NAIL in the range of 0~45 m from the boundary reached the level of moderate salinity (0.4~0.8 dS/m), especially in the surface soil, which reached the level of severe salinity (0.8~1.6 dS/m). Compared with that before autumn irrigation, the difference in soil salinity in the horizontal and vertical directions of the AIL decreased after irrigation by 72.7% and 37.5% and by 67.1% and 71.9% in the two years, while the difference in soil salinity in the NAIL increased by 66.7% and 55.6% and by 28.1% and 43.9%, respectively. It can be observed that the distribution uniformity of soil salts in the horizontal and vertical directions of AIL improved after irrigation, while the distribution of soil salts in the horizontal and vertical directions of the NAIL worsened.

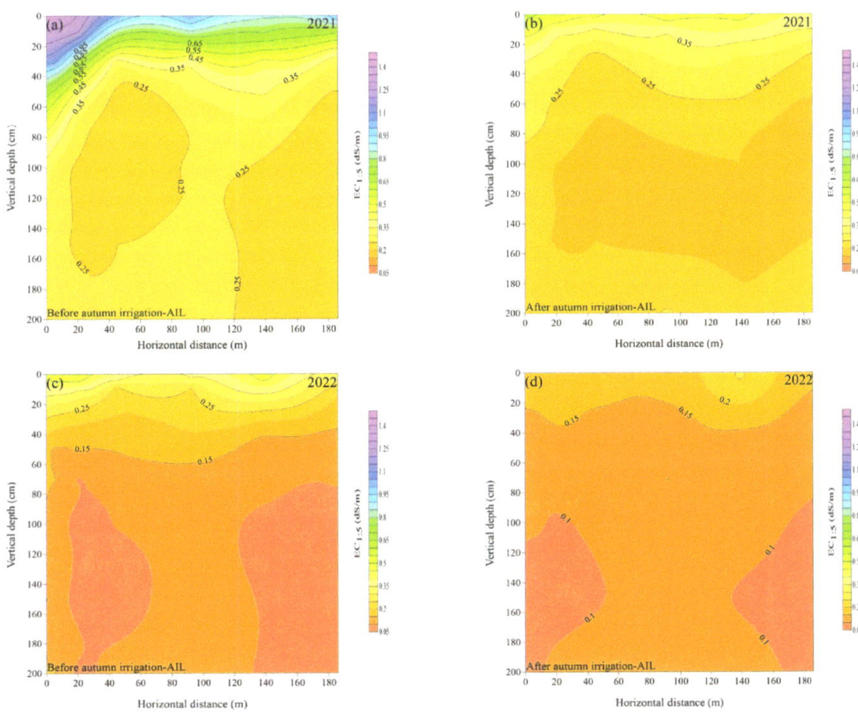


**Figure 5.** *Cont.*

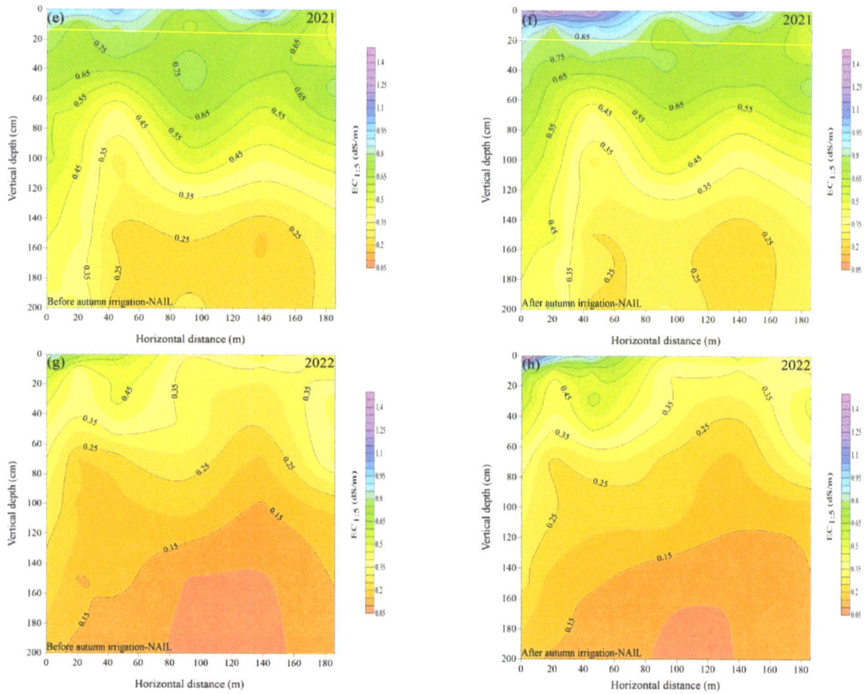

**Figure 5.** Distribution of soil salinity ($EC_{1:5}$) in autumn-irrigated land (AIL, **a–d**) and non-autumn-irrigated land (NAIL, **e–h**) in 2021 and 2022.

### *3.2. Changes in Soil Profile Water Salinity*

#### 3.2.1. Changes in Soil Profile Water Content

The changes in soil profile water content at each sampling point in AIL and NAIL before and after autumn irrigation are shown in Figure 6. After autumn irrigation, the average SWC in the root zone (0~50 cm) of the AIL increased by 33.1% and 29.2% in the two-year test period, which was 58.7% lower (Figure 6a) and 37.7% higher (Figure 6b) than that in the deeper layer (50~200 cm), respectively. In the second year, the increase in SWC in the root zone was greater than that in the deep layer. The main reason for this was that irrigation water did not fully infiltrate and froze in the top layer of the soil (0~10 cm). In the NAIL, the average SWC in the root zone (0~50 cm) increased by 17% (Figure 6c) and 9.9% (Figure 6d) in two years, respectively, and this was smaller than the changes in the deep SWC (42% and 14.8%), indicating that the water in the NAIL was more concentrated in the deeper soil layers after irrigation. In the horizontal direction, the largest increase in SWC among the AIL was in Q92, with increases of 67.7% and 30.2% in two years, and the smallest increase was in Q5, with increases of 38.5% and 15.5% in two years, respectively. The rest of the sampling sites changed in between. Under the influence of AIL irrigation, the SWC of each sampling point in NAIL increased by 18.3~54.5% and 8.3~19.7% in the two years, respectively, and the closer the sampling point to the boundary, the greater the increase in SWC.

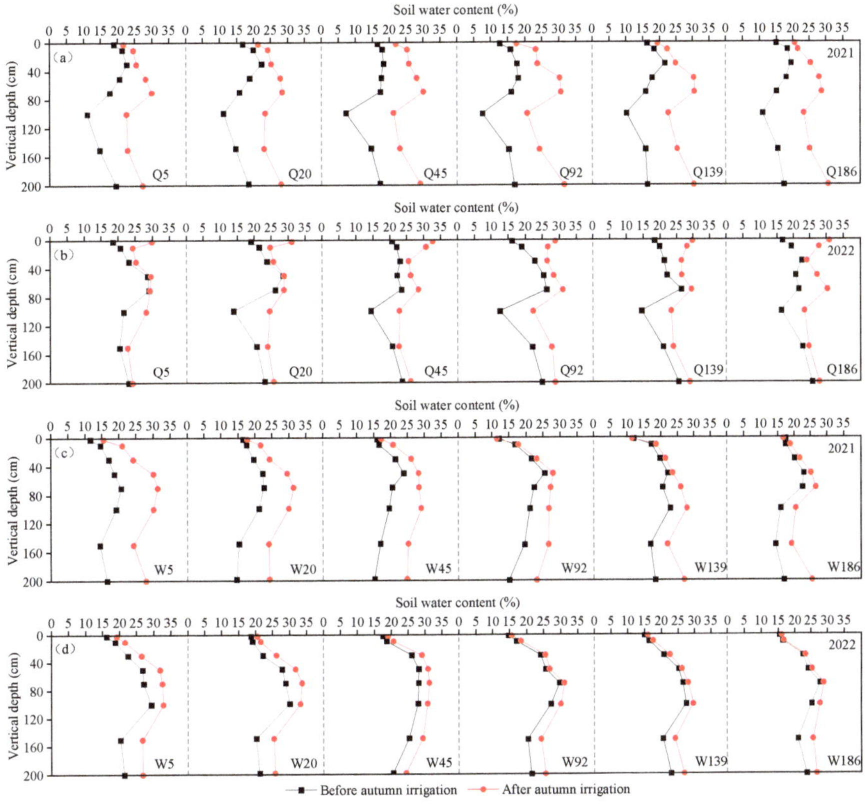


**Figure 6.** Changes in soil water content at each sampling point in autumn-irrigated land (**a**,**b**) and non-autumn-irrigated land (**c**,**d**) before and after autumn irrigation in 2021 and 2022.

### 3.2.2. Changes in Soil Profile Salinity

The changes in soil salinity in the 0~200 cm soil layer before and after autumn irrigation at each sampling point of AIL and NAIL are shown in Figure 7. As observed in Figure 7a,b, the degree of soil desalination at each sampling point in AIL decreased with an increase in vertical depth. The closer the sampling point to the boundary (Q5), the greater the desalination rate of the soil profile, which was 54.5% and 45.3% in the two years, respectively, while the desalination rates of the remaining sampling points were between 34.5~51.8% and 21.6~33.2%, respectively. Salts washed from the AIL entered the groundwater and migrated to the NAIL through lateral flow. Soil salts in the root zone (0~50 cm) of the NAIL accumulated due to evaporation. The salt accumulation rates in two years were 10.3% and 17.9%, respectively, which were about 2.9 and 2.8 times that of the deeper soil (Figure 7c,d). All sampling points were affected by fluctuations in the water table: the greater the depth of fluctuation, the greater the salt accumulation rates in the soil profile. For example, in 2021, the W5, W20, W45, W92, W139, and W186 sampling sites accumulated 14.8, 14, 11.7, 3.1, 3.1, and 1.2% salt after irrigation and, in 2022, they accumulated 25, 24.4, 23.9, 6.3, 3.1, and 1.7% salt, respectively. Further analyses revealed that the salt accumulation rate in the soil profile of the sampling points within 45 m from the boundary was significantly higher than that of the sampling points beyond 92 m ($p < 0.5$), which was about 5.5 and 6.6 times higher than the average salt accumulation rate of the sampling points beyond 92 m. Therefore, there was a range limit for the effect of autumn irrigation on soil salinity in the adjacent NAIL, which was between 45 and 92 m.

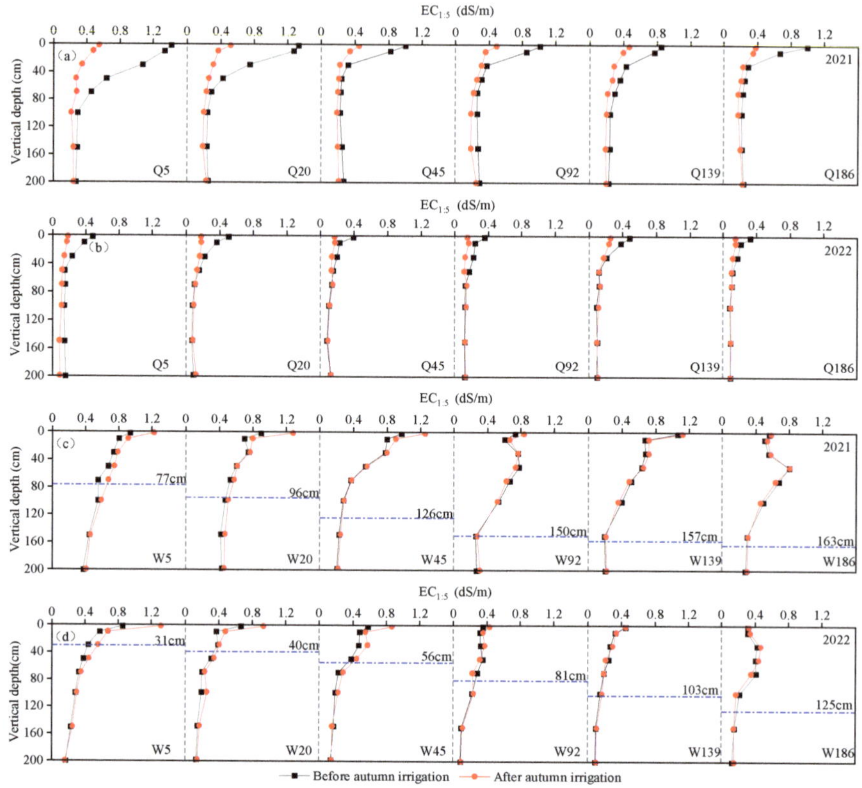

**Figure 7.** Changes in soil salinity ($EC_{1:5}$) at each sampling point in autumn-irrigated land (**a**,**b**) and non-autumn-irrigated land (**c**,**d**) before and after autumn irrigation in 2021 and 2022 (the blue dotted line is the highest influence depth line of groundwater levels after autumn irrigation).

## 3.3. *Changes in Groundwater Levels and Salinity*

### 3.3.1. Changes in Groundwater Levels

As illustrated in Figure 8, the change process of groundwater level in each observation well during the two-year test period was the same, exhibiting a first increasing and then decreasing change. Before autumn irrigation, the depths of the groundwater level in each observation well were substantial and similar to one another, and they varied between 2.72–2.81 m and 2.42–2.59 m in the two years. The hydraulic gradient between the observation wells in AIL and NAIL was small and groundwater flow was weak (Figure 9). This was mainly due to the study area not being irrigated since spring irrigation, long-term diving evaporation, and the adjustment to the groundwater's constant movement. After autumn irrigation, the infiltration of irrigation water resulted in a rapid increase in groundwater levels in AIL and reached a peak on the third day after irrigation, but, due to the obvious water level differences between AIL and NAIL, the larger hydraulic gradient promoted the groundwater's continuous flow from AIL to NAIL. The groundwater level of AIL decreased while the groundwater level of NAIL increased (Figure 9). On day 10, after irrigation, in 2021, the groundwater level of AIL and NAIL was between 0.01~0.42 m and 0.75~1.96 m, respectively. In contrast, in 2022, it was between 0.01~0.28 m and 0.43~1.17 m (before soil freezing), respectively. With time, the hydraulic gradient between the observation wells became smaller, and the groundwater level decreased synchronously. Until the 30th day after irrigation, the groundwater level varied between 1.52~1.6 m and 1.72~2 m (before soil freezing) and between 1.91~1.94 m and 2~2.08 m for the two years in the AIL and NAIL, respectively (Figure 8). The following could be observed: NAIL received the drainage of the AIL, which could cause the groundwater level after the rise of the AIL to quickly decrease, but earlier autumn irrigation was more capable of ensuring that the groundwater level of AIL fell below the designed critical depth of the HID before freezing (about 1.5) [35].

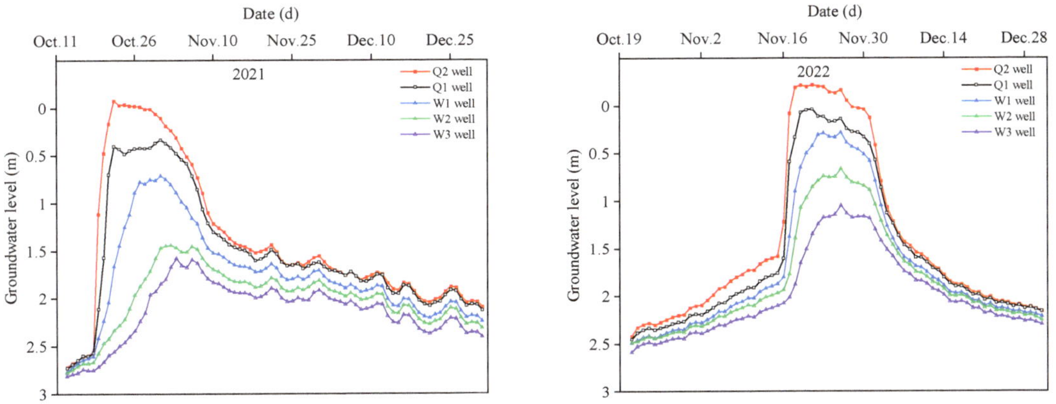


**Figure 8.** The change process of groundwater levels at each observation well in autumn-irrigated land and non-autumn-irrigated land before and after autumn irrigation in 2021 and 2022.

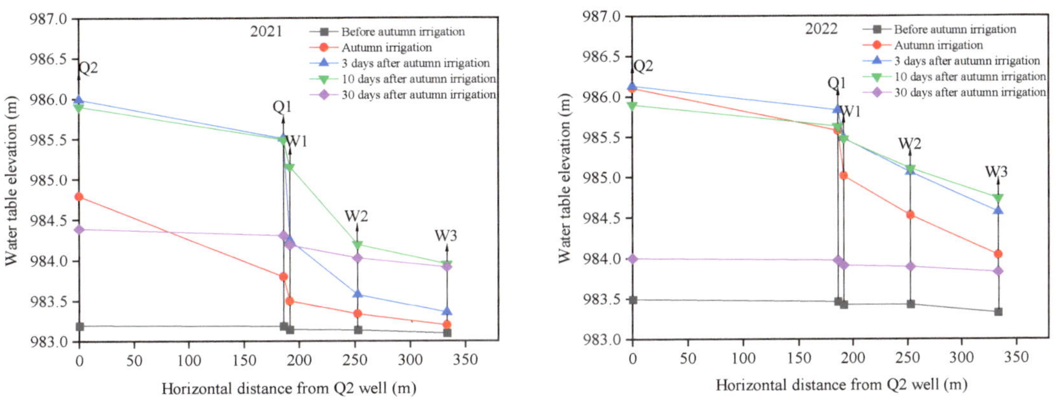


**Figure 9.** The change process of the hydraulic gradient between observation wells in autumn-irrigated land and non-autumn-irrigated land before and after autumn irrigation in 2021 and 2022.

### 3.3.2. Changes in Groundwater Salinity

As observed in Figure 10, the groundwater salinity of each observation well in the two years had the same change rule, increasing at the beginning of autumn irrigation and decreasing at the end of the autumn irrigation process. Before autumn irrigation, groundwater salinity in each observation well for two years was between 0.75~0.89 dS/m and 0.83~1.15 dS/m. After autumn irrigation, groundwater salinity increased in all observation wells, and the increase in groundwater salinity in NAIL was greater than that in AIL; this may have been caused by the leaching of salts from AIL into the groundwater and their flow toward NAIL through the horizontal flow and the further dissolution of salts in the soil during the rise of the groundwater.

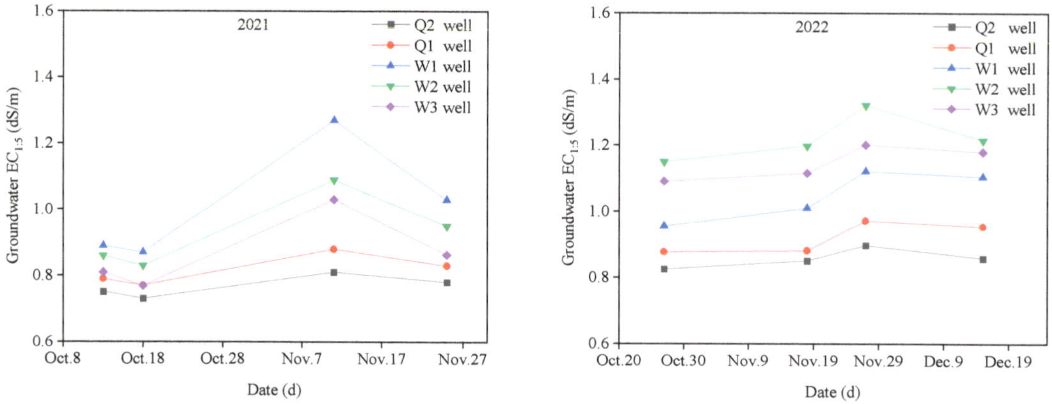


**Figure 10.** The change in groundwater salinity (EC1:5) at each observation well in autumn-irrigated land and non-autumn-irrigated land before and after autumn irrigation in 2021 and 2022.

## 4. Discussion

### *4.1. Effect of Autumn Irrigation on Soil Moisture*

In the study area, where the water table was shallow, soil moisture conditions in different horizontal areas were influenced by irrigation, precipitation, evapotranspiration, groundwater movement, and freeze–thaw cycles, with irrigation being the main driver affecting the distribution and change in soil moisture in AIL and NAIL [36,37]. In this study, after autumn irrigation, the infiltration of irrigation water resulted in smaller differences in soil moisture between horizontal and vertical directions in AIL, which concurs with the findings of Feng et al. [38]. The closer the sampling point to the boundary (Q5), the stronger the hydrodynamic conditions (Figure 9), the faster the water movement, and the greater the drainage (salt discharge). However, in NAIL, the continuous evaporation of topsoil and the increase in deeper soil water increased the difference in soil moisture in the vertical direction [30]. With the increase in the distance from the boundary, the water content of the soil profile gradually decreased. For example, the SWC of W5, W20, and W45 after irrigation was significantly higher than that of the other three positions (there was no significant difference between the other three positions) ($p < 0.05$), similar to the results of Yin et al. [39]. The autumn irrigation period in HID is short and is followed by a freezing period. The freezing effect prompts the water in the deep soil to carry salt upward, and the water moves and amasses within the frozen soil layer [14]. Following autumn irrigation, the deeper soils of NAIL received increased leaching water recharge, which increased the flow of water from deeper to shallower layers throughout the freezing period and aggravated the salinization process of surface soil when the soil melted the next year.

### *4.2. Effect of Autumn Irrigation on Soil Salinity*

Irrigation not only affects the distribution of water in soil but also changes the distribution of salt in soil, especially in cold-arid regions [40,41]. In addition, irrigation water may also introduce new salts to the soil [42]. In this study, the soil desalination rate of the soil layer above 50 cm was significantly higher than that of the soil layer below 50 cm in the AIL after irrigation, and the difference in soil salinity in the vertical direction was decreased; in contrast, the further accumulation of salts in the surface soil of NAIL increased the difference compared to that of the deeper soil layer. In terms of the entire soil profile, the AIL as a whole exhibited desalination, while the NAIL as a whole exhibited an accumulation of salts, which was consistent with the findings of Peng et al. [30]. Over the two-year experimental period, although the time of soil sampling was different, the precipitation and reference evapotranspiration caused by temporal difference was small (Figure 3), and the groundwater level in both years had already dropped to 1.7~2.0 m and

1.8~2.0 m by post-irrigation sampling (Figure 8). Therefore, different soil sampling times had little effect on the change of water–salt in the soil profile (0~200 cm). In the horizontal direction, there was no significant difference in soil profile salinity at different locations in the AIL after autumn irrigation ($p > 0.5$), but soil salinity was significantly greater in the proximal zone of NAIL (distance from the boundary of the NAIL was less than 45 m) than in the distal zone (distance from the boundary of NAIL was more than 92 m). This is a different conclusion compared to the study of Yuan et al. [43], which noted that the soil profile's salt in the near-distance zone of non-irrigated land after irrigation was less than that in the far-distance zone. The cause for this distinction was the different topography of the two test areas. Yuan et al.'s test area had a relatively large elevation difference, and the larger elevation difference tended to make it easier for soil profiles with low topography to become drenched with lateral water flow seepage from areas of high topography. It has been shown [23,41,44] that there exists a critical value for the dry drainage control salt effect of non-irrigated land when the critical value is exceeded: the drainage salt control ability of some non-irrigated land is reduced or even has no effect. The results of the current study are consistent with the above studies. This study observed the following: the influence distance of AIL irrigation on adjacent NAIL was between 45 and 92 m from the boundary of NAIL. As for the specific impact distance, further research is needed.

### 4.3. Effects of Autumn Irrigation on Groundwater

Groundwater levels and salinity are closely linked to soil water–salt conditions. When the groundwater level and groundwater mineralization are more substantial, soil salinity in the root zone is more severe [45,46]. Thus, controlling the groundwater table is essential in order to prevent and control soil salinization. Substantial water infiltration after irrigation will cause the groundwater level to increase sharply. Surface soil salts migrate to deeper soils and groundwater due to leaching [12], while drainage is an important way to control the dynamics and changes in the groundwater levels [47]. It has been shown that in HID, the groundwater level is below the designed critical depth (about 1.5 m) before the freezing period, which can prevent the salt in the surface soil from exceeding the crop salt tolerance standard in the next spring period [35]. In this study, it was observed that after autumn irrigation, NAIL received the drainage of AIL, which caused the groundwater level of AIL to rapidly decrease below the groundwater-designed critical depth in the irrigation area before the freezing period (in 2021); this is consistent with an earlier study [30]. However, in the 2022 test results, it was further observed that the groundwater levels of AIL did not decrease below the irrigation area designed critical depth before the freezing period. The primary causes of this phenomenon are diverse. On the one hand, the larger autumn irrigation quota and the proximity of the autumn irrigation time to the freezing period resulted in the groundwater levels being higher in AIL before the freezing period, and the groundwater did not have sufficient time to discharge; on the other hand, relatively low air temperatures after autumn irrigation (Figure 3) resulted in low-temperature soil water, high soil water viscosity, and slow movement, which resulted in a slow groundwater level decrease rate. Therefore, in order to reduce the secondary salinization of soil, which is caused by high groundwater levels before the freezing period, under the conditions of dry drainage, determining a reasonable autumn irrigation system is particularly important for controlling the depth of the water table in AIL before the freezing period, and methods for determining this system form a new research area.

## 5. Conclusions

(1) Autumn irrigation improved the uniformity of soil water–salt distributions in AIL but reduced the uniformity of soil water–salt distributions in adjacent NAIL. After autumn irrigation, the maximum difference between the average values of two-year soil water content (salinity) in the horizontal and vertical directions decreased by 75.9% and 25.2% (55.1% and 69.5%) for AIL, respectively, while it increased by 49.4% and 36.9% (61.2% and 36%) for NAIL, respectively.

(2) During the two-year experimental period, autumn irrigation increased the water content of soil profiles in AIL by 38.5~67.7% and 15.5~30.2% and it decreased the salt content by 34.5~54.5% and 21.6~45.3%, respectively. The closer the sampling point to the boundary (Q5), the greater the drainage and salt discharge. In contrast, in NAIL, the soil profile's water (salt) content increased by 18.3~54.5% and 8.3~19.7% (1.2~14.8% and 1.7%~25%) in the two years, respectively. The closer the sampling point to the boundary, the greater the increase in water–salt content in the soil profile.

(3) After autumn irrigation, the W5, W20, and W45 sampling sites all exhibited large increases in soil profile salinity, with increases of 11.7~14.8% and 23.9~25% within two years, respectively. The average salt accumulation rate was about 5.5 and 6.6 times that of the average salt accumulation rate of sampling points W92, W139, and W186. It can be observed that there was a range limit of the effect of AIL irrigation on the salinity of the soil in NAIL, which was between approximately 45 and 92 m from the boundary. If the limit distance was exceeded, the dry salt discharge effect of the increased NAIL was weakened.

(4) NAIL received drainage from AIL, causing post-irrigation AIL's groundwater level to decrease rapidly, especially from earlier autumn irrigation, which could reduce the depth of AIL's groundwater level before freezing to less than 1.5 m. This meets the requirements for the groundwater-designed critical depth in HID. When the autumn irrigation time is delayed to the eve of soil freezing, the depth of the groundwater table in AIL is still above 1.5 m, which will aggravate the risk of salt presence in surface soil during the freeze–thaw period.

**Author Contributions:** Writing–original draft, Y.Z.; methodology, Y.Z. and W.F.; funding acquisition, R.L.; resources, Q.M., M.S., X.Y., W.W. and Y.H.; project administration, R.L.; writing—review and editing, Y.Z., R.L. and W.F.; formal analysis, Y.Z. and Q.M.; visualization Y.Z. and M.S.; data curation, X.Y., W.W. and Y.H. All authors have read and agreed to the published version of the manuscript.

**Funding:** This research was funded by the National Natural Science Foundation of China (52069021).

**Data Availability Statement:** The original contributions presented in the study are included in the article, further inquiries can be directed to the corresponding author.

**Acknowledgments:** Thanks to the National Natural Science Foundation of China for supporting this work.

**Conflicts of Interest:** The authors declare no conflicts of interest.

## References

1. Li, P.; Ren, L. Evaluating the saline water irrigation schemes using a distributed agro-hydrological model. *J. Hydrol.* **2021**, *594*, 125688. [CrossRef]
2. Ramos, T.B.; Darouich, H.; Oliveira, A.R.; Farzamian, M.; Monteiro, T.; Castanheira, N.; Paz, A.; Alexandre, C.; Gonçalves, M.C.; Pereira, L.S. Water use, soil water balance and soil salinization risks of Mediterranean tree orchards in southern Portugal under current climate variability: Issues for salinity control and irrigation management. *Agric. Water Manag.* **2023**, *283*, 108319. [CrossRef]
3. Cao, Z.; Zhu, T.; Cai, X. Hydro-agro-economic optimization for irrigated farming in an arid region: The Hetao Irrigation District, Inner Mongolia. *Agric. Water Manag.* **2023**, *277*, 108095. [CrossRef]
4. Rosegrant, M.W.; Ringler, C.; Zhu, T. Water for Agriculture: Maintaining Food Security under Growing Scarcity. *Annu. Rev. Environ. Resour.* **2009**, *34*, 205–222. [CrossRef]
5. Ringler, C.; Agbonlahor, M.; Barron, J.; Baye, K.; Meenakshi, J.; Mekonnen, D.K.; Uhlenbrook, S. The role of water in transforming food systems. *Glob. Food Secur.* **2022**, *33*, 100639. [CrossRef]
6. Devkota, K.P.; Devkota, M.; Rezaei, M.; Oosterbaan, R. Managing salinity for sustainable agricultural production in salt-affected soils of irrigated drylands. *Agric. Syst.* **2022**, *198*, 103390. [CrossRef]
7. Feng, W.; Wang, T.; Zhu, Y.; Sun, F.; Giesy, J.P.; Wu, F. Chemical composition, sources, and ecological effect of organic phosphorus in water ecosystems: A review. *Carbon Res.* **2023**, *2*, 12. [CrossRef]
8. Li, L.; Liu, H.; Gong, P.; Lin, E.; Bai, Z.; Li, P.; Wang, C.; Li, J. Multi-objective optimization of winter irrigation for cotton fields in salinized freeze-thaw areas. *Eur. J. Agron.* **2023**, *143*, 126715. [CrossRef]
9. Qadir, M.; Quillérou, E.; Nangia, V.; Murtaza, G.; Singh, M.; Thomas, R.J.; Drechsel, P.; Noble, A.D. Economics of salt-induced land degradation and restoration. *Nat. Resour. Forum* **2014**, *38*, 282–295. [CrossRef]

10. Singh, A. Soil salinization management for sustainable development: A review. *J. Environ. Manag.* **2021**, *277*, 111383. [CrossRef]
11. Chang, X.; Gao, Z.; Wang, S.; Chen, H. Modelling long-term soil salinity dynamics using SaltMod in Hetao Irrigation District, China. *Comput. Electron. Agric.* **2019**, *156*, 447–458. [CrossRef]
12. Ramos, T.B.; Liu, M.; Paredes, P.; Shi, H.; Feng, Z.; Lei, H.; Pereira, L.S. Salts dynamics in maize irrigation in the Hetao plateau using static water table lysimeters and HYDRUS-1D with focus on the autumn leaching irrigation. *Agric. Water Manag.* **2023**, *283*, 108306. [CrossRef]
13. Tan, X.; Wu, J.; Wu, M.; Huang, J.; Tan, B.; Li, L. Effects of ice cover on soil water, heat, and solute movement: An experimental study. *Geoderma* **2021**, *403*, 115209. [CrossRef]
14. Chen, H.; Peng, Z.; Zeng, W.; Wu, J. Salt Movement during Soil Freezing Events in Inner Mongolia, China. *J. Coast. Res.* **2018**, *82*, 55–63. [CrossRef]
15. Liu, J.; Huang, Q.; Li, Z.; Liu, N.; Li, J.; Huang, G. Effect of Autumn Irrigation on Salt Leaching under Subsurface Drainage in an Arid Irrigation District. *Water* **2023**, *15*, 2296. [CrossRef]
16. Lu, X.; Li, R.; Shi, H.; Liang, J.; Miao, Q.; Fan, L. Successive simulations of soil water-heat-salt transport in one whole year of agriculture after different mulching treatments and autumn irrigation. *Geoderma* **2019**, *344*, 99–107. [CrossRef]
17. Ansari, S.; Abedi-Koupai, J.; Mostafazadeh-Fard, B.; Shayannejad, M.; Mosaddeghi, M.R. Assessment of solute transport and distribution under dry drainage conditions using a physical model. *Irrig. Drain.* **2019**, *68*, 797–807. [CrossRef]
18. Wichelns, D.; Oster, J. Sustainable irrigation is necessary and achievable, but direct costs and environmental impacts can be substantial. *Agric. Water Manag.* **2006**, *86*, 114–127. [CrossRef]
19. Abedi-Koupai, J.; Ansari, S.; Mostafazadeh-Fard, B.; Shayannejad, M.; Mosaddeghi, M.R. Experimental study and numerical simulation of soil water and salt transport under dry drainage conditions. *Environ. Earth Sci.* **2020**, *79*, 89. [CrossRef]
20. Konukcu, F.; Gowing, J.; Rose, D. Dry drainage: A sustainable solution to waterlogging and salinity problems in irrigation areas? *Agric. Water Manag.* **2006**, *83*, 1–12. [CrossRef]
21. Zhang, W.; Shi, H.; Li, Z.; Wang, W.; Fu, X.; Li, Z. Redistribution Mechanism for Irrigation Water and Salinity in Typical Irrigation and Drainage Unit in the Hetao Irrigation District, China. *J. Irrig. Drain. Eng.* **2022**, *148*, 04022021. [CrossRef]
22. Wu, J.; Zhao, L.; Huang, J.; Yang, J.; Vincent, B.; Bouarfa, S.; Vidal, A. On the effectiveness of dry drainage in soil salinity control. *Sci. China Technol. Sci.* **2009**, *52*, 3328–3334. [CrossRef]
23. Wei, F.; Shen, C.; Liu, J.; Wu, J. Analysis Based on Numerical Simulation on the Influencing Factors of Salinity Control Effect of Dry Drainage. *China Rural Water Hydropower* **2015**, *5*, 85–90.
24. Yang, X.; Zhu, Q.; Ma, M.; Wu, J. Research on the Evaporation Capacity and Water—Salt VariationCharacteristics of Different Types of Saline Wasteland. *China Rural Water Hydropower* **2019**, *9*, 49–53.
25. Wang, C.; Wu, J.; Zeng, W.; Zhu, Y.; Huang, J. Five-year experimental study on effectiveness and sustainability of a dry drainage system for controlling soil salinity. *Water* **2019**, *11*, 111. [CrossRef]
26. Liu, G.; Wang, C.; Wang, X.; Huo, Z.; Liu, J. Growing season water and salt migration between abandoned lands and adjacent croplands in arid and semi-arid irrigation areas in shallow water table environments. *Agric. Water Manag.* **2022**, *274*, 107968. [CrossRef]
27. Mao, W.; Zhu, Y.; Wu, J.; Ye, M.; Yang, J. Evaluation of effects of limited irrigation on regional-scale water movement and salt accumulation in arid agricultural areas. *Agric. Water Manag.* **2022**, *262*, 107398. [CrossRef]
28. Shi, H.; Yang, S.; Li, R.; Li, X.; Li, W.; Yan, J.; Miao, Q.; Li, Z. Soil Water and Salt Movement and Soil Salinization Control in Hetao Irrigation District: Current State and Future Prospect. *J. Irrig. Drain. Eng.* **2020**, *39*, 1–17. [CrossRef]
29. Fan, L.; Shi, H.; Yan, J.; Li, X.; Dou, X.; Qi, Q.; Li, H. Analysis of Soil Water and Salt Redistribution during the Freeze-thaw Periodin "Autumn Watering-spring Irrigation". *J. Irrig. Drain. Eng.* **2023**, *42*, 90–97+113. [CrossRef]
30. Peng, Z.; Wu, J.; Huang, J. Water and salt movement under partial irrigation in Hetao Irrigation District, Inner Mongolia. *J. Hydraul. Eng.* **2016**, *47*, 110–118. [CrossRef]
31. Dou, X.; Shi, H.; Li, R.; Miao, Q.; Tian, F.; Yu, D. Distribution Characteristics of Salinity and Nutrients in Salinized Soil. *Trans. Chin. Soc. Agric. Mach.* **2022**, *53*, 279–290+330. [CrossRef]
32. Wang, R.; Wan, H.; Chen, S.; Xia, Y.; Bai, Q.; Peng, Z.; Shang, S. Spatial distribution pattern of soil salinization in farmland of Wulate sub-irrigation areas in Hetao Irrigation District in Inner Mongolia in spring. *Trans. Chin. Soc. Agric. Eng.* **2021**, *37*, 105–113.
33. Wu, K.; Zhao, R. Soil Texture Classification and Its Application in China. *Acta Pedol. Sin.* **2019**, *56*, 227–241. [CrossRef]
34. Allen, R.G.; Pereira, L.S.; Raes, D.; Smith, M. *Crop Evapotranspiration (Guidelines for Computing Crop Water Requirements)*; FAO Irrig. Drain. Paper No. 56; FAO: Rome, Italy, 1998.
35. Guo, S. Influence of Autumn Irrigation time upon Moisture-Salt Movement and Agricultural Environment in Hetao Irrigation Area of Inner Mongolia. *J. China Agric. Univ.* **1997**, *S1*, 147–150.
36. Wen, Y.; Wan, H.; Shang, S. A monthly distributed water and salt balance model in irrigated and non-irrigated lands of arid irrigation district with shallow groundwater table. *J. Hydrol.* **2023**, *616*, 128811. [CrossRef]
37. Wu, T.; Li, H.; Lyu, H. Effect of freeze–thaw process on heat transfer and water migration between soil water and groundwater. *J. Hydrol.* **2023**, *617*, 128987. [CrossRef]
38. Feng, Z.-Z.; Wang, X.-K.; Feng, Z.-W. Soil N and salinity leaching after the autumn irrigation and its impact on groundwater in Hetao Irrigation District, China. *Agric. Water Manag.* **2005**, *71*, 131–143. [CrossRef]

39. Yin, X.; Feng, Q.; Zheng, X.; Zhu, M.; Wu, X.; Guo, Y.; Wu, M.; Li, Y. Spatio-temporal dynamics and eco-hydrological controls of water and salt migration within and among different land uses in an oasis-desert system. *Sci. Total Environ.* **2021**, *772*, 145572. [CrossRef] [PubMed]
40. Wang, T.; Feng, W.; Liu, J.; Fan, W.; Li, T.; Song, F.; Yang, F.; Liao, H.; Leppäranta, M. Eutrophication in cold-arid lakes: Molecular characteristics and transformation mechanism of DOM under microbial action at the ice-water interface. *Carbon Res.* **2024**, *3*, 42. [CrossRef]
41. Soltani, M.; Rahimikhoob, A.; Sotoodehnia, A.; Mendicino, G.; Akram, M.; Senatore, A. Numerical evaluation of the effects of increasing ratio of cropped to uncropped width on dry drainage efficiency in salty soils. *Irrig. Drain.* **2018**, *67*, 91–100. [CrossRef]
42. Malakar, A.; Snow, D.; Ray, C. Irrigation Water Quality-A Contemporary Perspective. *Water* **2019**, *11*, 1482. [CrossRef]
43. Yuan, C.; Feng, S.; Zhuang, X.; Qian, Z. Analyzing water-salt dynamics in typical cultivated and wasteland in Hetao irrigation district of Inner Mongolia. *Agric. Res. Arid. Areas.* **2022**, *40*, 76–85. [CrossRef]
44. Huang, Y.; Ma, Y.; Zhang, S.; Li, Z.; Huang, Y. Optimum allocation of salt discharge areas in land consolidation for irrigation districts by SahysMod. *Agric. Water Manag.* **2021**, *256*, 107060. [CrossRef]
45. Khasanov, S.; Li, F.; Kulmatov, R.; Zhang, Q.; Qiao, Y.; Odilov, S.; Yu, P.; Leng, P.; Hirwa, H.; Tian, C. Evaluation of the perennial spatio-temporal changes in the groundwater level and mineralization, and soil salinity in irrigated lands of arid zone: As an example of Syrdarya Province, Uzbekistan. *Agric. Water Manag.* **2022**, *263*, 107444. [CrossRef]
46. Shokri-Kuehni, S.M.; Raaijmakers, B.; Kurz, T.; Or, D.; Helmig, R.; Shokri, N. Water table depth and soil salinization: From pore-scale processes to field-scale responses. *Water Resour. Res.* **2020**, *56*, e2019WR026707. [CrossRef]
47. Askri, B.; Khodmi, S.; Bouhlila, R. Impact of subsurface drainage system on waterlogged and saline soils in a Saharan palm grove. *Catena* **2022**, *212*, 106070. [CrossRef]

land

Review

# Assessment and Mapping of Soil Salinity Using the EM38 and EM38MK2 Sensors: A Focus on the Modeling Approaches

**Panagiota Antonia Petsetidi and George Kargas ***

Laboratory of Agricultural Hydraulics, Department of Natural Resources Development and Agricultural Engineering, Agricultural University of Athens, Iera Odos 75, GR11855 Athens, Greece; tpetsetidi@aua.gr
* Correspondence: kargas@aua.gr

**Abstract:** Soil salinization and its detrimental agricultural, environmental, and socioeconomic impact over extended regions represent a major global concern that needs to be addressed. The sustainability of agricultural lands and the development of proper mitigation strategies require effective monitoring and mapping of the saline areas of the world. Therefore, robust modeling techniques and efficient sensors that assess and monitor the spatial and temporal variations in soil salinity within an area, promptly and accurately, are essential. The aim of this paper is to provide a comprehensive and up-to-date review of the modeling approaches for the assessment and mapping of saline soils using data collected by the EM38 and EM38MK2 (MK2) sensors at different scales. By examining the current and latest approaches and highlighting the most noteworthy considerations related to their accuracy and reliability, the intention of this review is to elucidate and underline the role of the EM38 and the MK2 type in the recent needs of detecting and interpreting soil salinity. Another aim is to assist researchers and users in selecting the optimal approach for future surveys and making well-informed decisions for the implementation of precise management practices. The study's findings revealed that the integration of the EM38 and MK2 sensors with remote sensing data and advanced methods like machine learning and inversion is a promising approach to the accurate prediction and mapping of the spatiotemporal variations in soil salinity. Therefore, future research focused on validating and expanding such sophisticated modeling applications to regional and global scales should be increased.

**Keywords:** soil salinity; assessment; EM38; EM38MK2; ECe; models; remote sensing; monitoring; mapping

**Citation:** Petsetidi, P.A.; Kargas, G. Assessment and Mapping of Soil Salinity Using the EM38 and EM38MK2 Sensors: A Focus on the Modeling Approaches. *Land* **2023**, *12*, 1932. https://doi.org/10.3390/land12101932

Academic Editors: Tiago Brito Ramos, Maria da Conceição Gonçalves and Mohammad Farzamian

Received: 19 September 2023
Revised: 10 October 2023
Accepted: 11 October 2023
Published: 17 October 2023



## 1. Introduction

Soil salinization has been considered one of the most challenging global threats, affecting large cultivated and irrigated areas all over the world. Its detrimental impacts on environmental quality, agricultural productivity, and socioeconomic stability are about to become even more pronounced in the coming years due to climate change [1]. The extent of soil salinity along with the frequency of floods and duration of droughts, as a consequence of climate change, are expected to be more intense and exacerbated in the arid and semi-arid regions where the sustainability of natural resources is imperative [2]. Considering the increasing demands for global food supplies and arable land, these effects necessitate urgent control and mitigation. For this purpose, regular and accurate monitoring of soil salinity distribution and its spatial variations across multiple scales is crucial for preventing soil salinization hazards and preserving the long-term sustainability of agricultural and environmental systems.

Successful monitoring of soil salinization requires rigorous modeling techniques and advanced tools to reliably assess the soil salinity levels and interpret its severity in different areas of the world. Recently, satellite remote sensing technology has been widely applied as an effective tool for identifying and mapping the soil salinity of large-scale areas [3].

However, the sensors are incapable of detecting the subsurface distribution of the soluble salts and the highly spatial heterogeneities in the soil profile [4]. Thus, their implementation is usually combined with other sensors or data for more accurate results [5].

On the other hand, proximal sensing (PSS) technology with ground-based, electromagnetic induction (EMI) sensors can quantify and characterize the spatial patterns of soil salinity within the soil profile by measuring the soil's apparent electrical conductivity (ECa). Besides the popular EM38 sensor, a variety of commercial EMI devices have also been developed, enabling the investigation of the solute's variability at different soil layers and allowing soil salinity mapping, particularly on the field scale [6].

Despite the accessibility of improved proximal sensing devices such as the EM38MK2 (MK2), an upgraded type of EM38, and the range of studies that evaluate various EMI data conversion techniques for the soil salinity assessment [7–10], the choice of the most suitable approach for each survey is still a challenging task. The uncertainties that emerge with their application depend on a number of site-specific environmental factors, including the complex interactions of soil properties that affect the ECa and distinct data processing requirements, which can significantly impact the credibility of the results. Furthermore, while the technical guidelines and considerations regarding the employment of EM38 in soil salinization surveying have been well documented by existing reviews and scientific publications [11–13], a concise compilation of the currently available approaches that convert ECa measurements by the EM38- and the MK2-type sensors into soil salinity as expressed in ECe has not yet been attempted.

In this respect, the objective of this paper is to provide a comprehensive and up-to-date review of the modeling approaches for monitoring and mapping the saline soils using the EM38 and the MK2 sensors. Through an examination of the approaches and techniques that have been applied for ECe assessment and mapping using the obtained ECa measurements, the aim of this review is to foster a deeper understanding of the sensor's efficiency in the recent and constantly rising demands of detection and monitoring soil salinity.

Specifically, in the following sections:

- The fundamental principles underlying the EM38 and MK2 probes are described, offering thorough insight into their operational mechanisms, capabilities, and constraints.
- Subsequently, the modeling approaches that utilize the EM38 and the MK2 data, for the estimation, prediction, and interpretation of the ECe at different scales, are extensively discussed. The various models and methods that have been developed and convert the sensor's ECa values into ECe are explored, highlighting the most noteworthy considerations regarding their accuracy and reliability.
- Finally, the fusion of the EM38, MK2, and remote sensing data for monitoring and mapping the saline soils is overviewed and followed by a brief summary of conclusions and future directions.

By addressing these aspects, the aim of this review is to enrich the current field of soil salinization research and elaborate on the potential of EM38 and MK2 sensors in the future modeling and assessment of soil salinity and eventually in developing precise management practices that will prevent land degradation and protect resources.

## 2. Materials and Methods

With the aim of obtaining a comprehensive and up-to-date overview of the modeling approaches that utilize the EM38 and MK2 data, we adopted an in-depth research methodology. Initially, for the collection and the thorough examination of the literature related to our objective, an online search using academic databases and search engines was employed. These included Elsevier Scopus, ScienceDirect, MDPI, as well as Google Scholar. The research criteria and the keywords that were applied in the selected electronic resources were: ("EM38" OR "EM38MK2" OR "EMI") AND ("salinity models" OR "soil salinity") OR ("Remote Sensors" AND "soil salinity"). The types of publications scoped for our review were restricted to those written in English. This encompassed journal articles, research papers, review papers, book chapters, conference papers, theses, and technical notes related to the principles and characteristics of the EM38 and MK2 sensors.

From searching all the databases, we retrieved a total of more than 300 articles. In order to identify and acquire the most relevant literature from the gathered publications, we reviewed their titles and abstracts. After excluding duplicates and less relevant publications, the remaining articles were full-text reviewed. Ultimately, a total of 170 documents were selected, analyzed, and cited in the present study.

## 3. Results

### 3.1. Fundamental Principles and Considerations of the EM38 and MK2 Sensors

#### 3.1.1. Basic Operational Features of EM38 and MK2 Sensors

The deployment of electromagnetic induction (EMI) instruments has been consolidated in agricultural science and soil surveying since the 1970s [14–16]. Owing to their low cost and their capability to detect the spatial variations in edaphic properties and heterogeneities within the field and at larger scales, in real time, and non-destructively [17,18], they have been studied and used for numerous environmental and geophysical applications [12].

Unlike other geophysical methods such as TDR and GPR, the quantification of soil salinity's spatial variations by electrical resistivity (ER) and EMI devices has become a vital component of precision management implementations [11,19]. This is mainly amplified by the measurements of apparent electrical conductivity (ECa), which have been found to be correlated with soil salinity estimates and can be used as an indirect indicator for many soil properties [20]. Frequency domain reflectometry (FDR) technology, including WET sensors, has also been successful in the appraisal of soil salinity in the laboratory [21–23] and in situ [19,24] using ECa measurements. Nevertheless, the single utilization of these probes might be exacting and locally restricted since they have a substantially smaller measurement volume and are invasive, as their operation is based on contact with the soil and its sublayers [24].

Over the years, commercial EMI sensors, such as the DUALEM (Dualem Inc., Milton, ON, Canada), EM31, EM34 (Geonics Ltd., Mississauga, ON, Canada), and Profiler EMP-400 (Geophysical Survey Systems, Inc., Salem, NH, USA) have been investigated for the assessment of various soil properties, including soil salinity, and their performance has been documented [25–27].

The ground conductivity meter EM38 (Geonics Ltd., Mississauga, ON, Canada), introduced in 1980, was revolutionary in soil salinization surveys due to its light weight, portability, and the large volume of ECa measurements taken in various types of soils and fields, e.g., stony, which until then were difficult to acquire with electrode-based devices [28,29]. Thus, it quickly gained the attention of the agricultural community and became the most frequently applied tool for monitoring and mapping soil salinity [20]. The adaptation of EM38 can also be attributed to the fact that it was intentionally designed to support the assessment of near-surface variations in soil properties and specifically of soluble salts that affect crops within the rooting zone [30,31].

The EM8 sensor is constructed with two coils, one transmitter and one receiver coil, which are installed at the opposite ends of the instrument with a fixed spacing of 1 m, and it operates at a 14.6 kHz frequency. The orientation of the coils determines the cumulative depth response of the instrument associated with the ECa measurements. When located in the horizontal configuration (EMh), the device's signal corresponds to a depth range of roughly up to 0.75 m, whereas in the vertical mode (EMv) the penetration depth is approximately up to 1.5 m. The depth range of the instrument sufficiently covers the root volume of most plants [32]. These depth-weighted responses of ECa, however, are theoretical measuring depths that rely on a non-linear function in homogeneous soils. As a result, the absolute depth values cannot be easily defined [33].

Since its release, the EM38 has undergone several modifications, updates, and technical improvements, including the addition of a GPS receiver, which allows for accurate georeferencing of the data, and the development of user-friendly software for data analysis and visualization. The dual-dipole EM38 DD sensor is an example of these modifications. This version consists of two EM38 units attached together and placed horizontally and vertically for recording simultaneous EMh and EMv measurements [34,35].

In 2008, the EM38MK2 (Geonics Ltd., Mississauga, ON, Canada) was launched as an updated version of the original EM38 instrument [30]. The MK2 type encompasses new attributes and enhancements regarding the depth range response, stability, and facilitations concerning the field survey and data acquisition. Compared to its predecessor, the MK2 has hardware and software that offer more automation in operation, easier processing, and better interpretation of the raw data. In particular, it can simultaneously measure both soil conductivity (Q/P) and magnetic susceptibility (I/P) within two discrete depth ranges. It entails temperature compensation circuitry, which reduces the occurring temperature drifts during the survey, and it supports automatic calibration without laborious adjustments. This can be achieved through a wireless Bluetooth data logger, which enables the collection of the data and the communication with the instrument conveniently from a relative distance. Alternatively, data recording can be performed through a serial port. The duration of field operations has also been enhanced with the addition of a power connector, which allows for the use of an external rechargeable battery [36]. Both instruments are presented in Figure 1.

**Figure 1.** The EM38-MK2 and EM38 devices. (**a**) The EM38 in horizontal mode. (**b**) The EM38 in vertical mode (from Geonics Ltd., Catalano [37] and Siddique [38]).

Besides the technical advancements, the main fundamental difference between the two sensor types lies in the second receiver coil of the MK2, which corresponds to an additional depth range of measurements. The MK2 consists of one transmitter coil and two receiver coils that are positioned at two fixed distances of 1 m and 0.5 m from the transmitter coil, respectively. Hence, in the MK2, the effective depth range is determined by both the coil separation and the dipole modes of horizontal and vertical orientation. Consequently, with the new coil (0.5 m coil separation), measurements of ECa can be additionally taken at two distinct depths: at 0.375 m depth when the device is placed in horizontal mode (EMh) and at 0.75 m in the vertical mode (EMv) (Figure 2). This version allows users to detect and investigate variations in shallower layers, which may be optimal for precise agriculture practices [26,33]. Moreover, along with the rest of the depth ranges of the sensor, the profile of soil salinity distribution up to a 1.5 m depth can be promptly acquired and evaluated.

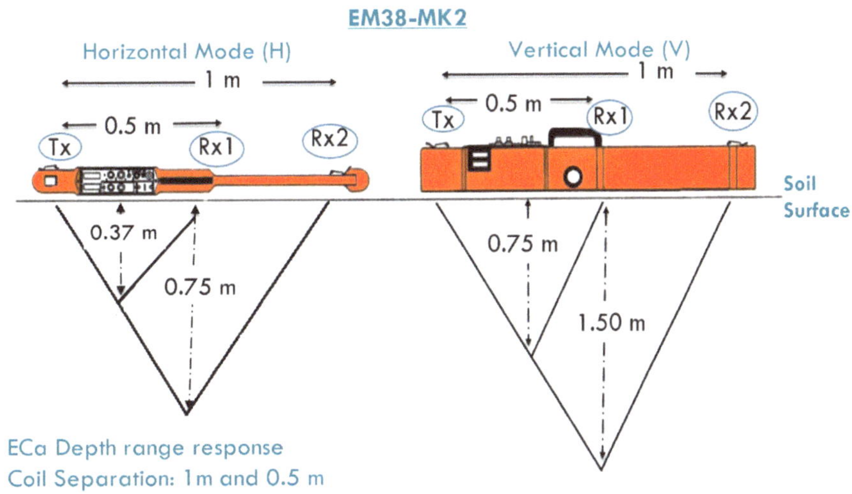


**Figure 2.** The EM38MK2 sensor in horizontal (H) and vertical (V) mode with the effective depth range responses of ECa for coil separation of 1 m and 0.5 m when placed on the soil surface. T× refers to the transmitter and R × 1, R × 2 to the two receiver coils.

The relative differences, advantages, and applicability of the EM38 and MK2 sensors over various geophysical instruments in mapping of soil properties have been discussed in a few studies [12,26,39,40]. Gebbers et al. [26], by comparing a variety of EMI and other technology devices (ARP03, CM-138, EM38, EM38-DD, MK2, OhmMapper, Veris 3100), concluded that the main disadvantage of EM38, EM38-DD, and CM138 sensors is their sensitivity to deeper soil layers, which is irrelevant to the crop's rootzone. On the other hand, MK2, ARP03, OhmMapper, and Veris3100 were found to be more effective in detecting shallower soil variations that are important for precision agriculture. Likewise, the EM31 and EM34 sensors, which have exploration depths of up to 6 and 60 m, respectively, may also be considered inappropriate for detecting the variability in shallower soil layers [12]. In a study conducted by Dooltitle et al. [39], the use of EM38 and the multifrequency device GEM 300 was investigated, revealing that both sensors provide reasonable estimates of soil salinity. Moreover, Urdanoz et al. [40], by comparing the EM38 and the DUALEM sensor, indicated that although EM38 tends to produce slightly higher horizontal ECa readings than the DUALEM, both sensors can be used interchangeably. Generally, the EMI sensors exhibit close similarities in their collected data, with the main differences attributed to the different operational modes and sensing depths [30].

#### 3.1.2. Principles of the EM38 and MK2 Operation

The operation of the EM38 and MK2 instruments is based on the principles of EMI and has been established by McNeil [41,42]. Once the sensors are turned on and properly calibrated for recording ECa measurements, the transmitter coil sends, at a frequency of 14.6 kHz, an alternating electrical current to the soil, generating a primary magnetic field ($H_p$). When the primary field interacts with the subsurface, it induces electrical currents (eddy currents) that, in turn, produce secondary magnetic fields ($H_s$). These secondary fields interact with the receiver coils by inducing alternating currents in the coils. The sum of the amplitude and phase of the induced voltages from the primary and secondary fields is amplified in an output voltage, which is read by the user.

Accordingly, under low induction number (LIN) conditions, where Nb << 1 and assuming homogeneity in the depth profile, the apparent electrical conductivity, ECa, is sensed and expressed as the ratio of the primary ($H_p$) and the secondary magnetic fields

($H_s$) (Equation (1)), where f is the operating frequency (Hz), $\mu_o$, the magnetic permeability of free space ($4\pi 10^{-7}$ H $m^{-1}$), s is the intercoil spacing (m), and $\omega = 2\pi f$ [41].

$$ECa = \frac{4}{2\pi f \mu_0 s^2}\left(\frac{H_s}{H_P}\right) \tag{1}$$

Besides other factors, ECa readings by the EM38 and MK2 sensors for an investigated depth range are influenced by the orientation and coil spacing of the instruments. The relative ranging depths for the horizontal and vertical modes have been determined by McNeil [41] in homogeneous soils as non-linear functions that describe the relative contribution to the secondary magnetic fields in respect to normalized depth z.

Consequently, the depth-weighted response, which indicates the cumulative depth response R(z) of the sensors, is a non-linear function that represents the relative contributions of all soil electrical conductivities from a soil volume below a normalized depth z. The R(z) equations, based on the horizontal and vertical orientation and expressed as a percentage (%) of the measured signal, have been defined for 1 m (Equations (2) and (3)) and 0.5 m (Equations (4) and (5)) coil separation [33,41]:

$$R_v(z) = \frac{1}{\left(4z^2+1\right)^{1/2}} \tag{2}$$

$$R_H(z) = \frac{1}{\left(4z^2+1\right)^{1/2}} - 2z \tag{3}$$

$$R_v(z) = \frac{1}{\left(4\left(\frac{z}{0.5}\right)^2+1\right)^{1/2}} \tag{4}$$

$$R_H(z) = \left(4\left(\frac{z}{0.5}\right)^2+1\right)^{1/2} - 2\left(\frac{z}{0.5}\right) \tag{5}$$

where z (m) is the depth and $R_H(z)$ and $R_v(z)$ are the cumulative relative ECa for horizontal and vertical mode, respectively.

From the derived cumulative functions, the depth of investigation (DOI), which refers to the depth from which more than 70% of the signal response derives, can be determined for each sensor. Heil et al. [33] compared the two instruments and examined the effective depth responses for each orientation and coil distance. The coil spacing of 0.5 and the horizontal mode are generally influenced by near-surface variability, making them more suitable for shallower depths. Instead, the 1 m spacing coil and the vertical mode seem to have an increased sensitivity along with the depth. It is noteworthy that the EM38's vertical response decreases drastically at depths above 90 cm, in contrast to MK2. Practically, the DOI of the sensors may vary under natural soil conditions due to existing heterogeneities and the interrelations of ECa with subsurface soil features that affect the signal.

In addition to the effective depth ranges when placed on the ground surface, both devices can be lifted at different heights above the soil surface to investigate interval depth variations and model the distribution of salt content in the soil layers [43–45]. Also, they are designed for handheld measurements or can be mounted on non-metallic sleds and attached to vehicles for mobile measurements. The mobile and real-time collection of ECa data by the EM38 and MK2 sensors is a simple process owing to their software and the direct connection to the GPS. Thus, they can be an ideal option for monitoring and mapping soil salinization at field scales [46–48].

### 3.1.3. Considerations in the EM38 and EMK2 Applications

One of the key factors in the employment of the EM38 and MK2 sensors is that on all occasions of soil salinity surveying, either at field or larger scales, site-specific calibration is required. Therefore, soil sampling for ground-truth data cannot be omitted [30].

Another important consideration when using these probes for the collection of ECa measurements is their susceptibility to metal and electrical interference, like fences and power lines. In comparison to other technologies, such as capacitance sensors, the presence of metallic objects in the study area can affect the signal, especially in the horizontal configuration [36]. Although the detection of metals may be beneficial for archaeological prospecting [49], for efficient soil salinity estimation and mapping, uniform, metal-free soils are a prerequisite. Furthermore, as the manufacturer recommends, in the automatic mode of ECa recording, more frequent calibrations might be needed to minimize any potential effects from the drifts on the accuracy [50]. The drifts by temperature are stronger in the original EM38 [20,33], while for the MK2, they are considered insignificant due to the internal enhancements. An exception might arise in the case of near-surface measurements with the 0.5 m spacing coil, where the effects from the drifts need to be managed [50].

Finally, one of the most concerning and constraining aspects of EM38's utility for determining solute distribution within the soil is its application under dry moisture conditions or in fields where there is insufficient moisture through the penetration depths. Conducting ECa surveys in fields where soil water content levels are less than those of field capacity and reportedly under 50% [51] can lead to unreliable and biased results. Likewise, ECa measurements in shallow and moderately deep soils above bedrock should be avoided [30,52].

### 3.2. *Modeling Approaches for the Assessment and Mapping of ECe Using the EM38 and MK2 Data*

The conventional strategies for the assessment of soil salinity consist of soil sampling and laboratory estimation of the electrical conductivity of saturated paste extract, ECe, which is the standard method. These laborious, costly, and time-consuming methods tend to be impractical for fields and large-scale areas since they are point-based and cannot provide a sufficient number of measurements for extended monitoring [30,53]. The non-invasive, cost-effective, and rapid measurements of ECa by the EM38 and the latest MK2 have become one of the most widely accepted and reliable alternatives for determining the spatiotemporal variation in soil salinity in arid and semi-arid regions [54]. In practice, the main benefit of these EMI instruments is that they allow quick and large numbers of georeferenced ECa measurements, which can be significantly correlated to the spatial variability of soil properties and especially soil salinity, providing information on the soil quality of the croplands [30,55,56]. In addition, the obtained data can be efficiently utilized to generate detailed maps of subsurface property spatial patterns and processes. These high-resolution maps enable the design of field and large-scale sustainable management decisions [10].

ECa is the weighted average of the vertical electrical conductivity distribution within the soil volume as depicted in a one-dimensional (1D) earth model [57]. It is influenced by various physical and chemical soil properties and their interrelations [19]. In this sense, factors such as soil moisture, soil salinity, texture, mineralogy, and temperature affect the EM38 signal and need to be considered when interpreting geospatial ECa measurements with respect to a particular property survey [51,58].

In salt-affected areas where ECa values are higher than 2 ds $m^{-1}$, the spatial variability of solute concentration has been proven to be the dominant factor contributing to ECa, and the provided soil moisture is close to field capacity across the research district [30,59]. In this instance, EMI measurements can most likely be directly correlated with soil salinity mapping [48]. In non-saline soils and soils with relatively low conductivity levels, however, ECa is strongly related to a function of soil properties, which include soil water content, the amount and type of clay, cation exchange capacity (CEC), organic matter (OM), and soil temperature [20,60–64]. Among them, volumetric water content and clay proportion have been reported as the major factors affecting the values of ECa [61–63,65,66]. In most situations of ECa surveying, soil moisture and soil salinity are considered the most influential factors, whereas the effect of other factors, like soil temperature, is weaker. It is worth mentioning that the strength of the effect of each soil property on ECa varies depending on specific soil conditions.

The complex relationships between ECa and soil properties have been examined in numerous site-specific management studies and described by various models. The volumetric water content of the soil is the primary pathway for electric current flow [60,67], thus exhibiting a strong positive relationship with ECa. When water content drops below a threshold value, ECa decreases significantly and becomes negligible since the conductance path ceases to exist. Moreover, like soil salinity, water content is considered a dynamic property, meaning that after certain circumstances (e.g., rainfall, nonuniform irrigation applications), it can gradually change across the field and soil profile, complicating the interpretation of its effects on ECa [11]. Similarly, the high concentration of the dissolved salts in the soil solution leads to an increase in ECa readings [60,62]. In non-saline soils, the increase in soil moisture enhances the pathway for current to flow through the soil, resulting in higher values of ECa. The contribution of soil texture to the spatial patterns of ECa varies according to the particle sizes and the charge density of their surface area. ECa values tend to be higher in fine-textured and clay-rich soils due to their larger surface area, which allows them to absorb and retain more ions [62,68]. Conversely, in coarser-textured soils, ECa tends to decrease and exhibit larger variations. Soil texture can influence the water-holding capacity of the soil. Therefore, in uniform cropping systems, soils with high clay and water content are likely to lead to an increase in ECa [60,62]. In conditions with highly spatial texture variations, however, characterizing soil ECa variability becomes difficult [69]. In addition, temperature fluctuations during the survey have a substantial impact on field measurements. Due to the positive temperature dependency of ECa, which increases by almost 2% for every 1 °C temperature rise, ECa readings are recommended to be referenced at 25 °C using corresponding equations [41,64]. Finally, cation exchange capacity (CEC) and organic matter (OM) also show a positive correlation with ECa [62,70]. Overall, various soil properties and their influence on ECa are summarized in Table 1.

**Table 1.** Different soil factors and their effect on ECa.

| Factor | Effect on ECa Values |
|---|---|
| Water Content | Higher moisture levels increase ECa<br>Dry soils have lower ECa |
| Soil Salinity | High salinity levels increase ECa |
| Texture | Clay content: Higher proportion increases ECa<br>Silt, Sand Content: Higher proportion decreases ECa |
| Temperature | Increasing temperature increases ECa |
| Cation Exchange Capacity (CEC) | Higher CEC increases ECa |
| Organic Matter (OM) | Higher percentage increases ECa |

The depth-weighted average conductivity of ECa does not indicate the distribution of the actual salt concentration with depth in the soil profile but rather reflects the average cumulative response of the sensor, weighted according to respective soil depths [41,42]. Thus, a modeling approach to establish a relationship between ECa measurements and true salinity levels like ECe, or EC1:5, at various depths is necessary for the prediction and monitoring of soil salinization across different spatial and temporal scales. The conversion of the EM38 data for soil salinity analysis encompasses a variety of simple or more advanced statistical and mathematical procedures, integration with data and sources from other technologies, as well as spatial modeling techniques. Among these, the preferred modeling approach can be applied to estimate, predict, and map the soil salinity profile, leading to precise rootzone management and mitigation of salinization impacts.

Based on the land use, the size of the study area (e.g., plot [71], field [7], landscape [72], regional [10], urban greenery [73]), and the purpose of the salinity survey, several approaches applying the EM38 and MK2 sensors have been reported for assessing soil salinity in terms of ECe and displaying its spatiotemporal characteristics.

### 3.2.1. Deterministic and Stochastic Conversion of ECa

One of the initial considerations in the assessment of ECe using an EM38 sensor depends upon the conversion of ECa through a stochastic or deterministic approach [70,74,75].

ECe values can be deterministically predicted from the in situ ECa data using geophysical models that have been developed based on the laws of EMI response to saturated or unsaturated conditions [12,76]. In this non-geostatistical approach, ECa readings are converted into ECe through theoretically or empirically determined formulas that incorporate physicochemical soil characteristics estimated or measured within the study area [35]. Well-established models such as Archie's 1942 model [77] and the dual pathway parallel conductance (DPPC) of Rhoades et al. [67,78] describe the ECa as a multiplicative function of soil parameters under different porous media and soil circumstances. These and other similar models [79] have been broadly examined for the detection and estimation of water and solute concentrations in the soil profile.

While deterministic models acknowledge the significance of the complex interactions between ECa and soil properties, their application in soil salinization monitoring is limited by their static nature. The requirement for precise information on additional soil properties such as soil water content or texture and the dynamic process of soil salinity, varying over space and time, makes this approach suitable mostly for local and short-term simulations [80,81].

An alternative method to overcome the challenges in interpreting the spatiotemporal variations in ECe with EM38 measurements is the stochastic or geostatistical approach [35]. This modeling technique relies on the correlation of ECa data with ECe for the direct prediction of soil salinity. This involves using soil sampling and geostatistical or spatial regression models. During the ECa survey, a number of soil samples are collected from the measurement points and analyzed to determine the corresponding ECe. The paired set of ECe and ECa values is then used to establish an ECe = f (ECa) relationship with the aid of regression and geostatistical analysis. Subsequently, this developed calibration equation is applied to predict the ECe values from the remaining non-sampled ECa measurements [7,82].

The site-specific calibration between ECe and the simultaneous EM38 or MK2 measurements, as a necessity for the accurate appraisal and mapping of soil salinity, has been the center of attention in numerous studies [7,8,83–86]. Over the past decades, the calibration of ECa data has been extensively explored and improved at field and local scales due to advancements in geostatistics and data processing tools. However, uncertainties in the generated maps of soil salinity still emerge from the employment of these techniques [87]. Furthermore, monitoring the spatial and temporal trends of soil salinity at a regional scale demands continuous data and calibration parameters from different fields, which may not always be available or easily accessible [88,89].

Geostatistical methods have been indispensable tools in soil salinity monitoring and mapping, as they are applied to model and predict the spatiotemporal variability in large salt-affected areas. Based on the spatial dependence and structure of the georeferenced variables, techniques like kriging and variogram modeling offer the advantage of predicting ECe values at unsampled locations [90], with relatively high accuracy [91]. They require dense soil sampling with approximately more than 50 sampling points to ensure reliable calibration with minimum errors [92]. Therefore, their application is not recommended for field surveys [91].

Spatially referenced regression models, on the other hand, have gained great recognition due to their simplicity and the reduced need for soil sampling [53]. They are regression equations with optional trend surface variables that assume an independent underlying error structure. This error is related to the variations between ECa and estimated ECe values. Despite their benefits, as regards their predictive efficacy, they cannot reach the same degree of accuracy as the geostatistical models [7,93]. Nonetheless, the regression modeling approach yields viable results in most cases of regional scales [94,95]; hence, it remains one of the most appealing and preferable calibration techniques.

These approaches are also adopted when applying ECa-directed soil sampling for soil salinity mapping [11,35,51]. According to this concept, when a correlation between measurements and soil salinity exists, geospatial ECa measurements can be used to construct a directed sampling plan for the selection of the optimal soil sampling sites. This method is accomplished either deterministically, through design-based sampling schemes, or stochastically with model-based sampling designs.

Based on ECa-directed soil sampling, in fields where salinity is not the dominant soil property influencing the EM38 measurements, the variations in ECa can act as a surrogate for identifying the site locations that depict an adequate range and variability of the soil salinity. The soil samples collected from these particular sites can then serve as ground-truth data for the accurate calibration of ECa to ECe. The calibration model, in turn, results in producing spatially reliable maps of soil salinity across multiple scales [48]. In saline soils with uniform texture and water content conditions, however, the ECa variation is mainly attributed to salt concentration; thus, it can be used as a soil salinity indicator.

The overall process of ECa-directed sampling for mapping the spatial distribution of soil salinity may be expedited by the available commercial software packages. The ECe Sampling Assessment and Prediction (ESAP) developed in USDA [96] is a conductivity modeling software, which based on a response-surface sampling design, may generate the minimum set of sampling points needed for the calibration of the EM38 measurements to ECe. Additionally, it embraces both deterministic and stochastic methods for the prediction of the spatial ECe values from the ECa survey data [71]. To date, several researchers have used the ESAP programs with EM38 readings as a tool to delineate and map ECe distribution at irrigated fields [71,97–99] and district scales [100]. Amezketa [71], using customized ESAP software and EM38 data, assessed and displayed the spatial patterns of soil salinity at an irrigated plot in Spain. The accuracy of the developed models for the multiple-depth ECe values ranged from $R^2 = 0.38$ in the topsoil (0–30 cm) to $R^2 = 0.90$ in the subsurface (30–60 cm). Also, based on the selected calibration models, the average ECe profile (0–90 cm) was mapped. Slimane et al. [100], after obtaining EM38 readings from a 240 ha region of Tunisia, imported the data into the ESAP to determine the appropriate soil sampling locations and to estimate and map the spatial variability of the area's ECe. According to the study's results, the average $R^2$ of the prediction models for different soil depths was approximately 0.78, ($0.6 \leq R^2 \leq 0.8$) and the ECe variation maps showed that salinity increased with depth.

Besides this general classification, the most common and current modeling approaches for the conversion of EM38 and MK2 data into soil salinity are listed below. Each of these models includes techniques that can be applied individually or in combination to achieve the specific objectives they are intended for. Essentially, simple regression techniques could be appropriate for ECe estimation in the field, but for mapping the spatiotemporal variability of ECe in irrigated regional districts, more sophisticated procedures might be necessary. The purpose of conducting a salinity survey with EM38-type sensors, whether qualitative or quantitative, is of major importance in defining the level of accuracy and the modeling approach to be utilized [101]. The following section focuses on the diverse approaches that address the conversion of raw EM38 and MK2 data into ultimate ECe values and their representations across a variety of scales and applications, as documented in the existing literature. The categorization of the available modeling approaches serves as a means to comprehend their fundamental characteristics and techniques and explore the range of their consistency and weaknesses. It is important to note that the types of models presented in the study are not rigidly distinct but may share common attributes. The various categories are designed to provide guidance and facilitate the identification of areas where the methods of soil salinity appraisal and mapping can be enhanced.

#### 3.2.2. Regression-Based Models (Linear, MLR, Simple Depth-Weighted Coefficients, Established Coefficients, Modeled Coefficients, Mathematical Coefficients)

The earliest studies that were carried out for soil salinity assessment using the EM38 probe were entirely based on producing sets of regression equations and searching for calibration coefficients for a specified range of depths at the site of interest. In order to detect the salinity distribution within the field, the weighted ECa data are transformed to determine the ECe profile by fitting linear or non-linear regression models at depth intervals and calculating the coefficients for each depth range. This framework involves the development of empirical calibrations relating the instrument's horizontal and vertical observations, either separately or in combination, to ECa measured with a probe at known depth intervals [102–104]. The predicted equations of the probe's ECa can then be reconstructed to estimate the soil ECe. Alternatively, equations are developed to directly relate the sensor's measurements to a single weighted ECe value [83,105] or ECe at different depth increments. Typically, in these approaches, the vertical profiling of ECe is derived through a one- or two-step process that leverages the operational dipole modes of the sensor and depends on the establishment of regression relationships between the obtained and the ground-truth conductivity data. They utilize the field EM38 readings, taken on the soil surface or at distinct heights above it, to define the ECa layering and the collection of soil samples from corresponding depth increments for the construction of calibration models [106].

The various published calibration model functions that have been retrieved mainly through linear (LR) [83] and multiple linear (MLR) regression analysis [71,84,85,107] include known approaches such as depth-weighted coefficients [105]; established coefficients, which are empirical-mathematical coefficients [108,109]; modeled coefficients [103]; mathematical coefficients based on the theoretical EMI depth response function [110]; and the logistic profile model [85]. Johnston et al. [111] compared several of the developed calibration approaches to evaluate their performance in soil salinity estimation. As they reported, the established coefficients of Corwin and Rhoades [108,109] and the modeled coefficients of Slavich [103], which predict the probe ECa, exhibited results with low bias but with significant errors. Also, the depth-weighted coefficients model of Wollenhaupt et al. [105], which determines the weighted ECe, was observed to perform poorly with high error values (RMSE = 5.33). Furthermore, Triantafilis et al. [85] indicated that the logistic profile, which consists of a mixed nonlinear model, can provide smoother and less erratic prediction profiles than the established coefficients and the multiple-regression coefficients model [104].

The common trait of the proposed models and approaches is that they all use regression parameters that are site-specific and time-dependent to a significant degree [112]. Hence, they will not perform at the same level of accuracy when applied in dissimilar soil conditions to those they have been developed in. The fluctuations in the prediction results can be attributed to the variations in soil volumetric water content and soil texture across different locations [111,113] and to misleading homogeneous assumptions [85]. This arises from the fact that the calibrations of ECa measurements are being established, assuming that soil salinity is the only soil property affecting the response of the EM38, thus excluding the influence of other soil factors in the equation. In these instances, a strong relationship between EM38 data and ECe values can be confirmed [12]. Additionally, since most of these calibration models rely on linear equations, they use a best fit line through the data, usually an ordinary least-squares (OLS) regression, a technique that is accompanied by certain attributes. This prediction method is associated with key assumptions such as the normality and independence of errors and homoscedasticity for obtaining valid coefficients [93,114]. Therefore, deviations from meeting these criteria can compromise the reliability of the coefficients and result in inaccurate predictions and potentially biased conclusions for the salt-affected area.

### 3.2.3. Geostatistical Models

The comprehension and characterization of soil salinity spatial distribution constitutes a fundamental element of the sustainable management practices and prevention of soil salinization [7,115,116]. Due to the need for high spatial resolution data and expanding coverage, determining and mapping the spatial variations in soil salinity at the land and regional scales by soil sampling appears unfeasible [30,117].

In contrast, this process can be accomplished by using the EM38 and MK2 instruments and performing spatial predictions at the field and land scales [30,55]. The promptly collected ECa data can be employed with geostatistical methods, enabling the quantification and interpretation of the spatial variability of ECe over large areas [118,119]. Soil salinity severity maps may also be generated to support the proper decision-making for crop and land management [120].

The geostatistical models, in the context of conversion, refer to a set of geostatistical methods and tools, such as interpolation and variogram analysis, which are implemented as a means of modeling, predicting, and mapping the spatial variations [92] of soil salinity from the sensor's data. The geostatistical processes can be carried out by appropriate software packages.

More specifically, the models are constructed to analyze the spatial patterns of the data and also predict values for the variable at unsampled locations, diminishing the weakness of point measuring from proximal sensing and creating a constant spatial coverage [73,121]. Considering the scientific concern and practical constraints in characterizing the distribution of soil salt in a continuous space, the contribution of geostatistics, which identifies the spatial variation structure and predicts unsampled data, is significant.

The analysis, definition, and quantification of the variance structure of soil salinity can be achieved with variogram calculations, whereas the prediction of the spatial variability of ECe for generating informative salinity maps is obtained with an interpolation method.

The spatial prediction and mapping of ECe, based on geostatistics, can be potentially adopted in two different ways. The ECa measurements can be initially interpolated for the prediction of unsampled ECa values, and the amount of spatial ECa data can then be calibrated for the estimation of ECe and eventually analyzed with a variogram. This approach has proven to have controversial results depending on the interpolation method and the density of the EM38 data [8,119].

The second and most established method prioritizes the prediction of ECe by developing a relationship between ECa measurements and ECe values, usually with linear regression analysis [94,122,123]. Thereafter, variogram analysis and interpolation techniques can be applied to the predicted ECe for mapping and evaluating the spatial patterns of soil salinity across the study area. The accuracy of this modeling approach depends on various factors that need to be taken into account. Aside from ensuring the quality of the EM38 data, the primary aspect is associated with the selection of the interpolation technique and the distinct sampling design that they require [7]. In a similar sense, the sampling scheme is essential for the reliability of the variograms [92]. Moreover, different and more sophisticated approaches than linear regression for calibrating ECa data might perform with higher accuracy [120].

To date, there is a diversity of geostatistical interpolation methods available, several of which have been examined and compared using the EM38 and MK2 sensors for the prediction and mapping of the spatial variability of ECe at the landscape [124], district [8,120], and regional scales [94,123]. They extend from basic kriging to more contemporary tools, such as ordinary kriging [123], universal kriging, and hybrid interpolation techniques that integrate different technologies, like regression kriging, co-kriging [7], indicator kriging [122], or 3D kriging [94]. All geospatial procedures, including variogram analysis, interpolation methods, and spatial data visualization, can be executed with commercially available software packages, such as Geostatistics Software GS+ [123], Golden Software Surfer [97], ESRI ArcGIS [122] and ArcMap [120], and Geo R [82]. Yet, despite the variety of these spatial estimation tools, there has not been a single optimal method reported for the interpolation of the data [73]. For example, Jantaravikorn et al. [120], after identifying a strong ECa–ECe correlation for both modes of EM38 (r = 0.86 and r = 0.87), examined the accuracy of four distinct spatial interpolation methods for soil salinity prediction on categorized validation datasets using ESRI ArcMap and SAGA software. The predicted horizontal and vertical ECa data were then classified to create soil salinity severity maps. ordinary co-kriging was found to be an accurate interpolation method for predicting soil salinity when using the horizontal data of the sensor ($R^2 = 0.85$), while the deterministic inverse distance weighting (IDW) was found to be more suitable for the vertical mode ($R^2 = 0.83$).

The applicability of these methods has been investigated mostly for the prediction and interpretation of the spatial distribution of soil salinity at a point in time. Recently, Xie et al. [123], based on ECa data using the MK2 and the geostatistical methods of semivariograms and ordinary kriging, which were conducted with the aid of Geostatistics Software (GS + 7.0) and ArcMap 10.2, respectively, quantified and interpreted the spatiotemporal distribution and variations in regional soil salinity across several years. The satisfactory results of the linear prediction model, which was conducted with ECa measurements for the topsoil ($R^2 > 0.90$), indicate the use of the MK2 in the assessment and mapping of the spatiotemporal variability of soil salinity and foster the potential for further applications under similar soil conditions.

In the field, changes in solute distribution on a spatial and temporal scale have been characterized and evaluated in three dimensions using the MK2 and the non-geostatistical interpolation method of inverse distance weighting (IDW) [125]. In this work, multivariate linear models were established for the relationship of ECe with ECa in various soil layers and times. The predicted ECe values were then interpolated using the deterministic 3D IDW technique. Besides the high model reliability ($0.82 < R^2 < 0.99$), the 3D IDW was also proven to predict the three-dimensional spatiotemporal variations in soil salinity with good accuracy and $R^2$ values ranging from 0.76 to 0.77.

Regardless of the option of a deterministic or geostatistical interpolation method, mapping the spatiotemporal characteristics of solute distribution by the geostatistical modeling approach comes with certain limitations. The existence of a strong relationship between ECe and ECa measurements is a basic requirement [11]. In addition, the accumulation of high-quality data following specific distributional assumptions for the prediction of unsampled values is necessary. Finally, since these methods are based on the spatial dependence of the variables, they may not be suitable for processing and predicting more complex and non-spatial relationships.

#### 3.2.4. Inversion Models

Monitoring and mapping of the vertical extent of salinity within the soil profile using the EM38 probes has been remarkably improved by the inversion approach [126]. Inversion modeling is an evolving and steadily increasing applied process for the estimation and mapping of depth-specific soil salinity from EMI data [48,127]. It consists of various complex algorithms and calculation methods that enable the conversion of the recorded ECa data to depth-specific estimates of electrical conductivity. The inverted data are then modeled using calibration techniques for the prediction of ECe at any depth [128,129] or depth increments [130], and the production of multidimensional (1D, 2D, 3D) maps of the salinity profile for the investigated district [131]. The rapid and efficient assessment of solute distribution with depth is essential for acquiring an accurate and real-time quantitative interpretation of the salt dynamics in the profile and especially in the rootzone [112,132].

Recently, the growing interest in inversion procedures for the spatiotemporal analysis of soil salinity has been associated with the development of multi-coil (e.g., CMD-Mini Explorer) and multi-frequency (e.g., GEM 2) EMI sensors, which are designed to take simultaneous measurements at multiple depth ranges [132–134]. However, the cost and limited access to these instruments [126] make the application of EM38 and MK2 an attractive alternative.

Several inversion approaches that employ EM38 data have been explored for modeling the vertical patterns of solutes. Some of the initial attempts were linear [135] and non-linear models [136], which included the Tikhonov regularization to invert the ECa data. The data in these cases were obtained by taking multi-height EM38 measurements in the horizontal and vertical orientations at different sites. Second-order Tikhonov regularization is a mathematical method suggested for reducing possible data errors and stabilizing the inversion process. It is applied as a technique to overcome the "ill-posedness" and the "non-uniqueness" that are encountered in the inversion. These problems entail measurement or data errors that can induce significant changes in the outcome and the fact that there might be more than one solution for different ECa profiles [117,137]. In addition to these issues, the 1D inversion models, as mentioned above, though applicable, characterize only the vertical distribution with depth, thus providing limited information on the actual transfer of salt within the soil profile [131,138].

Other approaches, in order to predict the estimates of soil salinity in discrete soil layers, joined the EM38 measurements with ECa data from different EMI sensors, like the EM31 [139] or EM34 [140], and inverted them with a 1D algorithm with 2D smoothness constraints to display the vertical and lateral variations in soil salinity in the subsurface along transects [139]. Moghadas et al. [141] proposed a joint inversion of the horizontal and vertical EM38 data using a probabilistic optimization algorithm. The derived one-dimensional inversion models were merged to generate a 3D image of the subsurface distribution of soil salinity at a regional scale. To evaluate the robustness of the models, the inverted data were calibrated to ground-truth ECe values using linear regression for certain depths. As observed, the models exhibited a good prediction ($R^2 = 0.67$) for the shallow layers (30 cm), while for the deeper soil layers (60, 90 cm) the discrepancies were attributed to the high clay content of the study area.

The efficient models that have been developed during the last decade rely on advanced inversion strategies that can extract 2D or 3D electromagnetic conductivity images (EMCIs) of the spatial distribution of soil salinity from the ECa data [131]. Software packages like EM4Soil have been released to assist in the process of inversion of ECa data directly in 1D [72], 2D [128,142,143], or quasi-3D [126,129] layered conductivity values and generate EMCIs by applying inversion parameters. The prediction of ECe from the inverted EM38 readings can be achieved with high accuracy by establishing simple calibration equations, such as linear regression (LR) [128–130]. Farzamian et al. [126], addressing the need for affordable and easily accessible monitoring tools as an alternative to non-available multi-coil instruments, proposed the use of multi-height EM38 data and a quasi-3D inversion algorithm for the development of ECe maps in the landscape. They indicated that by collecting multiple ECa measurements at different heights, a single regional calibration equation (LR) instead of discrete depth-specific calibrations may predict the ECe at any desired depth.

By these means, detailed maps of the spatial patterns of ECe across various depths in large and landscape-irrigated areas can be efficiently produced. Also, the quasi-3D inversion models of ECe can be combined with remote sensing to create a 3D map and illustrate the quantitative and qualitative spatial distribution of soil salinity in the survey area [130]. The application of inverse modeling to the three-dimensional distribution of soil salinity is an important evolutionary step for characterizing and interpreting the lateral and vertical variations within the soil profile. This contribution might be particularly prominent in mapping the local 3D patterns of the solutes variability in more complicated irrigation schemes, such as micro-irrigation systems [99].

Moreover, the temporal distribution of soil salinity with depth has been examined most recently by using time-lapse inversion of ECa measurements with multiconfiguration systems [132,144,145]. Time-lapse inversion of ECa data by EM38 and MK2 instruments is a challenging approach that has not been investigated for soil salinity assessment [146,147] and only with a strictly limited scope for moisture content distribution, which involves measurements from the electrical resistivity tomography (ERT) [145,146,148]. The difficulty is ascribed to the complex dynamics of solutes, which can vary significantly within the soil profile. Therefore, the method requires multiple local soil data and measurements to be repeated over time for the determination of the solute's temporal trend [48]. A recent attempt has been made by Paz et al. [127], who collected repeated measurements by the EM38 sensor at specific locations and dates and inverted them to obtain time-lapse EMCIs of the vertical profiling of the layered estimates of electrical conductivity. By using a pre-constructed calibration model of ECe for the same area, the layered conductivity estimates were converted to predict ECe. The prediction model was found to be adequately precise ($R^2 = 0.88$) in depicting the spatiotemporal variations in soil salinity with depth and across the study area.

While the inversion modeling approach yields positive outcomes in the spatiotemporal mapping of soil salinity, it should be mentioned that the credibility and the accuracy of its application depend on the quality of the acquired data and regularization factors [148]. Therefore, thorough validation using datasets independent of those employed in the prediction process is pivotal for minimizing uncertainties in the results [127] and ensuring the credibility of the survey. Furthermore, when using time-lapse inversion at regional scales, the establishment of precise local-specific calibrations needs to be taken into account. High temporal variations in dynamic soil properties, such as water content, can have a substantial impact on the assessment of the solute variations and consequently misdirect soil treatment decisions [127].

#### 3.2.5. Machine-Learning-Based Models

Lately, a new modeling approach has been introduced for characterizing and quantifying the spatial variations and depth distribution of soil salinity with data from EM38 and MK2 devices. This encompasses the integration of ECa measurements, remote sensing data, and environmental variables with machine-learning (ML) technology to generate digital soil maps (DSMs) of ECe in large arid zones [4,52,149,150]. According to this approach, the prediction and mapping of salt-affected areas can be substantially improved when proximal sensed data and multi-spectral information from remote sensing imagery are incorporated [54], and advanced machine-learning algorithms are utilized to assess and model the relationship of ECe with these multiple variables [4,150].

Plenty of modeling methods have been elaborated and extensively used in DSM applications for various soil properties [151], including soil salinity. However, advanced methods that utilize various machine-learning algorithms with EMI and the EM38 and MK2 probes in particular for the prediction of soil salinity are currently being explored. The growing interest in machine-learning modeling is attributed to its ability to process huge volumes of datasets and identify the complex and non-linear interactions between soil properties and various environmental features [152]. For soil salinity assessment, the models use a composite of statistical and mathematical formulas through which they capture and "learn" the relationship between the spatial variations in ECa data and input variables. After successful training, they can predict the unknown spatial patterns of soil salinity at large scales [4].

Contrary to geostatistical models, machine-learning models employ a more computational operation for the prediction of soil salinity's spatial variability. While geostatistical modeling relies on the spatial dependency structure of the data, machine-learning-based models use algorithms that detect the relations and patterns between soil properties and the variables.

There are many machine-learning algorithms available for the prediction and digital mapping of soil properties, with varying principles, complexity, and overall performance. Some of them include decision trees (DTs), random forest (RF), support vector machines (SVMs), genetic programming (GP), and artificial neural networks (ANNs). Among them, random forest has been proven to be a reliable and robust option for developing prediction models of soil salinity and producing explicit digital soil maps, particularly in arid areas [10]. Ding et al. [4], based on the dipole mode of the portable MK2, established a random forest model with ECa measurements, auxiliary variables from remote sensing, and environmental products for predicting the spatial variations in different land types of the oasis agroecosystems in China. The good relationship between the horizontal and vertical ECa data with ECe at all depth increments indicated ECa as a valid predictor for modeling the spatial variations in soil salinity. The application of random forest models with ECa measurements and auxiliary data was found to have high accuracy, with $R^2$ ranging from 0.77 to 0.84 for all coil configurations. Furthermore, according to the results for various land uses, the derived RF prediction models of ECa seemed to perform better for the deeper soil layers and the bare lands, with an $R^2$ range from 0.84 to 0.91. Slightly different results for all depth intervals ($0.61 < R^2 < 0.65$) were demonstrated in the same regional area when RF models of ECe were constructed with the involvement of MK2 measurements and remote sensing imagery as covariates [150].

Owing to the implementation of machine-learning-based models with the use of the MK2 probe, fine-resolution maps of soil salinity in large areas may be produced [150,153]. Moreover, the magnitude of influence the diverse soil, environmental, or other site-related variables have in the prediction of soil salinity can easily be exhibited and evaluated [149]. Their capability to provide complex non-linear interrelations between soil properties and a vast amount of existing data without making any distributional assumptions may offer great opportunities in future surveys of soil salinity mapping.

Despite the inspiring benefits, their application is associated with practical weaknesses. The prediction of subsurface soil salinity using machine-learning techniques is primarily driven by the diverse variables that are utilized as inputs for their training. Thus, collecting and employing accurate and relevant attributes from the available resources require some level of data expertise. Also, given that the efficiency of these models depends on the quality of the input data, there cannot be a unique algorithm for assessing the spatial and temporal variability of soil salinity in all soil and site circumstances with the same accuracy.

#### 3.2.6. Hybrid Models

The nature of soil salinity comprises alterations over time and space. The diverse changes, which extend from the complex distribution in the rootzone to spatial variations across the field and larger scales, are crucial to determine in order to interpret and control the sources of soil salinization. These issues, however, need to be reconciled with rapid and resilient methods, as precise irrigation systems and soil sustainability strategies at the regional and global levels are imperative. The application of individual modeling techniques like simple regression-based models may fail to adapt to all these requirements. To this end, hybrid models have emerged, which combine different methods and data sources for estimating and mapping the spatial variations in soil salinity at different scales [101].

The hybrid modeling approach is presented as a broader category of models that encompasses the integration of multiple techniques with data obtained by proximal EM38 and MK2 devices and additional data mainly derived from remote sensing.

These models incorporate geostatistical, machine learning, and empirical methods to leverage the strengths and advantages of each approach and overcome the constraints or weaknesses that they might have as individual models in soil salinity assessment [154]. Particularly for the detection and mapping of the spatial distribution of solutes within the profile and across a large area, the hybrid models based on EM38, MK2, and remote sensing data integrate advanced hybrid geostatistical techniques like regression kriging with machine-learning algorithms, such as random forest or Cubist. This fusion of approaches and data sources can identify and predict the complex relationships of solutes with different variables within the profile and, at the same time, determine the soil salinity's spatial variations in the study area.

This innovative and promising modeling approach could potentially lead to more accurate predictions of spatiotemporal patterns of soil salt content and eventually to a better understanding and representation of soil salinity dynamics at a regional level. To date, however, few studies have examined its efficiency in monitoring and mapping soil salinity using the EM38 and MK2 probes.

The hybrid modeling approach of Taghizadeh-Mehrjardi et al. [149] combined remote sensing and MK2 data with hybrid geostatistical methods and machine learning. They applied regression kriging for the spatial prediction and digital mapping of horizontal and vertical ECa using auxiliary data at certain depth intervals. Then, the advanced machine-learning algorithm Cubist (regression tree) was applied to the set of variables to model the spatial distribution of ECe values at the standard depths. As they documented, the prediction models of ECe in the arid region of Arkadan had varying accuracy ($0.11 < R^2 < 0.78$), with the spatial distribution performing better in the upper soil (0–30 cm).

Another hybrid approach for modeling and mapping the spatial variations in soil salinity is the combination of multi-spectral data from high-resolution remote imagery with measurements from the MK2 and the use of partial least-squares regression (PLSR) [101]. As was reported, the constructed spectral-PLSR-based prediction models could reliably ($R^2 = 0.67$) detect and monitor the variations in soil salinity in the oasis region of the Keriya River in China.

In an attempt to explore methods for integrating new sources of soil data as inputs in DSMs, Zare et al. [52] compared three approaches using MK2 measurements and ECe data collected from a region near the saline Maharlu Lake in Iran. They concluded that the approach involving the combination of machine learning, quantile random forest model, and regression co-kriging on the residuals, with $R^2$ values up to 0.79, can be effectively used for the prediction of ECe. Also, the MK2 data were found to be a reliable and meaningful input for digitally mapping the soil salinity variations, especially in cases where there is a lack of appropriate remote sensing data.

Nonetheless, the hybrid modeling approach, in order to be applicable, requires a correlation between the input variables. Specifically for generating soil salinization maps, a strong relationship between ECa measurements and ground-truth data of ECe needs to be valid. In addition, the selection of the appropriate modeling techniques according to the existing survey data plays a significant role in the reliability and accuracy of the results [154].

A summary of the documented modeling approaches for the assessment and mapping of soil salinity by the EM38 and MK2 measurements is presented in Table 2. Each modeling approach is accompanied by the various techniques or methods that have been employed within it. Referenced studies for each case are also shown for further exploration.

### 3.3. *Fusion of the EM38, MK2, and Remote Sensing Data for Soil Salinity Monitoring and Mapping*

To mitigate the global impacts of soil salinization, there is an urgent need to retrieve accurate information on its status in arid and semi-arid regions quickly and consistently. This has prompted researchers to gradually develop more robust approaches that integrate multiple data and technologies to interpret the changes in soil salt content and preserve the agricultural sustainability of the irrigated systems in these areas [155,156]. In the last few years, these efforts have focused on the combination of remote sensing data with ground-based EMI sensors for complementary and potentially high-precision monitoring and mapping of soil salinity variability at different scales [1,54,56,149,157].

Remote sensing, by providing repetitive, current, and prompt high-resolution images like those of WorldView 2, has become a valuable tool for identifying and mapping the solutes' spatiotemporal variations across the surface of large and severely saline areas [158,159]. The capability, however, of the available multi-spectral satellite data is restricted to the detection and assessment of surface soil salinity [160], disregarding the three-dimensional spatial distribution of salt content in the soil profile. Moreover, spectral reflectance can be problematic in regions with slightly low to moderate salinity levels and limited or no visible salt features due to interference from other site factors, such as vegetation cover or soil type [159,161].

**Table 2.** Short summary of the modeling approaches and methods that have been used for the assessment and mapping of soil salinity in terms of ECe using EM38 or EM38 MK2 measurements, at different scales.

| Model | Methods | Reference Studies |
|---|---|---|
| Regression-Based (Calibration Models) | Linear, multiple linear | [104,107] |
| | Simple depth-weighted coefficients | [105] |
| | Established coefficients | [108,109] |
| | Modeled coefficients | [103] |
| | Mathematical coefficients | [110] |
| Geostatistical | Interpolation methods (OK [1], CO-K [2], OCK [3], universal kriging, indicator kriging, 3D kriging) Semi/variogram analysis | [8,94,98,120,122,123] |
| Deterministic Spatial Interpolation | 3D IDW | [125] |
| Inversion | Tikhonov regularization | [135,136] |
| | Joint inversion | [72,139,141] |
| | 2D algorithms | [128,142,143] |
| | Quasi-3D algorithms | [126,129] |
| Machine-Learning-Based | Random forest | [4,150] |
| Hybrid | Cubist (ML), regression kriging | [149] |
| | Partial least-squares regression, spectral index | [101] |
| | Quantile random forest (ML), regression co-kriging | [52] |

[1] OK: ordinary kriging, [2] CO-K: co-kriging, [3] OCK: ordinary co-kriging.

On the contrary, the EM38 and MK2 sensors have been broadly employed and evaluated for the quantification and characterization of soluble salt distribution within the subsurface and the rootzone. The fusion of EM38 and MK2 field measurements with multi-temporal [162,163] and multi-spectral [73,101,124,150] remote sensing datasets has been documented in a few soil salinity surveys, which vary from agricultural and urban greenery systems to different soil crops, irrigation schemes, and scales. In most circumstances, particularly in bare lands and large arid or semi-arid areas, it has been suggested as an effective, inexpensive, and time-efficient method for predicting and monitoring the spatiotemporal variations in soil salinity with relative accuracy [54,101,149,163].

For instance, Wu et al. [162], based on acquired MK2 field measurements and ECe from soil sampling, used multi-temporal remote sensing (vegetation indices) data to develop salinity models of the severely salinized region of Dujaila in Iraq. The derived multi-year maxima-based models, which were applied for mapping and tracking the spatiotemporal changes in the salt-affected areas, achieved high accuracy in predicting soil salinity, with $R^2$ reaching a value of 82.5. Vegetation indices such as the NDVI that are utilized in remote sensing tend to become a significant innovative tool for mapping the entire rootzone salinity at the regional scale and potentially in the field and landscape as well [48]. However, they need to be calibrated. Thus, even though the process of ground-truth data collection can be considerably reduced owing to the MK2 measurements [52], it remains a necessity for appropriate field calibration [160]. Ding and Yu [54] constructed regression models using EM38 readings and various spectral indices from satellite images for the prediction and evaluation of seasonal and spatial variations in soil salinity in the Delta Oasis of the Tarim Basin, China. They also examined three interpolation techniques for assessing the distribution patterns of salt concentration within the region. According to their findings, the fusion of EM38 data with salinity indices from remote sensing images can provide the assessment of salinity at a regional scale for both dry and wet seasons with fairly high accuracy. Moreover, the results indicate the importance of EM38 measurements in designing rapidly and simply the most suitable soil sampling strategy for the survey. Additional soil sampling is needed for the interpolation's accuracy. Among universal kriging, spectral index regression, and regression kriging, the regression kriging with nested spherical model was found to have the closest fit to the measured ECa ($R^2 > 0.90$).

On other occasions, however, the combination of the sensors seems to need further research. By employing an EM38 sensor and using data from remote sensing, Nouri et al. [73] investigated the soil salinity in urban greenery spaces of the Adelaide Parklands in Australia. In order to predict the variations in solutes at different spatial and temporal scales, regression models for the EM38 were developed with various vegetation and salinity indices from high-resolution satellite images. The EM38 data were validated with ECe values from soil samples. As the authors observed, although the soil-adjusted vegetation index (SAVI) was found to be a good predictor, the proximal sensor was considered more efficient than the spectral indices in predicting soil salinity in urban landscapes in semi-arid climates. Additionally, Gharsallah et al. [124] observed that, in large arid lands that are densely planted, such as an irrigated olive orchard, multivariate models of EM38 data with soil spectral indices from low-resolution images may exhibit poor accuracy in predicting and mapping the soil salinity distribution ($R^2 < 0.12$). Nonetheless, in these models, the integration of vegetation indices and high-resolution images may resolve the influence of the dense tree canopy on the remote sensing reflectance and improve the overall reliability [124].

The fusion of field EM38 and MK2 data with remote sensing techniques, including high-quality satellite data and selected indicators like vegetation indices, can contribute to more thorough and concordant maps of soil salinity. Monitoring the spatial and temporal distribution of solutes through field EMI surveys and remote sensing can be sufficiently accomplished without excessive soil sampling. This can accelerate the understanding of soil salinization severity at multiple scales and lead to rapid remediation measures.

## 4. Discussion

From the results of the research, it was observed that the approaches utilizing the EM38 and MK2 data for soil salinity assessment and mapping have been centered around regression-based (calibration) models, geostatistical methods, deterministic interpolation methods, inversion models, machine-learning algorithms, and hybrid modeling.

According to the reviewed literature, the developed calibration models exhibit limited applicability since they use entirely site-specific equations and regression coefficients that cannot be accurately applied in different locations [48,85,111,112]. Moreover, these models rely on distributional and subsurface homogeneous assumptions, which exclude from their equations the impact of influential soil properties, such as soil texture and soil water content, on ECa measurements [63]. Nevertheless, linear regression can be employed as a viable, easy, and simple calibration technique in conjunction with the geostatistical [94,122,123] and the inversion [126–130] modeling approaches. The geostatistical models have contributed significantly to the prediction and interpretation of soil salinity spatial variations over the years. With a diversity of classic and more advanced interpolation methods available, they have been proven to provide relatively accurate estimations of ECe on a range of soil scales [94,120,123]. However, since their application depends on the spatial structure of the collected data, their solitary use is not suggested in circumstances where more complex and non-spatial relationships among the variables need to be assessed. As was observed from the findings, the inversion modeling approach has gained wide acceptance in the scientific community, and its applications in salinity surveys using the EM38 and MK2 sensors are increasing and evolving. The benefit of these methods lies in the easy conversion of multiple ECa measurements through software to determine ECe at different depths and generate detailed multidimensional (2D, 3D) images of the soil salinity profile with high accuracy [118–120,126,128,142,143]. Besides the variety of studies that characterize the spatial variations using the EM38 and MK2, recent studies have made efforts to assess the temporal variability of solutes, yielding good results [127]. In all cases, the inversion prediction models require rigorous validation with independent datasets to be credible [127]. Machine-learning-based models have also emerged as a powerful approach in the current field of soil salinity monitoring and mapping. Their capability to process immense amounts of data and capture complex, non-linear interrelations between the sensor's ECa data and a plethora of input variables is considered their major advantage over other modeling techniques. The findings from the literature demonstrated that the implementation of machine-learning techniques combined with MK2 data, auxiliary data from remote sensing, and environmental variables can predict with significant accuracy the spatial and temporal variability of soluble salts on the large scale of arid zones and generate fine-resolution maps [4,52,149,150]. Recently, an increasing number of studies have focused on developing such integrating approaches and coupling them with hybrid techniques of geostatistical interpolation and machine learning. These approaches have been found to produce reliable soil salinity predictions and DSMs, particularly in the arid regions of the world [52,149], but caution is needed in the selection of the proper techniques. In a similar sense, the review's findings highlighted the fusion of the EM38 and MK2 data with high-quality satellite data and vegetation indices from remote sensing as an effective, rapid, and promising method for the prediction and mapping of the spatiotemporal variations in soil salinity in large arid areas [54,101,149,162,163].

## 5. Conclusions and Future Directions

By exploring the latest modeling approaches for the assessment and mapping of soil salinity using the EM38 and MK2 sensors, we attempted to enrich and enhance the current research field of soil salinization monitoring and mitigation through this review. Specifically, through a comprehensive summary of the applied approaches and methods, the paper intended to elucidate and underline the role of EM38 and MK2 in the detection and interpretation of soil salinity and facilitate researchers and users in making well-informed decisions concerning future salinity surveys and agricultural treatments.

The advantage of the EM38 and MK2 sensors in soil salinity surveys is their capability to collect rapid and effortlessly numerous field ECa measurements that can be conveniently related to soil salinity, especially in high-salt-affected areas. By applying simple or more sophisticated modeling approaches, the ECa data can be converted to salinity estimates of ECe, revealing valuable insights into the spatiotemporal patterns and depth distribution of solutes at various scales and land types. Due to the complicated interactions of soil salinity with various soil properties and factors and its continuous changes over time and space, the measurements by EM38 and MK2 sensors are pivotal for the subsurface mapping of saline soils in a timely manner. The enhanced ground conductivity meter MK2, which is easy to operate and cost-effective compared to the latest multi-coil sensors, can be significantly useful in subsurface salinity mapping, especially in cases where other soil data sources, such as remote sensing, are unavailable.

To date, despite the research and technological progress, there is not a universally approved and applicable modeling approach for the assessment and monitoring of soil salinization. Therefore, the construction of efficient salinity models using the EM38 and MK2 probes should be carried out considering the available approaches and their adaptation under different soil and environmental conditions. It is essential to have a full understanding of the current modeling options and their techniques so the optimal method can be selected and the spatiotemporal characteristics of the salt-affected areas can be accurately identified and interpreted.

In this sense, by using the sensor's field readings at various depths and employing an inverse modeling technique, the spatial and temporal variability of soil salinity at large irrigated farms can be predicted and quantified in 3D maps. This enables the detailed monitoring of soil salt variations within the field and the design of proper soil management practices. Moreover, advanced approaches like hybrid models, which combine EM38 and MK2 readings with multiple site-specific datasets and methods like machine learning and interpolation, may overcome the limitations of single model applications and increase the accuracy of ECe predictions. Subsequently, the interpretation of the complex soil salinity dynamics across the study area can be fulfilled. Finally, the recent trend of integrating georeferenced EM38 and MK2 measurements with remote sensing data indicates great opportunities in delineating the soil salinity distribution within the rooting zone and across large arid regions. This approach can improve the implementation of precise and timely sustainable strategies and substantially reduce the need for excessive soil sampling.

Consequently, further exploration of such promising methods at multiple scales is crucial for generating reliable and up-to-date salinity maps. More attention should be directed toward extending and validating the inverse modeling at a regional scale and in distinct irrigation systems to create consistent 3D maps of the spatiotemporal variability of soil salinity. In addition, efforts to incorporate machine-learning algorithms with multi-period remote sensing data, auxiliary variables, and EMI data should be intensified to track and map the long-term variations in soluble salt depositions at the regional level, refine the method, and expand it toward global-scale predictions. Furthermore, a deeper investigation of the diverse environmental, topographic data, and spectral indices as covariates is needed to identify and optimize the predictor variables of soil salinity mapping in different soil conditions and cultivation systems.

Ultimately, the validation of robust, sophisticated approaches that combine field measurements using sensors like the EM38MK2 and contemporary technologies can potentially minimize the time, labor, and cost of ground-truth data and render a fundamental basis for the efficient prediction and real-time monitoring of soil salinization at the field, regional, and global scales.

**Author Contributions:** Conceptualization, G.K. and P.A.P.; methodology, P.A.P.; validation, G.K. and P.A.P.; formal analysis, P.A.P.; investigation, P.A.P.; resources, P.A.P.; data curation, P.A.P.; writing—original draft preparation, P.A.P.; writing—review and editing, P.A.P.; visualization, G.K. and P.A.P.; supervision, G.K.; project administration, G.K. and P.A.P. All authors have read and agreed to the published version of the manuscript.

**Funding:** This research received no external funding.

**Data Availability Statement:** Not applicable.

**Conflicts of Interest:** The authors declare no conflict of interest.

## References


1. Corwin, D.L. Climate Change Impacts on Soil Salinity in Agricultural Areas. *Eur. J. Soil Sci.* **2021**, *72*, 842–862. [CrossRef]
2. Besma, Z.; Christian, W.; Didier, M.; Pierre, M.J.; Mohamed, H. Soil Salinization Monitoring Method Evolution at Various Spatial and Temporal Scales in Arid Context: A Review. *Arab. J. Geosci.* **2021**, *14*, 283. [CrossRef]
3. Scudiero, E.; Corwin, D.; Anderson, R.; Yemoto, K.; Clary, W.; Wang, Z.; Skaggs, T. Remote Sensing Is a Viable Tool for Mapping Soil Salinity in Agricultural Lands. *Calif. Agric.* **2017**, *71*, 231–238. [CrossRef]
4. Ding, J.; Yang, S.; Shi, Q.; Wei, Y.; Wang, F. Using Apparent Electrical Conductivity as Indicator for Investigating Potential Spatial Variation of Soil Salinity across Seven Oases along Tarim River in Southern Xinjiang, China. *Remote Sens.* **2020**, *12*, 2601. [CrossRef]
5. Goldshleger, N.; Ben-Dor, E.; Lugassi, R.; Eshel, G. Soil Degradation Monitoring by Remote Sensing: Examples with Three Degradation Processes. *Soil Sci. Soc. Am. J.* **2010**, *74*, 1433–1445. [CrossRef]
6. Rodrigues, H.M.; Vasques, G.M.; Oliveira, R.P.; Tavares, S.R.L.; Ceddia, M.B.; Hernani, L.C. Finding Suitable Transect Spacing and Sampling Designs for Accurate Soil ECa Mapping from EM38-MK2. *Soil Syst.* **2020**, *4*, 56. [CrossRef]
7. Lesch, S.M.; Strauss, D.J.; Rhoades, J.D. Spatial Prediction of Soil Salinity Using Electromagnetic Induction Techniques: 1. Statistical Prediction Models: A Comparison of Multiple Linear Regression and Cokriging. *Water Resour. Res.* **1995**, *31*, 373–386. [CrossRef]
8. Triantafilis, J.; Odeh, I.O.A.; McBratney, A.B. Five Geostatistical Models to Predict Soil Salinity from Electromagnetic Induction Data Across Irrigated Cotton. *Soil Sci. Soc. Am. J.* **2001**, *65*, 869–878. [CrossRef]
9. Aboelsoud, H.; Abdel-Rahman, M. Rapid Field Technique for Soil Salinity Appraisal in North Nile Delta Using EM38 through Some Empirical Relations. *Int. J. Plant Soil Sci.* **2017**, *14*, 1–9. [CrossRef]
10. Wang, F.; Shi, Z.; Biswas, A.; Yang, S.; Ding, J. Multi-Algorithm Comparison for Predicting Soil Salinity. *Geoderma* **2020**, *365*, 114211. [CrossRef]
11. Corwin, D.L.; Scudiero, E. Field-scale Apparent Soil Electrical Conductivity. *Soil Sci. Soc. Am. J.* **2020**, *84*, 1405–1441. [CrossRef]

12. Heil, K.; Schmidhalter, U. The Application of EM38: Determination of Soil Parameters, Selection of Soil Sampling Points and Use in Agriculture and Archaeology. *Sensors* **2017**, *17*, 2540. [CrossRef] [PubMed]
13. Heil, K.; Schmidhalter, U. Theory and Guidelines for the Application of the Geophysical Sensor EM38. *Sensors* **2019**, *19*, 4293. [CrossRef] [PubMed]
14. Williams, B.; Baker, G. An Electromagnetic Induction Technique for Reconnaissance Surveys of Soil Salinity Hazards. *Soil Res.* **1982**, *20*, 107. [CrossRef]
15. De Jong, E.; Ballantyne, A.K.; Cameron, D.R.; Read, D.W.L. Measurement of Apparent Electrical Conductivity of Soils by an Electromagnetic Induction Probe to Aid Salinity Surveys. *Soil Sci. Soc. Am. J.* **1979**, *43*, 810–812. [CrossRef]
16. Nogués, J.; Robinson, D.A.; Herrero, J. Incorporating Electromagnetic Induction Methods into Regional Soil Salinity Survey of Irrigation Districts. *Soil Sci. Soc. Am. J.* **2006**, *70*, 2075–2085. [CrossRef]
17. Abdu, H.; Robinson, D.A.; Jones, S.B. Comparing Bulk Soil Electrical Conductivity Determination Using the DUALEM-1S and EM38-DD Electromagnetic Induction Instruments. *Soil Sci. Soc. Am. J.* **2007**, *71*, 189–196. [CrossRef]
18. De Benedetto, D.; Castrignanò, A.; Rinaldi, M.; Ruggieri, S.; Santoro, F.; Figorito, B.; Gualano, S.; Diacono, M.; Tamborrino, R. An Approach for Delineating Homogeneous Zones by Using Multi-Sensor Data. *Geoderma* **2013**, *199*, 117–127. [CrossRef]
19. Kargas, G.; Londra, P.A.; Sotirakoglou, K. Evaluation of Soil Salinity Using the Dielectric Sensor WET-2. *Soil Res.* **2022**, *61*, 397–409. [CrossRef]
20. Sudduth, K.A.; Drummond, S.T.; Kitchen, N.R. Accuracy Issues in Electromagnetic Induction Sensing of Soil Electrical Conductivity for Precision Agriculture. *Comput. Electron. Agric.* **2001**, *31*, 239–264. [CrossRef]
21. Krishan, G.; Kumar, R.; Suresh; Gopal, K.; Saha, S.K. Measuring Soil Salinity with WET Sensor and Characterization of Salt Affected Soils. *Agropedology* **2008**, *18*, 124–128.
22. Kargas, G.; Persson, M.; Kanelis, G.; Markopoulou, I.; Kerkides, P. Prediction of Soil Solution Electrical Conductivity by the Permittivity Corrected Linear Model Using a Dielectric Sensor. *J. Irrig. Drain. Eng.* **2017**, *143*, 04017030. [CrossRef]
23. Kargas, G.; Kerkides, P. Determination of Soil Salinity Based on WET Measurements Using the Concept of Salinity Index. *J. Plant Nutr. Soil Sci.* **2018**, *181*, 600–605. [CrossRef]
24. Visconti, F.; De Paz, J.M. Field Comparison of Electrical Resistance, Electromagnetic Induction, and Frequency Domain Reflectometry for Soil Salinity Appraisal. *Soil Syst.* **2020**, *4*, 61. [CrossRef]
25. Huang, J.; Koganti, T.; Santos, F.A.M.; Triantafilis, J. Mapping Soil Salinity and a Fresh-Water Intrusion in Three-Dimensions Using a Quasi-3d Joint-Inversion of DUALEM-421S and EM34 Data. *Sci. Total Environ.* **2017**, *577*, 395–404. [CrossRef] [PubMed]
26. Gebbers, R.; Lück, E.; Dabas, M.; Domsch, H. Comparison of Instruments for Geoelectrical Soil Mapping at the Field Scale. *Near Surf. Geophys.* **2009**, *7*, 179–190. [CrossRef]
27. Khongnawang, T.; Zare, E.; Srihabun, P.; Triantafilis, J. Comparing Electromagnetic Induction Instruments to Map Soil Salinity in Two-Dimensional Cross-Sections along the Kham-Rean Canal Using EM Inversion Software. *Geoderma* **2020**, *377*, 114611. [CrossRef]
28. Rhoades, J.D. Electrical Conductivity Methods for Measuring and Mapping Soil Salinity. In *Advances in Agronomy*; Sparks, D.L., Ed.; Academic Press: Cambridge, MA, USA, 1993; Volume 49, pp. 201–251. [CrossRef]
29. Hendrickx, J.M.H.; Baerends, B.; Raza, Z.I.; Sadig, M.; Chaudhry, M.A. Soil Salinity Assessment by Electromagnetic Induction of Irrigated Land. *Soil Sci. Soc. Am. J.* **1992**, *56*, 1933–1941. [CrossRef]
30. Doolittle, J.A.; Brevik, E.C. The Use of Electromagnetic Induction Techniques in Soils Studies. *Geoderma* **2014**, *223–225*, 33–45. [CrossRef]
31. Bennett, D.L.; George, R.J. Using the EM38 to Measure the Effect of Soil Salinity on Eucalyptus Globulus in South-Western Australia. *Agric. Water Manag.* **1995**, *27*, 69–85. [CrossRef]
32. Domsch, H.; Giebel, A. Estimation of Soil Textural Features from Soil Electrical Conductivity Recorded Using the EM38. *Precis. Agric.* **2004**, *5*, 389–409. [CrossRef]
33. Heil, K.; Schmidhalter, U. Comparison of the EM38 and EM38-MK2 Electromagnetic Induction-Based Sensors for Spatial Soil Analysis at Field Scale. *Comput. Electron. Agric.* **2015**, *110*, 267–280. [CrossRef]
34. Saey, T.; Van Meirvenne, M.; Vermeersch, H.; Ameloot, N.; Cockx, L. A Pedotransfer Function to Evaluate the Soil Profile Textural Heterogeneity Using Proximally Sensed Apparent Electrical Conductivity. *Geoderma* **2009**, *150*, 389–395. [CrossRef]
35. Corwin, D.L.; Lesch, S.M. Application of Soil Electrical Conductivity to Precision Agriculture: Theory, Principles, and Guidelines. *J. Agron.* **2003**, *95*, 455–471. [CrossRef]
36. EM38-MK2-Operating-Manual.Pdf. Available online: https://geophysicalequipmentrental.com/files/2020/01/EM38-MK2-Operating-Manual.pdf (accessed on 30 August 2023).
37. Catalano-Measuring Soil Conductivity with Geonics Limited E.Pdf. Available online: https://adamchukpa.mcgill.ca/gwpss/Presentations/GWPSS_2011_Catalano.pdf (accessed on 1 October 2023).
38. Siddique, M.N.A. Potential of Soil Sensor (EM38) Measurements for Soil Fertility Mapping. Master's Thesis, Ghent University, Brussel, Belgium, 2020.
39. Doolittle, J.; Petersen, M.; Wheeler, T. Comparison of Two Electromagnetic Induction Tools in Salinity Appraisals. *J. Soil Water Conserv.* **2001**, *56*, 257–262.

40. Urdanoz, V.; Aragüés, R. Comparison of Geonics EM38 and Dualem 1S Electromagnetic Induction Sensors for the Measurement of Salinity and Other Soil Properties: Comparison of Geonics EM38 and Dualem 1S Sensors. *Soil Use Manag.* **2012**, *28*, 108–112. [CrossRef]
41. McNeill, J.D. *Electromagnetic Terrain Conductivity Measurement at Low Induction Numbers: Mississauga. Ontario, Canada*; Technical Note TN-6; Geonics Ltd.: Mississauga, ON, Canada, 1980; Volume 6.
42. Bonomi, E.; Manzi, C.; Pieroni, E.; Deidda, G.P. *Inversion of EM38 Electrical Conductivity Data*; II International Workshop on Geo-Electro-Magnetism (Proccedings), Lerici, Italy, 26–28 September 2001; Accademia Lunigianese di Scienze G. Capellini: La Spezia, Italy, 2001.
43. Yao, R.-J.; Yang, J.-S.; Wu, D.-H.; Xie, W.-P.; Gao, P.; Wang, X.-P. Geostatistical Monitoring of Soil Salinity for Precision Management Using Proximally Sensed Electromagnetic Induction (EMI) Method. *Environ. Earth Sci.* **2016**, *75*, 1362. [CrossRef]
44. Korsaeth, A. Height above Ground Corrections of EM38 Readings of Soil Apparent Electrical Conductivity. *Acta Agric. Scand. B Soil Plant Sci.* **2006**, *56*, 333–336. [CrossRef]
45. Song, J.H. Assessment of the Accuracy of a Soil Salinity Model for Shallow Groundwater Areas in Xinjang Based on Electromagnetic Induction. *Appl. Ecol. Environ. Res.* **2019**, *17*, 10037–10057. [CrossRef]
46. Kitchen, N.R.; Sudduth, K.A.; Drummond, S.T. Mapping of Sand Deposition from 1993 Midwest Floods with Electromagnetic Induction. *J. Soil Water Conserv.* **1996**, *51*, 336–340. Available online: https://www.ars.usda.gov/ARSUserFiles/50701000/cswq-0111-kitchen.pdf (accessed on 31 August 2023).
47. Urdanoz, V.; Amezketa Lizarraga, E.; Clavería, I.; Ochoa, V.; Aragües Lafarga, R. Mobile and Georeferenced Electromagnetic Sensors and Applications for Salinity Assessment. *Span. J. Agric. Res.* **2008**, *6*, 469. [CrossRef]
48. Corwin, D.L.; Scudiero, E. Review of Soil Salinity Assessment for Agriculture across Multiple Scales Using Proximal and/or Remote Sensors. *Adv. Agron.* **2019**, *158*, 1–130. [CrossRef]
49. Dalan, R.A. Defining Archaeological Features with Electromagnetic Surveys at the Cahokia Mounds State Historic Site. *Geophysics* **1991**, *56*, 1280–1287. [CrossRef]
50. Grellier, S.; Florsch, N.; Camerlynck, C.; Janeau, J.L.; Podwojewski, P.; Lorentz, S. The Use of Slingram EM38 Data for Topsoil and Subsoil Geoelectrical Characterization with a Bayesian Inversion. *Geoderma* **2013**, *200–201*, 140–155. [CrossRef]
51. Corwin, D.L.; Lesch, S.M. Protocols and Guidelines for Field-Scale Measurement of Soil Salinity Distribution with $EC_a$-Directed Soil Sampling. *J. Environ. Eng. Geophys.* **2013**, *18*, 1–25. [CrossRef]
52. Zare, S.; Abtahi, A.; Fallah Shamsi, S.R.; Lagacherie, P. Combining Laboratory Measurements and Proximal Soil Sensing Data in Digital Soil Mapping Approaches. *Catena* **2021**, *207*, 105702. [CrossRef]
53. Brevik, E.C.; Fenton, T.E.; Lazari, A. Soil Electrical Conductivity as a Function of Soil Water Content and Implications for Soil Mapping. *Precis. Agric.* **2006**, *7*, 393–404. [CrossRef]
54. Ding, J.; Yu, D. Monitoring and Evaluating Spatial Variability of Soil Salinity in Dry and Wet Seasons in the Werigan–Kuqa Oasis, China, Using Remote Sensing and Electromagnetic Induction Instruments. *Geoderma* **2014**, *235–236*, 316–322. [CrossRef]
55. Corwin, D.L. Past, Present, and Future Trends of Soil Electrical Conductivity Measurement Using Geophysical Methods. In *Handbook of Agricultural Geophysics*; CRC Press: Boca Raton, FL, USA, 2008.
56. Ren, D.; Wei, B.; Xu, X.; Engel, B.; Li, G.; Huang, Q.; Xiong, Y.; Huang, G. Analyzing Spatiotemporal Characteristics of Soil Salinity in Arid Irrigated Agro-Ecosystems Using Integrated Approaches. *Geoderma* **2019**, *356*, 113935. [CrossRef]
57. Moghadas, D.; Behroozmand, A.A.; Christiansen, A.V. Soil Electrical Conductivity Imaging Using a Neural Network-Based Forward Solver: Applied to Large-Scale Bayesian Electromagnetic Inversion. *Appl. Geophys.* **2020**, *176*, 104012. [CrossRef]
58. Friedman, S.P. Soil Properties Influencing Apparent Electrical Conductivity: A Review. *Comput. Electron. Agric.* **2005**, *46*, 45–70. [CrossRef]
59. Corwin, D.L.; Lesch, S.M. Apparent Soil Electrical Conductivity Measurements in Agriculture. *Comput. Electron. Agric.* **2005**, *46*, 11–43. [CrossRef]
60. Rhoades, J.D.; Raats, P.A.C.; Prather, R.J. Effects of Liquid-phase Electrical Conductivity, Water Content, and Surface Conductivity on Bulk Soil Electrical Conductivity. *Soil Sci. Soc. Am. J.* **1976**, *40*, 651–655. [CrossRef]
61. Brevik, E.C.; Fenton, T.E. Influence of Soil Water Content, Clay, Temperature, and Carbonate Minerals on Electrical Conductivity Readings Taken with an EM-38. *Soil Horiz.* **2002**, *43*, 9. [CrossRef]
62. Kachanoski, R.G.; Wesenbeeck, I.J.V.; Gregorich, E.G. Estimating Spatial Variations of Soil Water Content Using Non-Contacting Electromagnetic Inductive Methods. *Can. J. Soil Sci.* **1988**, *68*, 715–722. [CrossRef]
63. Visconti, F.; De Paz, J.M. Sensitivity of Soil Electromagnetic Induction Measurements to Salinity, Water Content, Clay, Organic Matter and Bulk Density. *Precis. Agric.* **2021**, *22*, 1559–1577. [CrossRef]
64. Sheets, K.R.; Hendrickx, J.M.H. Noninvasive Soil Water Content Measurement Using Electromagnetic Induction. *Water Resour. Res.* **1995**, *31*, 2401–2409. [CrossRef]
65. McCutcheon, M.C.; Farahani, H.J.; Stednick, J.D.; Buchleiter, G.W.; Green, T.R. Effect of Soil Water on Apparent Soil Electrical Conductivity and Texture Relationships in a Dryland Field. *Biosyst. Eng.* **2006**, *94*, 19–32. [CrossRef]
66. Farahani, H.J.; Buchleiter, G.W.; Brodahl, M.K. Characterization of Apparent Soil Electrical Conductivity Variability in Irrigated Sandy And Non-Saline Fields In Colorado. *Trans. ASABE* **2005**, *48*, 155–168. [CrossRef]
67. Rhoades, J.D.; Manteghi, N.A.; Shouse, P.J.; Alves, W.J. Soil Electrical Conductivity and Soil Salinity: New Formulations and Calibrations. *Soil Sci. Soc. Am. J.* **1989**, *53*, 433–439. [CrossRef]

68. McKenzie, R.C.; George, R.J.; Woods, S.A.; Cannon, M.E.; Bennett, D.L. Use of the Electromagnetic-Induction Meter (EM38) as a Tool in Managing Salinisation. *Hydrogeol. J.* **1997**, *5*, 37–50. [CrossRef]
69. Heil, K.; Schmidhalter, U. Characterisation of Soil Texture Variability Using the Apparent Electrical Conductivity at a Highly Variable Site. *Comput. Geosci.* **2012**, *39*, 98–110. [CrossRef]
70. Rhoades, J.D.; Corwin, D.L.; Lesch, S.M. Geospatial Measurements of Soil Electrical Conductivity to Assess Soil Salinity and Diffuse Salt Loading from Irrigation. In *Geophysical Monograph Series*; Corwin, L., Loague, K., Ellsworth, R., Eds.; American Geophysical Union: Washington, DC, USA, 1999; Volume 108, pp. 197–215. [CrossRef]
71. Amezketa Lizarraga, E. Soil Salinity Assessment Using Directed Soil Sampling from a Geophysical Survey with Electromagnetic Technology: A Case Study. *Span. J. Agric. Res.* **2007**, *5*, 91. [CrossRef]
72. Huang, J.; Mokhtari, A.R.; Cohen, D.R.; Monteiro Santos, F.A.; Triantafilis, J. Modelling Soil Salinity across a Gilgai Landscape by Inversion of EM38 and EM31 Data: Modelling Salinity in Gilgai by EM Inversion. *Eur. J. Soil Sci.* **2015**, *66*, 951–960. [CrossRef]
73. Nouri, H.; Chavoshi Borujeni, S.; Alaghmand, S.; Anderson, S.; Sutton, P.; Parvazian, S.; Beecham, S. Soil Salinity Mapping of Urban Greenery Using Remote Sensing and Proximal Sensing Techniques; The Case of Veale Gardens within the Adelaide Parklands. *Sustainability* **2018**, *10*, 2826. [CrossRef]
74. Lesch, S.M.; Corwin, D.L.; Robinson, D.A. Apparent Soil Electrical Conductivity Mapping as an Agricultural Management Tool in Arid Zone Soils. *Comput. Electron. Agric.* **2005**, *46*, 351–378. [CrossRef]
75. Edeh, J.A. Quantifying Spatio-Temporal Soil Water Content Using Electromagnetic Induction. Ph.D. Thesis, University of the Free State, Bloemfontein, South Africa, 2017. Available online: http://hdl.handle.net/11660/6471 (accessed on 22 July 2023).
76. Herrero, J.; Pabuayon, I.L. The Problem with "Apparent Electrical Conductivity" in Soil Electromagnetic Induction Studies. *Adv. Agron.* **2021**, *165*, 161–173. [CrossRef]
77. Archie, G.E. The Electrical Resistivity Log as an Aid in Determining Some Reservoir Characteristics. *Trans. Am. Inst. Min.* **1942**, *146*, 54–62. [CrossRef]
78. Rhoades, J.D.; Loveday, J. Salinity in Irrigated Agriculture. In *American Society of Civil Engineers, Irrigation of Agricultural Crops*; Stewart, B.A., Nielsen, D.R., Eds.; Monograph, American Society of Agronomists: Madison, WI, USA, 1990; Volume 30, pp. 1089–1142.
79. Mualem, Y.; Friedman, S.P. Theoretical Prediction of Electrical Conductivity in Saturated and Unsaturated Soil. *Water Resour. Res.* **1991**, *27*, 2771–2777. [CrossRef]
80. Lesch, S.M.; Corwin, D.L. Using the Dual-Pathway Parallel Conductance Model to Determine How Different Soil Properties Influence Conductivity Survey Data. *J. Agron.* **2003**, *95*, 365–379. [CrossRef]
81. Hassani, A.; Azapagic, A.; Shokri, N. Predicting Long-Term Dynamics of Soil Salinity and Sodicity on a Global Scale. *Proc. Natl. Acad. Sci. USA* **2020**, *117*, 33017–33027. [CrossRef] [PubMed]
82. Corwin, D.L.; Carrillo, M.L.K.; Vaughan, P.J.; Rhoades, J.D.; Cone, D.G. Evaluation of a GIS-Linked Model of Salt Loading to Groundwater. *J. Environ. Qual.* **1999**, *28*, 471–480. [CrossRef]
83. McKenzie, R.C.; Chomistek, W.; Clark, N.F. Conversion of Electromagnetic Inductance Readings to Saturated Paste Extract Values In Soils For Different Temperature, Texture And Moisture Conditions. *Can. J. Soil Sci.* **1989**, *69*, 25–32. [CrossRef]
84. Lesch, S.M.; Rhoades, J.D.; Lund, L.J.; Corwin, D.L. Mapping Soil Salinity Using Calibrated Electromagnetic Measurements. *Soil Sci. Soc. Am. J.* **1992**, *56*, 540–548. [CrossRef]
85. Triantafilis, J.; Laslett, G.M.; McBratney, A.B. Calibrating an Electromagnetic Induction Instrument to Measure Salinity in Soil under Irrigated Cotton. *Soil Sci. Soc. Am. J.* **2000**, *64*, 1009–1017. [CrossRef]
86. Bouksila, F.; Persson, M.; Bahri, A.; Berndtsson, R. Electromagnetic Induction Prediction of Soil Salinity and Groundwater Properties in a Tunisian Saharan Oasis. *Hydrol. Sci. J.* **2012**, *57*, 1473–1486. [CrossRef]
87. Akramkhanov, A.; Brus, D.J.; Walvoort, D.J.J. Geostatistical Monitoring of Soil Salinity in Uzbekistan by Repeated EMI Surveys. *Geoderma* **2014**, *213*, 600–607. [CrossRef]
88. Harvey, O.R.; Morgan, C.L.S. Predicting Regional-Scale Soil Variability Using a Single Calibrated Apparent Soil Electrical Conductivity Model. *Soil Sci. Soc. Am. J.* **2009**, *73*, 164–169. [CrossRef]
89. Corwin, D.L.; Lesch, S.M. Validation of the ANOCOVA Model for Regional-scale $EC_a$ to $EC_e$ Calibration. *Soil Use Manag.* **2017**, *33*, 178–190. [CrossRef]
90. Trangmar, B.B.; Yost, R.S.; Uehara, G. Application of Geostatistics to Spatial Studies of Soil Properties. *Adv. Agron.* **1986**, *38*, 45–94. [CrossRef]
91. Corwin, D.L.; Lesch, S.M. Characterizing Soil Spatial Variability with Apparent Soil Electrical Conductivity. *Comput. Electron. Agric.* **2005**, *46*, 103–133. [CrossRef]
92. Oliver, M.A.; Webster, R. *Basic Steps in Geostatistics: The Variogram and Kriging*; SpringerBriefs in Agriculture; Springer International Publishing: Cham, Switzerland, 2015. [CrossRef]
93. De Caires, S.A.; St Martin, C.; Wuddivira, M.N.; Farrick, K.K.; Zebarth, B.J. Predicting Soil Depth in a Humid Tropical Watershed: A Comparative Analysis of Best-Fit Regression and Geospatial Models. *Catena* **2023**, *222*, 106843. [CrossRef]
94. Liu, G.; Li, J.; Zhang, X.; Wang, X.; Lv, Z.; Yang, J.; Shao, H.; Yu, S. GIS-Mapping Spatial Distribution of Soil Salinity for Eco-Restoring the Yellow River Delta in Combination with Electromagnetic Induction. *Ecol. Eng.* **2016**, *94*, 306–314. [CrossRef]
95. Corwin, D.L.; Lesch, S.M. A Simplified Regional-Scale Electromagnetic Induction—Salinity Calibration Model Using ANOCOVA Modeling Techniques. *Geoderma* **2014**, *230–231*, 288–295. [CrossRef]

96. Lesch, S.M.; Rhoades, J.D.; Corwin, D.L. *ESAP-95 Version 2.01R: User Manual and Tutorial Guide*; USDA-ARS; George, E.B., Jr., Ed.; Salinity Laboratory: Riverside, CA, USA, 2000.
97. Ganjegunte, G.; Leinauer, B.; Schiavon, M.; Serena, M. Using Electro-Magnetic Induction to Determine Soil Salinity and Sodicity in Turf Root Zones. *J. Agron.* **2013**, *105*, 836–844. [CrossRef]
98. Narjary, B.; Meena, M.D.; Kumar, S.; Kamra, S.K.; Sharma, D.K.; Triantafilis, J. Digital Mapping of Soil Salinity at Various Depths Using an EM38. *Soil Use Manag.* **2019**, *35*, 232–244. [CrossRef]
99. Corwin, D.L.; Scudiero, E.; Zaccaria, D. Modified Eca–ECe Protocols for Mapping Soil Salinity under Micro-Irrigation. *Agric. Water Manag.* **2022**, *269*, 107640. [CrossRef]
100. Ben Slimane, A.; Bouksila, F.; Selim, T.; Joumada, F. Soil Salinity Assessment Using Electromagnetic Induction Method in a Semi-Arid Environment—A Case Study in Tunisia. *Arab. J. Geosci.* **2022**, *15*, 1031. [CrossRef]
101. Kasim, N.; Tiyip, T.; Abliz, A.; Nurmemet, I.; Sawut, R.; Maihemuti, B. Mapping and Modeling of Soil Salinity Using WorldView-2 Data and EM38-KM2 in an Arid Region of the Keriya River, China. *Photogramm. Eng. Remote Sens.* **2018**, *84*, 43–52. [CrossRef]
102. Corwin, D.L.; Rhoades, J.D. Establishing Soil Electrical Conductivity—Depth Relations from Electromagnetic Induction Measurements. *Commun. Soil Sci. Plant Anal.* **1990**, *21*, 861–901. [CrossRef]
103. Slavich, P. Determining ECa-Depth Profiles from Electromagnetic Induction Measurements. *Soil Res.* **1990**, *28*, 443. [CrossRef]
104. Rhoades, J.D.; Corwin, D.L. Determining Soil Electrical Conductivity-Depth Relations Using an Inductive Electromagnetic Soil Conductivity Meter. *Soil Sci. Soc. Am. J.* **1981**, *45*, 255–260. [CrossRef]
105. Wollenhaupt, N.C.; Richardson, J.L.; Foss, J.E.; Doll, E.C. A Rapid Method For Estimating Weighted Soil Salinity from Apparent Soil Electrical Conductivity measured with An Aboveground Electromagnetic Induction Meter. *Can. J. Soil Sci.* **1986**, *66*, 315–321. [CrossRef]
106. Vlotman, W.F.; Chaudhry, M.R.B.A. *Electromagnetic Induction Device (EM38) Calibration and Monitoring Soil Salinity/Environment (Pakistan)*; ILRI: New Delhi, India, 2000; pp. 37–48.
107. Slavich, P.; Petterson, G. Estimating Average Rootzone Salinity from Electromagnetic Induction (EM-38) Measurements. *Soil Res.* **1990**, *28*, 453. [CrossRef]
108. Corwin, D.L.; Rhoades, J.D. An Improved Technique for Determining Soil Electrical Conductivity-Depth Relations from Above-Ground Electromagnetic Measurements. *Soil Sci. Soc. Am. J.* **1982**, *46*, 517–520. [CrossRef]
109. Corwin, D.L.; Rhoades, J.D. Measurement of Inverted Electrical Conductivity Profiles Using Electromagnetic Induction. *Soil Sci. Soc. Am. J.* **1984**, *48*, 288–291. [CrossRef]
110. Cook, P.G.; Walker, G.R. Depth Profiles of Electrical Conductivity from Linear Combinations of Electromagnetic Induction Measurements. *Soil Sci. Soc. Am. J.* **1992**, *56*, 1015–1022. [CrossRef]
111. Johnston, M.A.; Savage, M.J.; Moolman, J.H.; Du Plessis, H.M. Evaluation of Calibration Methods for Interpreting Soil Salinity from Electromagnetic Induction Measurements. *Soil Sci. Soc. Am. J.* **1997**, *61*, 1627–1633. [CrossRef]
112. Coppola, A.; Smettem, K.; Ajeel, A.; Saeed, A.; Dragonetti, G.; Comegna, A.; Lamaddalena, N.; Vacca, A. Calibration of an Electromagnetic Induction Sensor with Time-Domain Reflectometry Data to Monitor Rootzone Electrical Conductivity under Saline Water Irrigation: The EMI Sensor for Salinity Management of the Rootzone. *Eur. J. Soil Sci.* **2016**, *67*, 737–748. [CrossRef]
113. Bennett, D.L.; George, R.J.; Whitfield, B. The Use of Ground EM Systems to Accurately Assess Salt Store and Help Define Land Management Options for Salinity Management. *Explor. Geophys.* **2000**, *31*, 249–254. [CrossRef]
114. Lesch, S. Statistical Models for the Prediction of Field-Scale and Spatial Salinity Patterns from Soil Conductivity Survey Data. In *Agricultural Salinity Assessment and Management*, 2nd ed.; ASCE: Reston, VA, USA, 2011; pp. 461–482. [CrossRef]
115. Corwin, D.L. Geospatial Measurements of Apparent Soil Electrical Conductivity for Characterizing Soil Spatial Variability. In *Soil-Water-Solute Process Characterization—An Integrated Approach*; Alvarez-Benedi, J., Munoz-Carpena, R., Eds.; CRC Press: Boca Raton, FL, USA, 1993; Volume 18, pp. 639–672.
116. Lu, L.; Li, S.; Wu, R.; Shen, D. Study on the Scale Effect of Spatial Variation in Soil Salinity Based on Geostatistics: A Case Study of Yingdaya River Irrigation Area. *Land* **2022**, *11*, 1697. [CrossRef]
117. Killick, M. An Analysis of the Relationship of Apparent Electrical Conductivity to Soil Moisture in Alluvial Recent Soils, Lower North Island, New Zealand. Master's Thesis, Massey University, Palmerston North, New Zealand, 2013. Available online: http://hdl.handle.net/10179/4902 (accessed on 21 July 2023).
118. Guo, Y.; Zhou, Y.; Zhou, L.; Liu, T.; Wang, L.; Cheng, Y.; He, J.; Zheng, G. Using Proximal Sensor Data for Soil Salinity Management and Mapping. *J. Integr. Agric.* **2019**, *18*, 340–349. [CrossRef]
119. Yao, R.; Yang, J. Quantitative Evaluation of Soil Salinity and Its Spatial Distribution Using Electromagnetic Induction Method. *Agric. Water Manag.* **2010**, *97*, 1961–1970. [CrossRef]
120. Jantaravikorn, Y.; Ongsomwang, S. Soil Salinity Prediction and Its Severity Mapping Using a Suitable Interpolation Method on Data Collected by Electromagnetic Induction Method. *Appl. Sci.* **2022**, *12*, 10550. [CrossRef]
121. Islam, M.M.; Saey, T.; Meerschman, E.; De Smedt, P.; Meeuws, F.; Van De Vijver, E.; Meirvenne, M. Delineating Water Management Zones in a Paddy Rice Field Using a Floating Soil Sensing System. *Agric. Water Manag.* **2011**, *102*, 8–12. [CrossRef]
122. Dakak, H.; Benmohammadi, A.; Soudi, B.; Douaik, A.; Badraoui, M.; Zouahri, A. Mapping the Risk of Soil Salinization Using Electromagnetic Induction and Non-Parametric Geostatistics. In *Developments in Soil Salinity Assessment and Reclamation*; Shahid, S.A., Abdelfattah, M.A., Taha, F.K., Eds.; Springer Netherlands: Dordrecht, The Netherlands, 2013; pp. 155–166. [CrossRef]

123. Xie, W.; Yang, J.; Yao, R.; Wang, X. Spatial and Temporal Variability of Soil Salinity in the Yangtze River Estuary Using Electromagnetic Induction. *Remote Sens.* **2021**, *13*, 1875. [CrossRef]
124. Gharsallah, M.E.; Aichi, H.; Stambouli, T.; Ben Rabah, Z.; Ben Hassine, H. Assessment and Mapping of Soil Salinity Using Electromagnetic Induction and Landsat 8 OLI Remote Sensing Data in an Irrigated Olive Orchard under Semi-Arid Conditions. *Soil Water Res.* **2022**, *17*, 15–28. [CrossRef]
125. Li, H.; Liu, X.; Hu, B.; Biswas, A.; Jiang, Q.; Liu, W.; Wang, N.; Peng, J. Field-Scale Characterization of Spatio-Temporal Variability of Soil Salinity in Three Dimensions. *Remote Sens.* **2020**, *12*, 4043. [CrossRef]
126. Farzamian, M.; Bouksila, F.; Paz, A.M.; Santos, F.M.; Zemni, N.; Slama, F.; Ben Slimane, A.; Selim, T.; Triantafilis, J. Landscape-Scale Mapping of Soil Salinity with Multi-Height Electromagnetic Induction and Quasi-3d Inversion in Saharan Oasis, Tunisia. *Agric. Water Manag.* **2023**, *284*, 108330. [CrossRef]
127. Paz, M.C.; Farzamian, M.; Paz, A.M.; Castanheira, N.L.; Gonçalves, M.C.; Monteiro Santos, F. Assessing Soil Salinity Dynamics Using Time-Lapse Electromagnetic Conductivity Imaging. *Soil* **2020**, *6*, 499–511. [CrossRef]
128. Farzamian, M.; Paz, M.C.; Paz, A.M.; Castanheira, N.L.; Gonçalves, M.C.; Monteiro Santos, F.A.; Triantafilis, J. Mapping Soil Salinity Using Electromagnetic Conductivity Imaging—A Comparison of Regional and Location-specific Calibrations. *Land Degrad. Dev.* **2019**, *30*, 1393–1406. [CrossRef]
129. Dakak, H.; Huang, J.; Zouahri, A.; Douaik, A.; Triantafilis, J. Mapping Soil Salinity in 3-Dimensions Using an EM38 and EM4Soil Inversion Modelling at the Reconnaissance Scale in Central Morocco. *Soil Use Manag.* **2017**, *33*, 553–567. [CrossRef]
130. Khongnawang, T.; Zare, E.; Srihabun, P.; Khunthong, I.; Triantafilis, J. Digital Soil Mapping of Soil Salinity Using EM38 and Quasi-3-d Modelling Software (EM4Soil). *Soil Use Manag.* **2021**, *38*, 277–291. [CrossRef]
131. Guo, Y.; Shi, Z.; Huang, J.; Wang, L.; Cheng, Y.; Zheng, G. Mapping Horizontal and Vertical Spatial Variability of Soil Salinity in Reclaimed Areas. In *Digital Soil Mapping Across Paradigms, Scales and Boundaries*; Zhang, G.-L., Brus, D., Liu, F., Song, X.-D., Lagacherie, P., Eds.; Springer Environmental Science and Engineering; Springer: Singapore, 2016; pp. 33–45. [CrossRef]
132. Dragonetti, G.; Farzamian, M.; Basile, A.; Monteiro Santos, F.; Coppola, A. In Situ Estimation of Soil Hydraulic and Hydrodispersive Properties by Inversion of Electromagnetic Induction Measurements and Soil Hydrological Modeling. *Hydrol. Earth Syst. Sci.* **2022**, *26*, 5119–5136. [CrossRef]
133. Dragonetti, G.; Comegna, A.; Ajeel, A.; Deidda, G.P.; Lamaddalena, N.; Rodriguez, G.; Vignoli, G.; Coppola, A. Calibrating Electromagnetic Induction Conductivities with Time-Domain Reflectometry Measurements. *Hydrol. Earth Syst. Sci.* **2018**, *22*, 1509–1523. [CrossRef]
134. Flores, J.G.; Rodríguez, M.R.; Jiménez, A.G.; Farzamian, M.; Herencia, J.F.; Galán, B.; Sacristan, P.C.; Vanderlinden, K. Monitoring Tridimensional Soil Salinity Patterns at the Field Scale Using Electromagnetic Induction Sensing and Inversion. *Salt-Affect. Soils* **2022**, *XV*, 48.
135. Borchers, B.; Uram, T.; Hendrickx, J.M.H. Tikhonov Regularization of Electrical Conductivity Depth Profiles in Field Soils. *Soil Sci. Soc. Am. J.* **1997**, *61*, 1004–1009. [CrossRef]
136. Hendrickx, J.M.H.; Borchers, B.; Corwin, D.L.; Lesch, S.M.; Hilgendorf, A.C.; Schlue, J. Inversion of Soil Conductivity Profiles from Electromagnetic Induction Measurements: Theory and Experimental Verification. *Soil Sci. Soc. Am. J.* **2002**, *66*, 673–685. [CrossRef]
137. Gebbers, R.; Lück, E.; Heil, K. Depth Sounding with the EM38-Detection of Soil Layering by Inversion of Apparent Electrical Conductivity Measurements. *Precis. Agric.* **2007**, *7*, 95–102. [CrossRef]
138. Moghadas, D.; Jadoon, K.Z.; McCabe, M.F. Spatiotemporal Monitoring of Soil Moisture from EMI Data Using DCT-Based Bayesian Inference and Neural Network. *Appl. Geophys.* **2019**, *169*, 226–238. [CrossRef]
139. Triantafilis, J.; Santos, F.A.M.; Triantafilis, J.; Santos, F.A.M. Resolving the Spatial Distribution of the True Electrical Conductivity with Depth Using EM38 and EM31 Signal Data and a Laterally Constrained Inversion Model. *Soil Res.* **2010**, *48*, 434–446. [CrossRef]
140. Triantafilis, J.; Santos, F.A.M. 2-Dimensional Soil and Vadose-Zone Representation Using an EM38 and EM34 and a Laterally Constrained Inversion Model. *Soil Res.* **2009**, *47*, 809. [CrossRef]
141. Moghadas, D.; Taghizadeh-Mehrjardi, R.; Triantafilis, J. Probabilistic Inversion of EM38 Data for 3D Soil Mapping in Central Iran. *Geoderma Reg.* **2016**, *7*, 230–238. [CrossRef]
142. Paz, M.C.; Farzamian, M.; Santos, F.M.; Gonçalves, M.C.; Paz, A.M.; Castanheira, N.L.; Triantafilis, J. Potential to Map Soil Salinity Using Inversion Modelling of EM38 Sensor Data. *First Break* **2019**, *37*, 35–39. [CrossRef]
143. Paz, A.M.; Castanheira, N.; Farzamian, M.; Paz, M.C.; Gonçalves, M.C.; Monteiro Santos, F.A.; Triantafilis, J. Prediction of Soil Salinity and Sodicity Using Electromagnetic Conductivity Imaging. *Geoderma* **2020**, *361*, 114086. [CrossRef]
144. Jadoon, K.Z.; Moghadas, D.; Jadoon, A.; Missimer, T.M.; Al-Mashharawi, S.K.; McCabe, M.F. Estimation of Soil Salinity in a Drip Irrigation System by Using Joint Inversion of Multicoil Electromagnetic Induction Measurements. *Water Resour. Res.* **2015**, *51*, 3490–3504. [CrossRef]
145. Farzamian, M.; Monteiro Santos, F.A.; Khalil, M.A. Application of EM38 and ERT Methods in Estimation of Saturated Hydraulic Conductivity in Unsaturated Soil. *Appl. Geophys.* **2015**, *112*, 175–189. [CrossRef]
146. Martini, E.; Werban, U.; Zacharias, S.; Pohle, M.; Dietrich, P.; Wollschläger, U. Repeated Electromagnetic Induction Measurements for Mapping Soil Moisture at the Field Scale: Validation with Data from a Wireless Soil Moisture Monitoring Network. *Hydrol. Earth Syst. Sci.* **2017**, *21*, 495–513. [CrossRef]

147. Farzamian, M.; Autovino, D.; Basile, A.; De Mascellis, R.; Dragonetti, G.; Monteiro Santos, F.; Binley, A.; Coppola, A. Assessing the Dynamics of Soil Salinity with Time-Lapse Inversion of Electromagnetic Data Guided by Hydrological Modelling. *Hydrol. Earth Syst. Sci.* **2021**, *25*, 1509–1527. [CrossRef]
148. Von Hebel, C.; Rudolph, S.; Mester, A.; Huisman, J.A.; Kumbhar, P.; Vereecken, H.; Van Der Kruk, J. Three-Dimensional Imaging of Subsurface Structural Patterns Using Quantitative Large-Scale Multiconfiguration Electromagnetic Induction Data. *Water Resour. Res.* **2014**, *50*, 2732–2748. [CrossRef]
149. Taghizadeh-Mehrjardi, R.; Minasny, B.; Sarmadian, F.; Malone, B.P. Digital Mapping of Soil Salinity in Ardakan Region, Central Iran. *Geoderma* **2014**, *213*, 15–28. [CrossRef]
150. Wang, F.; Yang, S.; Wei, Y.; Shi, Q.; Ding, J. Characterizing Soil Salinity at Multiple Depth Using Electromagnetic Induction and Remote Sensing Data with Random Forests: A Case Study in Tarim River Basin of Southern Xinjiang, China. *Sci. Total Environ.* **2021**, *754*, 142030. [CrossRef] [PubMed]
151. McBratney, A.B.; Mendonça Santos, M.L.; Minasny, B. On Digital Soil Mapping. *Geoderma* **2003**, *117*, 3–52. [CrossRef]
152. Rentschler, T. Explainable Machine Learning in Soil Mapping: Peeking into the Black Box. Ph.D. Thesis, Universität Tübingen, Tübingen, Germany, 2021. [CrossRef]
153. Taghizadeh-Mehrjardi, R.; Ayoubi, S.; Namazi, Z.; Malone, B.P.; Zolfaghari, A.A.; Sadrabadi, F.R. Prediction of Soil Surface Salinity in Arid Region of Central Iran Using Auxiliary Variables and Genetic Programming. *Arid. Land Res. Manag.* **2016**, *30*, 49–64. [CrossRef]
154. Sahbeni, G.; Ngabire, M.; Musyimi, P.K.; Székely, B. Challenges and Opportunities in Remote Sensing for Soil Salinization Mapping and Monitoring: A Review. *Remote Sens.* **2023**, *15*, 2540. [CrossRef]
155. Farifteh, J.; Farshad, A.; George, R.J. Assessing Salt-Affected Soils Using Remote Sensing, Solute Modelling, and Geophysics. *Geoderma* **2006**, *130*, 191–206. [CrossRef]
156. Aldabaa, A.A.A.; Weindorf, D.C.; Chakraborty, S.; Sharma, A.; Li, B. Combination of Proximal and Remote Sensing Methods for Rapid Soil Salinity Quantification. *Geoderma* **2015**, *239–240*, 34–46. [CrossRef]
157. Casterad, M.A.; Herrero, J.; Betrán, J.A.; Ritchie, G. Sensor-Based Assessment of Soil Salinity during the First Years of Transition from Flood to Sprinkler Irrigation. *Sensors* **2018**, *18*, 616. [CrossRef] [PubMed]
158. Allbed, A.; Kumar, L. Soil Salinity Mapping and Monitoring in Arid and Semi-Arid Regions Using Remote Sensing Technology: A Review. *Adv. Remote Sens.* **2013**, *2*, 373–385. [CrossRef]
159. Metternicht, G.I.; Zinck, J.A. Remote Sensing of Soil Salinity: Potentials and Constraints. *Remote Sens. Environ.* **2003**, *85*, 1–20. [CrossRef]
160. Scudiero, E.; Skaggs, T.H.; Corwin, D.L. Comparative Regional-Scale Soil Salinity Assessment with near-Ground Apparent Electrical Conductivity and Remote Sensing Canopy Reflectance. *Ecol. Indic.* **2016**, *70*, 276–284. [CrossRef]
161. Rao, B.R.M.; Sharma, R.C.; Ravi Sankar, T.; Das, S.N.; Dwivedi, R.S.; Thammappa, S.S.; Venkataratnam, L. Spectral Behaviour of Salt-Affected Soils. *Int. J. Remote Sens.* **1995**, *16*, 2125–2136. [CrossRef]
162. Wu, W.; Mhaimeed, A.S.; Al-Shafie, W.M.; Ziadat, F.; Dhehibi, B.; Nangia, V.; De Pauw, E. Mapping Soil Salinity Changes Using Remote Sensing in Central Iraq. *Geoderma Reg.* **2014**, *2–3*, 21–31. [CrossRef]
163. Sultanov, M.; Ibrakhimov, M.; Akramkhanov, A.; Bauer, C.; Conrad, C. Modelling End-of-Season Soil Salinity in Irrigated Agriculture Through Multi-Temporal Optical Remote Sensing, Environmental Parameters, and In Situ Information. *PFG–J. Photogramm. Remote Sens. Geoinf. Sci.* **2018**, *86*, 221–233. [CrossRef]

*land*

MDPI

*Article*

# Comparison of Electromagnetic Induction and Electrical Resistivity Tomography in Assessing Soil Salinity: Insights from Four Plots with Distinct Soil Salinity Levels

**Maria Catarina Paz [1], Nádia Luísa Castanheira [2], Ana Marta Paz [2], Maria Conceição Gonçalves [2], Fernando Monteiro Santos [3] and Mohammad Farzamian [2,4,*]**

[1] RESILIENCE—Center for Regional Resilience and Sustainability, Escola Superior de Tecnologia do Barreiro, Instituto Politécnico de Setúbal, Rua Américo da Silva Marinho, 2839-001 Barreiro, Portugal; catarina.paz@estbarreiro.ips.pt

[2] Instituto Nacional de Investigação Agrária e Veterinária, Avenida da República, Quinta do Marquês (Edifício Sede), 2780-157 Oeiras, Portugal; nadia.castanheira@iniav.pt (N.L.C.); ana.paz@iniav.pt (A.M.P.); maria.goncalves@iniav.pt (M.C.G.)

[3] Instituto Dom Luiz, Faculdade de Ciências da Universidade de Lisboa, Campo Grande, Edifício C1, Piso 1, 1749-016 Lisboa, Portugal; fasantos@fc.ul.pt

[4] Centre of Geographical Studies (CEG), IGOT, Universidade de Lisboa, 1600-276 Lisbon, Portugal

* Correspondence: mohammad.farzamian@iniav.pt

**Abstract:** Electromagnetic induction (EMI) and electrical resistivity tomography (ERT) are geophysical techniques measuring soil electrical conductivity and providing insights into properties correlated with it to depths of several meters. EMI measures the apparent electrical conductivity ($EC_a$, dS $m^{-1}$) without physical contact, while ERT acquires apparent electrical resistivity ($ER_a$, ohm m) using electrodes. Both involve mathematical inversion to obtain models of spatial distribution for soil electrical conductivity ($\sigma$, mS $m^{-1}$) and electrical resistivity ($\rho$, ohm m), respectively, where $\rho$ is the reciprocal of $\sigma$. Soil salinity can be assessed from $\sigma$ over large areas using a calibration process consisting of a regression between $\sigma$ and the electrical conductivity of the saturated soil paste extract ($EC_e$, dS $m^{-1}$), used as a proxy for soil salinity. This research aims to compare the prediction abilities of the faster EMI to the more reliable ERT for estimating $\sigma$ and predicting soil salinity. The study conducted surveys and sampling at four locations with distinct salinity levels in Portugal, analysing the agreement between the techniques, and obtained 2D vertical soil salinity maps. In our case study, the agreement between EMI and ERT models was fairly good in three locations, with $\sigma$ varying between 50 and 500 mS $m^{-1}$. However, this was not the case at location 4, where $\sigma$ exceeded 1000 mS $m^{-1}$ and EMI significantly underestimated $\sigma$ when compared to ERT. As for soil salinity prediction, both techniques generally provided satisfactory and comparable regional-level predictions of $EC_e$, and the observed underestimation in EMI models did not significantly affect the overall estimation of soil salinity. Consequently, EMI demonstrated an acceptable level of accuracy in comparison to ERT in our case studies, supporting confidence in utilizing this faster and more practical technique for measuring soil salinity over large areas.

**Keywords:** electromagnetic induction; electrical resistivity tomography; soil salinity

**Citation:** Paz, M.C.; Castanheira, N.L.; Paz, A.M.; Gonçalves, M.C.; Monteiro Santos, F.; Farzamian, M. Comparison of Electromagnetic Induction and Electrical Resistivity Tomography in Assessing Soil Salinity: Insights from Four Plots with Distinct Soil Salinity Levels. *Land* **2024**, *13*, 295. https://doi.org/10.3390/land13030295

Academic Editor: Amrakh I. Mamedov

Received: 8 January 2024
Revised: 20 February 2024
Accepted: 22 February 2024
Published: 27 February 2024



## 1. Introduction

Electromagnetic induction (EMI) and electrical resistivity tomography (ERT) are two near-surface geophysical techniques that allow the electrical conductivity of soil to be measured and therefore for properties that are correlated with it to be monitored to depths that can reach up to several meters. These properties can be soil salinity [1–6], soil sodicity [7], soil water content [8–15], particle size distribution [16–19], soil cation exchange capacity [20–23], and organic matter [24].

EMI implies the acquisition of apparent electrical conductivity ($EC_a$, dS $m^{-1}$) using a device that does not touch the surface of the soil, while ERT involves the acquisition of apparent electrical resistivity ($ER_a$, ohm m), which is the reciprocal of $EC_a$, using a device that takes measurements in electrodes placed at the surface of the soil. Both EMI and ERT techniques involve the mathematical inversion of the apparent data [25] to obtain models of the spatial distribution of the soil electrical conductivity ($\sigma$, mS $m^{-1}$), and of the soil electrical resistivity ($\rho$, ohm m), respectively. Different inversion methods (e.g., [26–28]) and software (e.g., [29,30]) have been developed to estimate the distribution of $\sigma$ based on measured $EC_a$ data. Similarly, various inversion codes are available to estimate the distribution of $\rho$ based on measured $ER_a$ data (e.g., [31–33]). $\rho$ is the reciprocal of $\sigma$, so it can be easily converted to $\sigma$.

While the fundamental physical principles differ (induction versus galvanic phenomena), and the volume of ground investigated by the two techniques is also different, both could yield similar electrical conductivity values under specific assumptions. Theoretically, adopting low-frequency signals ($f < 105$ Hz) and the absence of metallic objects in the subsoil should result in comparable outcomes for ERT and EMI techniques. However, it is important to note that ERT is more sensitive to strong resistors, while EMI is more sensitive to strong conductors [34].

Soil salinization is a process of soil degradation that limits agricultural productivity and can lead to desertification and land abandonment. Salinization also decreases biodiversity, affects ground- and surface water, and degrades infrastructures. Such effects represent major negative economic, environmental, and social impacts. According to the Global Map of Salt-Affected Soils [35], salt-affected soils are distributed globally, but about two thirds of the area is located in arid and semi-arid climatic zones. FAO ([35]) estimates that 4.4% of the topsoil (0–30 cm) and more than 8.7% of soil at depths of 30–100 cm of the total land area is salt-affected. Given this threat, it is very important to be able to monitor soil salinity in agricultural areas. The monitoring of soil salinity along the soil profile is key to understanding the specific processes related to salinization and to defining and implementing measures to counter it and its impacts [36,37].

Soil salinity can be assessed from $\sigma$ over large areas through a calibration process consisting of a regression between $\sigma$ and the electrical conductivity of the saturated soil paste extract ($EC_e$, dS $m^{-1}$), used as a proxy for soil salinity, and the conversion of the $\sigma$ models into salinity maps using the obtained calibration equation (e.g., [21,38,39]). While both EMI and ERT offer non-invasive, rapid, and cost-effective analysis, EMI stands out for its capacity to cover extensive areas in a very short timeframe. However, several studies (e.g., [40–46]) have highlighted that EMI may not provide precise estimations of $\sigma$ distribution and may require the prior calibration of $EC_a$ data against ERT or time domain reflectometry (TDR) measurements to account for expected shifts and offsets to obtain more representative $EC_a$ measurements. However, the calibration process is site-specific, time-consuming, and may not be feasible in many cases.

In this study, we explore the prediction ability of EMI in assessing soil salinity without prior calibration, comparing the outcomes to equivalent results derived from the ERT data. The aim is to assess whether satisfactory prediction results can be achieved without a prior EMI calibration process. Given that many applications of EMI in soil salinity assessment often skip such a calibration step due to its time-consuming nature or the unavailability of necessary geophysical equipment, this case study offers insights into potential uncertainties associated with the absence of calibration. To achieve our goal, we conducted EMI and ERT surveys, along with soil sampling, at four locations with varying salinity levels from non-saline to severely saline soils in the Lezíria de Vila Franca de Xira, an alluvial agricultural area in Portugal. The selection of locations aimed to encompass a range of soil salinity levels, also ensuring a great variability of $\sigma$, expected when conducting EMI surveys over saline soil across the globe. To this aim, we generated 2D vertical $\sigma$ models and analysed the agreement between the two techniques in estimating $\sigma$ at the same subsurface points.

Then, we obtained one calibration equation for each technique, compared their abilities to predict $EC_e$ from the $\sigma$ models, and generated 2D vertical soil salinity maps.

## 2. Materials and Methods

### 2.1. Study Area

The investigation was conducted in Lezíria de Vila Franca, situated 10 km northeast of Lisbon, Portugal (see Figure 1). It is a 130-square-kilometer alluvial peninsula bounded by the Tejo and Sorraia rivers. The climate is classified as temperate with hot and dry summers, according to the Köppen classification. In the northern region, the soils exhibit a fine to very fine texture and are categorized as Fluvisols, and in the southern region they are categorized as Solonchaks, based on the Harmonized World Soil Database [47]. There is a gradient of soil salinity that increases from north to south that affects the land use distribution. This gradient is a combination of (1) primary salinization [48], attributed to the regional presence of marine sediments and the saline influence of the estuary on groundwater in the southern part of the study area; and (2) secondary salinization [48], attributed to the irrigated farming that, using good quality water, has washed the soil in the northern part of the region. In fact, in this region, land use primarily comprises mainly irrigated annual crops in the north and rainfed pastures in the south.

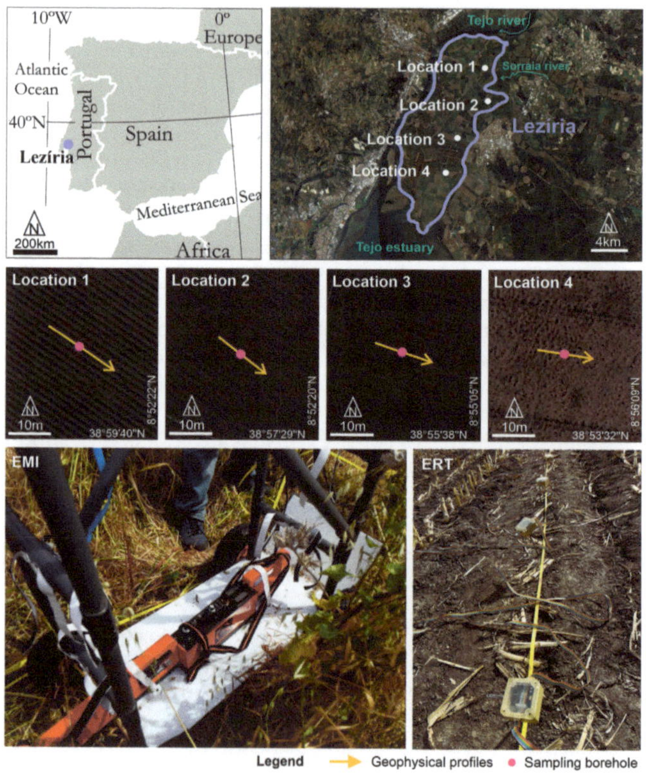


**Figure 1.** Location of Lezíria, the study area, details of the four locations with the geophysical transects and soil sampling sites, and images of the electromagnetic induction (EMI) and electrical resistivity tomography (ERT) instruments used in geophysical acquisition. © Google Earth.

For the comparative assessment, four distinct locations with varying salinity levels were selected within the study area (see Figure 1). According to the soil salinity classification defined by [49], location 1 exhibited non-saline conditions (ECe < 2 dS $m^{-1}$), location 2 was slightly saline (2–4 dS m$^{-1}$), location 3 was moderately to highly saline (4–16 dS m$^{-1}$), and location 4 was severely saline, surpassing 16 dS m$^{-1}$ at the time of the experiment [4,7,38].

Moreover, previous studies conducted by our team at this study area [4,7,38] have provided a detailed description of the soil's physico-chemical properties. We have determined the pH, cation exchange capacity, and exchangeable sodium percentage. From this analysis, we know that all locations have a pH above 8 at depths below 30 cm, and that the exchangeable sodium percentage is above 15% in the subsoil at locations 2, 3, and 4. The latter indicates a high concentration of sodium in the soil exchange complex in relation to those of calcium and potassium. The principal component analysis showed that the soil electrical conductivity in this study area is mainly influenced by soil salinity [7]. This correlation is attributed to the significant north–south gradient of soil salinity, which predominates in influencing soil salinity impact, coupled with the limited variability observed in other soil properties like soil texture.

During the surveys, location 1 featured drip-irrigated tomatoes, and locations 2 and 3 centred on pivot irrigated maize, with no irrigation occurring on survey days. Location 4 is a rainfed natural pasture that had not been ploughed for at least the past decade.

### 2.2. Electromagnetic Induction

Electromagnetic induction (EMI, Figure 1) data were gathered, employing the EM38 instrument. The instrument comprises two coils housed in a case—one for transmiting the electromagnetic signal and the other for receiving it—positioned 1 m apart. The instrument can be oriented vertically (horizontal dipole mode) for a maximum depth of investigation of 1.5 m or horizontally (vertical dipole mode) for a maximum depth of 0.75 m. EMI surveys were conducted in the dry seasons of 2017 (locations 1 and 4) and 2018 (locations 2 and 3). $EC_a$ measurements were collected at 1 m intervals along a 20 m transect at each location (refer to Figure 1) and at two heights from the soil surface (0.15 and 0.4 m) in both horizontal and vertical dipole orientations. This positioning was facilitated by placing the EM38 on a custom-built cart for precision [38].

The inversion of $EC_a$ data to derive $\sigma$ was executed using EM4SOIL software (V-3.05) [29]. The $EC_a$ responses in the model were determined through forward modelling based on the complete solution of the Maxwell equations [50]. The subsurface model employed in the inversion consisted of a series of 1-D models distributed according to the $EC_a$ measurement positions. Each subsurface model at a measurement position was influenced by neighbouring models, enabling algorithm use in regions characterized by high conductivity contrast. The inversion of $EC_a$ data employed an approach grounded in Occam regularization [51]. Data from all four locations were inverted, applying a five-layer earth initial model with an electrical conductivity of 100 mS m$^{-1}$ and a fixed layer thickness of 0.30 m. The algorithm's parameters, including the inversion algorithm type, number of iterations, and smoothing factor ($\lambda$) controlling model roughness, were chosen according to the methodology outlined in [38]. With the evenly distributed inverted data across the modelled area, 2D vertical $\sigma$ models were generated for each location utilizing triangulation with linear interpolation.

### 2.3. Electrical Resistivity Tomography

Electrical resistivity tomography (ERT, Figure 1) data were acquired using a 4-point light 10 W system (LIPPMANN, Schaufling, Germany). The technology of this system is based on the measurement of voltage between two reading electrodes installed on the soil surface, when direct current is injected into two other electrodes, to calculate subsurface electrical resistivity. In this system, the disposition of the electrodes changes according to the array used, so that a grid of subsoil $ER_a$ values is obtained. The maximum depth of investigation and resolution vary with electrode spacing and the array configuration. ERT surveys were carried out at the same transects as EMI surveys, on the same dates, with electrode spacings of 1 m at location 1 and 0.75 m at locations 2, 3, and 4. $ER_a$ data were collected using the Wenner array, known for effective mapping in areas with significant vertical gradients of $\sigma$ and a superior signal-to-noise ratio [31].

The inversion of $ER_a$ data to derive $\sigma$ was performed using RES2DINV software (V-3.71) (Geotomo Software, Penang, Malaysia). Given the pronounced vertical salinity gradients due to saline groundwater in the region, robust inversion and mesh refinement to half of the electrode spacing were implemented to address anticipated strong resistivity contrasts. In robust inversion, the objective function aimed to minimize the absolute change in resistivity values, yielding models with well-defined interfaces between regions with varying resistivity values [31]. Since the inverted data were evenly distributed over the modelled area, 2D vertical $\sigma$ models were produced for each location using triangulation with linear interpolation.

### 2.4. Soil Salinity

Soil samples were collected simultaneously as the geophysical surveys, in the medium point of each profile, as shown in Figure 1. There, five soil samples were collected between a depth of 0.15 m and 1.35 m representing topsoil (0–0.3 m), subsurface (0.3–0.6 m), upper subsoil (0.6–0.9 m), intermediate subsoil (0.9–1.2 m), and lower subsoil (1.2–1.5 m). It is noteworthy that the number of soil samples was limited to one borehole in each plot. This was because of the short length of the transects and the relatively small lateral variability of $\sigma$, suggesting that a single borehole could adequately represent soil properties. Furthermore, during the 18-month monitoring period at these four locations, a larger number of boreholes were drilled, and laboratory analysis confirmed the limited lateral variability of soil properties [4,7,38]. In the laboratory, $EC_e$ was determined using a conductivity meter (WTW 1C20-0211 inoLab) on liquid extracts obtained by suction filtering of the soil saturation paste derived from 300 g of air-dried and 2 mm sieved soil samples. The methodology employed for $EC_e$ measurement followed the procedures by [52], and soil samples' salinity was classified according to [49], as described in Section 2.1.

Two regional calibrations were developed to predict $EC_e$ from all locations together, one using $\sigma$ obtained from the inversion of EMI data and the other using $\sigma$ obtained from the inversion of ERT data. The prediction ability of these calibrations was analysed through cross-validation, using the leave-one-out cross-validation in R language [53] through the function train(). In this method, one sample is removed and a calibration is established based on the remaining samples to predict the $EC_e$ of the removed sample. This procedure is iteratively repeated for each sample, until all 19 samples have been removed. The prediction ability of the calibrations was evaluated by calculating the root mean square errors (RMSE), and the mean errors (ME). RMSE evaluates matching between measured and predicted data, indicating more precise predictions when closer to zero. ME evaluates whether the predicted data are overestimated (negative ME) or underestimated (positive ME). RMSE and ME were calculated according to the following equations:

$$\mathrm{RMSE} = \sqrt{\frac{\sum_{i=1}^{n}\left(\mathrm{mEC}_{e_i} - \mathrm{pEC}_{e_i}\right)^2}{n}} \tag{1}$$

$$\mathrm{ME} = \frac{\sum_{i=1}^{n}\left(\mathrm{mEC}_{e_i} - \mathrm{pEC}_{e_i}\right)}{n} \tag{2}$$

where $mEC_e$ indicates measured $EC_e$ and $pEC_e$ indicates predicted $EC_e$.

The two regional calibrations were then used to estimate $EC_e$ from $\sigma$ (obtained from either EMI or ERT). Since the estimated $EC_e$ data were evenly distributed over the mapping area, 2D vertical soil salinity classification maps were produced for each location using triangulation with linear interpolation.

*2.5. Agreement Analysis*

Agreement analysis was performed for EMI and ERT techniques in estimating $\sigma$ and $EC_e$ on the same points of the subsurface, based on the methodology described in [54]. In this methodology, the agreement is analysed using the differences between measurements made by two techniques on the same subject. In this study, we considered the subsurface position as the subject. However, both EMI and ERT data surveys and inversion processes produced subsurface $\sigma$ grids that were not entirely coincident in space (Figure 2). To overcome this issue, $\sigma$ grids obtained from EMI and ERT modelling results were ordered in data pairs selected for being distanced less than 0.22 m horizontally and 0.05 m vertically. This criterion was looser in the horizontal direction to maintain a statistically significant number of selected points, and it could be accepted because it was known from the previous studies [4,7,38] that the subsurface was highly horizontally stratified. The position attributed to each data pair was the position of the corresponding EMI measurement. These data pairs were then used for a comparison of EMI and ERT techniques in estimating $\sigma$, represented by ($\sigma_{ERT}$, $\sigma_{EMI}$).

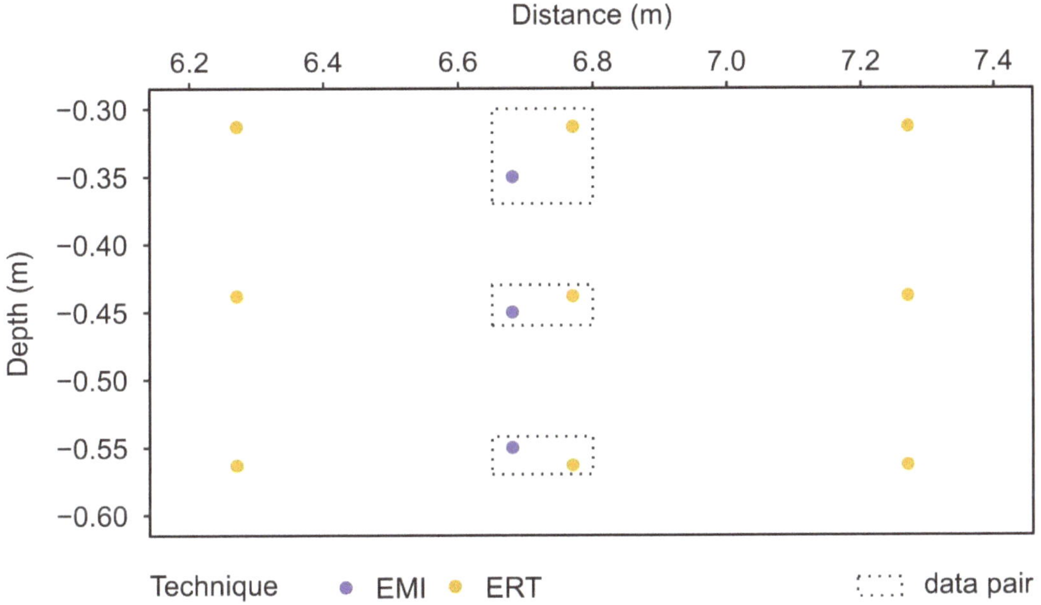


**Figure 2.** Comparison of electromagnetic induction (EMI) and electrical resistivity tomography (ERT) $\sigma$ grids for location 1, Lezíria, Portugal.

For the comparison of EMI and ERT techniques in predicting $EC_e$, the data pairs used were composed by predicted $EC_e$ using ERT and predicted $EC_e$ using EMI, ($pEC_{e\ ERT}$, $pEC_{e\ EMI}$), and were located at the position of the soil samples, as a result of the procedure for the calibration development.

To visualize the agreement of $\sigma$, data pairs were represented through $\sigma_{EMI}$ against $\sigma_{ERT}$ plots, and modified Bland–Altman plots [54]. In the latter type of plot, the difference between each data pair value is plotted against the mean between the two values, which represents the most approximate value to the true value being studied [54]. However, in this case, we used $\sigma_{ERT}$ in the y-axis, as we considered it to be the $\sigma$ reference and true value. A reference interval within which fall most differences between the data is also included, and it is called the 95% limit of agreement [54]. The 95% limit of agreement can be calculated in different ways, depending on the differences between the data pair values following a normal distribution or not [54], which was investigated using the qqPlot function of R language [53].

When the normal distribution of the differences between the data pair values was verified, the 95% limits of agreement were calculated using the mean difference (MD) and the standard deviation (SD) of the differences as MD ± 1.96SD agreement [54]. MD and SD were calculated according to the following equations:

$$MD = \frac{\sum_{i=1}^{n}\left(\sigma_{EMI_i} - \sigma_{ERT_i}\right)}{n} \tag{3}$$

$$SD = \sqrt{\frac{\sum_{i=1}^{n}\left(\sigma_{EMI_i} - \sigma_{ERT_i} - MD\right)^2}{n-1}} \tag{4}$$

When the normal distribution of the differences between the data pair values was not verified, the 95% limits of agreement were defined using the median (MeD), the 5 percentile (p5), and the 95 percentile (p95) of the differences [54].

To visualize the agreement between measured $EC_e$ and predicted $EC_e$ for each technique, data pairs were represented through another modification of the Bland–Altman plot, in which the difference between measured $EC_e$ and predicted $EC_e$ for each technique was plotted against the measured $EC_e$.

Spearman's rank correlation coefficient (RCC) between (i) the differences between $\sigma_{ERT}$ and $\sigma_{EMI}$, and $\sigma_{ERT}$, for $\sigma$ data, and (ii) the differences between measured $EC_e$ and predicted $EC_e$, and the measured $EC_e$, for $EC_e$ data, were calculated using the SpearmanRho() function of R language [53]. Spearman's RCC always returns a value placed between −1 and 1. In this case, it would indicate if the differences between the values being compared are related to the magnitude of the property being estimated. A Spearman's RCC closer to 1 means that the differences are closely related to the magnitude of the property being estimated [54].

## 3. Results and Discussion

### *3.1. Soil Electrical Conductivity Obtained from EMI vs. ERT*

Figure 3 shows the 2D vertical $\sigma$ models obtained by EMI and by ERT techniques for locations 1 to 4. For both techniques, a general increasing trend of $\sigma$ is evident from the north to the south of the peninsula, accompanying the known soil salinity gradient, illustrating the strong correlation of $\sigma$ with $EC_e$ in the region, as verified previously in [4,7,38]. Also, both techniques show that $\sigma$ increases with depth at locations 2, 3, and 4, which correlates well with groundwater depth and salinity in these locations. However, this trend is not observed at location 1, possibly due to irrigation and the deeper positioning of the saline groundwater at this site, making it less prone to capillary rise. Location 1 exhibits the lowest $\sigma$ values, ranging from 50 to 160 mS $m^{-1}$, followed by slightly higher $\sigma$ values at location 2 (i.e., 60–460 mS $m^{-1}$). Location 3 displays a greater variability in $\sigma$ compared to the other two locations, featuring a minimum $\sigma$ of 30 mS $m^{-1}$ and a higher maximum of 530 mS $m^{-1}$. The $\sigma$ gradient at location 4, however, is the most substantial, with an extreme maximum $\sigma$ of over 1600 mS $m^{-1}$.

To compare the EMI model with ERT, Figure 4 shows the plots that support the agreement analysis between EMI and ERT $\sigma$ estimations. At location 1, a normal distribution of the differences between $\sigma_{ERT}$ and $\sigma_{EMI}$ data was verified, so the 95% limits of agreement were calculated using the mean difference (MD) and the standard deviation of the differences (SD), as explained in the Materials and Methods section. It could be verified that most differences between $\sigma_{ERT}$ data and $\sigma_{EMI}$ data fell in the −14.98 mS $m^{-1}$ to 32.16 mS $m^{-1}$ interval. The mean difference was 8.59 mS $m^{-1}$, and, since it was a positive value, indicated that, at this location, the EMI model tended to underestimate $\sigma$ compared to the ERT model. This mean difference value was relatively low compared to the range of $\sigma_{ERT}$ data (82.20–143.10 mS $m^{-1}$, Table 1). In terms of depths, agreement was generally good in topsoil (0–0.3 m), subsurface (0.3–0.6 m), and upper subsoil (0.6–0.9 m). However, the agreement diminished in the intermediate subsoil (0.9–1.2 m) and lower subsoil (1.2–1.5 m), as evidenced by the distance of points to the 1:1 line in the $\sigma_{EMI}$ against $\sigma_{ERT}$ plot. Differences between $\sigma_{EMI}$ and $\sigma_{ERT}$ data in the topsoil, subsurface, and upper subsoil were the closest to the 0-horizontal line in the Bland–Altman plot. The Bland–Altman plot indicates that, at this location, agreement increased with σ and decreased with depth. In comparison to the ERT model, there was an obvious tendency for the EMI model to underestimate σ in the lower subsoil (1.2–1.5 m). Spearman's RCC of −0.34, as observed in location 1, indicates that generally the differences between $\sigma_{EMI}$ and $\sigma_{ERT}$ are not related to the magnitude of σ.

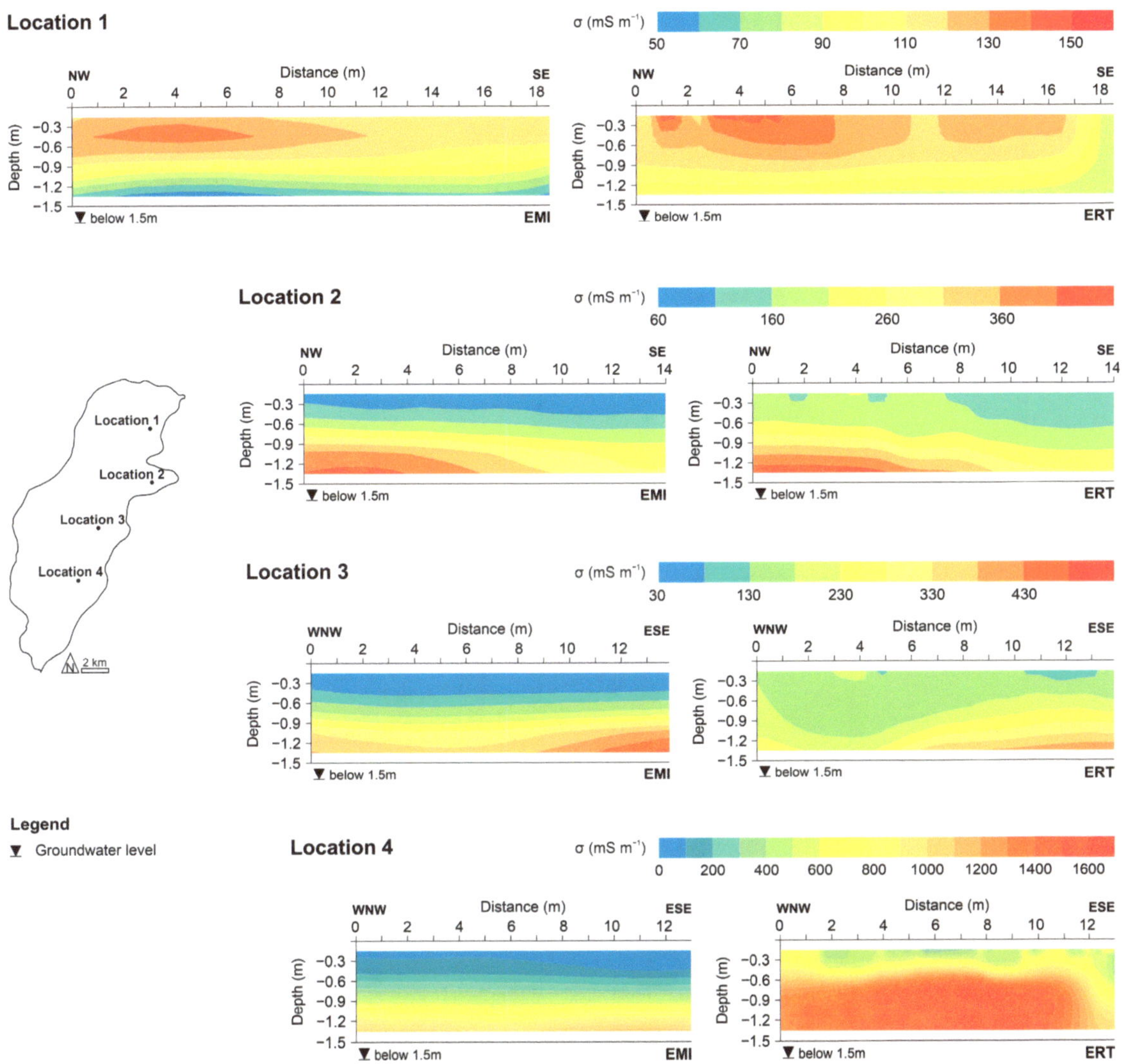


**Figure 3.** Two-dimensional vertical soil electrical conductivity (σ) models obtained from electromagnetic induction (EMI, **left**) and electrical resistivity tomography (ERT, **right**) techniques for locations 1 to 4.

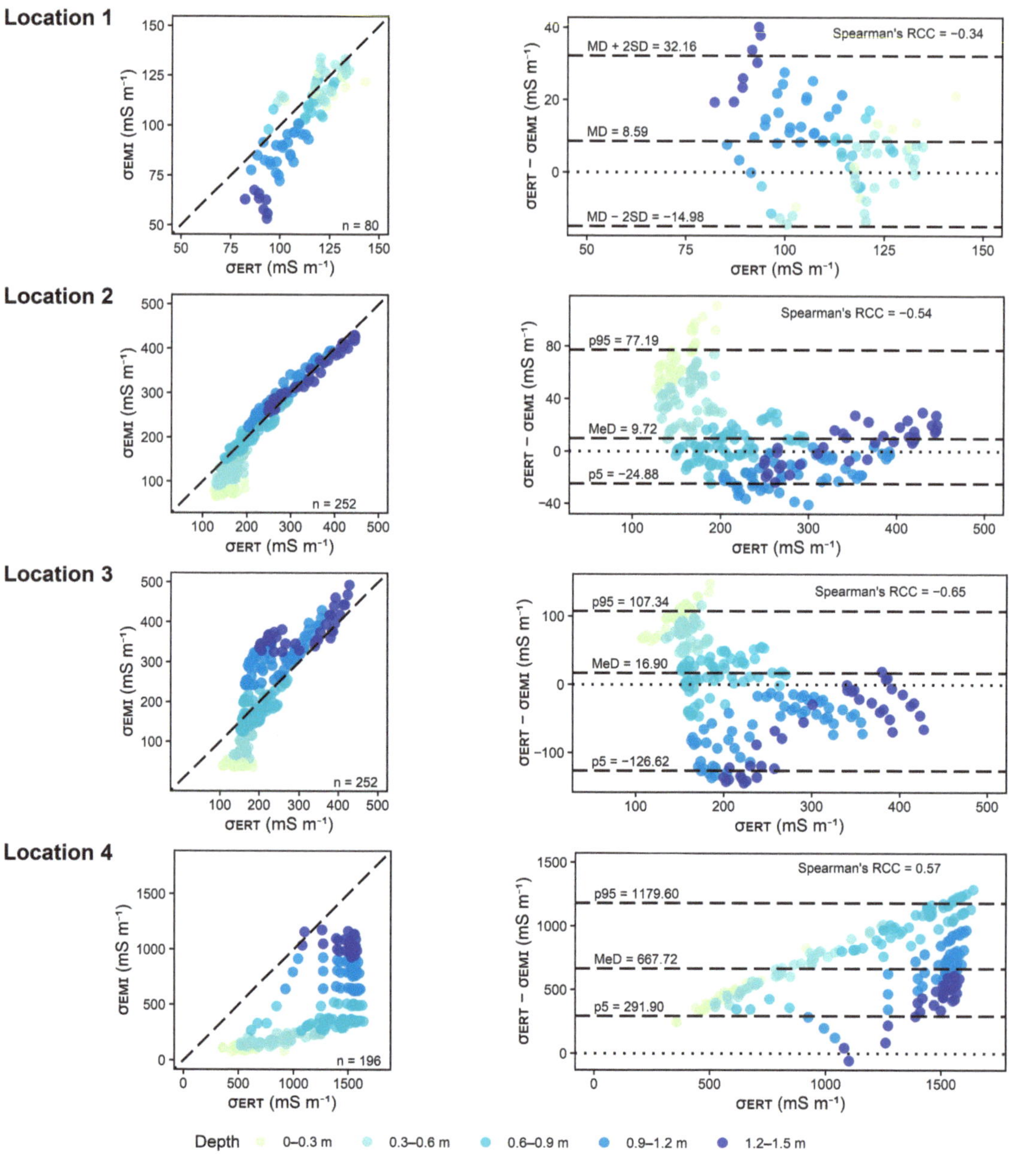


**Figure 4.** Agreement of soil electrical conductivity (σ) obtained from electromagnetic induction (EMI) and electrical resistivity tomography (ERT) techniques for locations 1 to 4: $\sigma_{EMI}$ against $\sigma_{ERT}$ plots with the 1:1 line (**left**), and modified Bland–Altman plots with the 95% limits of agreement (**right**). Plots include the number of data (n) (**left**), Spearman's rank correlation coefficient (RCC), mean difference (MD), standard deviation (SD), median (MeD), the 5 percentile (p5), and the 95 percentile (p95) (**right**).

**Table 1.** Statistics of soil's electrical conductivity obtained from electrical resistivity tomography ($\sigma_{ERT}$), and measured soil salinity ($mEC_e$): minimum, maximum, range, and amount of data, at locations 1 to 4 and at all locations together, respectively.

| | Unit | Location | Minimum | Maximum | Range | Amount of Data |
|---|---|---|---|---|---|---|
| $\sigma_{ERT}$ | mS m$^{-1}$ | 1 | 82.20 | 143.10 | 60.90 | 80 |
| | | 2 | 126.70 | 446.20 | 319.50 | 252 |
| | | 3 | 107.40 | 427.50 | 320.10 | 252 |
| | | 4 | 356.40 | 1640.00 | 1283.60 | 196 |
| $mEC_e$ | dS m$^{-1}$ | all | 0.75 | 37.10 | 36.75 | 19 |

$\sigma_{ERT}$ is soil electrical conductivity obtained by electrical resistivity tomography. $mEC_e$ is measured soil salinity obtained from saturated soil paste extract ($EC_e$).

At locations 2 to 4, a normal distribution of the differences between $\sigma_{EMI}$ and $\sigma_{ERT}$ data was not verified, so the 95% limits of agreement were calculated using the median (MeD), the 5 percentile (p5), and the 95 percentile (p95) of the differences (see Section 2.5).

At location 2, most differences between $\sigma_{ERT}$ and $\sigma_{EMI}$ data fell in the −24.88 mS m$^{-1}$ to 77.19 mS m$^{-1}$ interval. The median was 9.72 mS m$^{-1}$, which indicates that, in general, at this location the EMI model similarly tended to underestimate σ, compared to the ERT model. Also, this value was low compared to the range of $\sigma_{ERT}$ data (126.70–446.20 mS m$^{-1}$, Table 1). In terms of depth, agreement was generally good in the upper subsoil (0.6–0.9 m) and lower subsoil (1.2–1.5 m), which could be verified by the distance of points to the 1:1 line in the $\sigma_{EMI}$ against $\sigma_{ERT}$ plot. Also, differences between $\sigma_{EMI}$ data and $\sigma_{ERT}$ data in the upper and lower subsoil were the closest to the 0-horizontal line in the Bland–Altman plot. At this location, agreement increased with σ and with depth. This contrasted with the observations at location 1, where the most significant disagreement was noted in the lower subsoil (1.2–1.5 m). However, the distribution of σ at location 1 differed significantly from the variability observed in the other three locations, as σ decreased with depth at location 1. Spearman's RCC of −0.54 indicates that generally, the differences between $\sigma_{EMI}$ and $\sigma_{ERT}$ tend to slightly decrease when σ increases.

At location 3, most of the differences between $\sigma_{ERT}$ and $\sigma_{EMI}$ data fell in the −126.62 mS m$^{-1}$ to 107.34 mS m$^{-1}$ interval. The median was 16.90 mS m$^{-1}$, which indicates that, similar to locations 1 and 2, and in general, at this location the EMI model tended to underestimate σ, compared to the ERT model. Also, this value was low compared to the range of $\sigma_{ERT}$ data (107.40–427.50 mS m$^{-1}$, Table 1). In terms of depth, agreement was generally good in the subsurface (0.3–0.6 m) and upper subsoil (0.6–0.9 m), as evidenced by the proximity of points to the 1:1 line in the $\sigma_{EMI}$ against $\sigma_{ERT}$ plot and to the 0-horizontal line in the Bland–Altman plot. Spearman's RCC of −0.65 indicates that the differences between $\sigma_{EMI}$ and $\sigma_{ERT}$ slightly decrease when σ increases.

At location 4, most differences between $\sigma_{ERT}$ and $\sigma_{EMI}$ data fell in the 291.90 mS m$^{-1}$ to 1179.60 mS m$^{-1}$ interval. This interval was totally above the 0-horizontal line in the Bland–Altman plot, which, together with a median of 667.72 mS m$^{-1}$, indicated that at this location, the EMI model tended to drastically underestimate σ, compared to the ERT model. Also, this value was significantly high in the range of $\sigma_{ERT}$ data (356.40–1640.00 mS m$^{-1}$, Table 1). In terms of depth, there was no significant agreement between $\sigma_{ERT}$ and $\sigma_{EMI}$, as evidenced by the lack of proximity of points to the 1:1 line in the $\sigma_{EMI}$ against the $\sigma_{ERT}$ plot and to the 0-horizontal line in the Bland–Altman plot. Spearman's RCC of 0.57 indicates that the differences between $\sigma_{EMI}$ and $\sigma_{ERT}$ slightly increase when σ increases.

Comparing the results between locations suggests that the EMI models tend to underestimate σ when compared to the ERT models in all four locations. However, in locations 1, 2, and 3, with σ variability inferred from the ERT model in the 50–500 mS m$^{-1}$ range, the underestimation of the EMI model was not significant, suggesting that the obtained EMI models were in good and acceptable agreement with those inferred from detailed ERT investigation. This was not the case at location 4, as the underestimation tendency was quite drastic at this location, where the σ variability inferred from the ERT model fell in the range of 500–1600 mS m$^{-1}$. This is likely linked to the relationship between the quadrature component of the EMI signal and σ in superconductive soil at location 4, which may exhibit a non-monotonic behaviour [33]. In such situations, obtaining more representative $EC_a$ values may need the use of both the in-phase and quadrature components of the EMI signal [55]. However, the in-phase component has to be adequately calibrated, as it is susceptible to signal instability and offsets [56], which was not considered in this study. This presents a significant challenge when estimating σ in highly saline soil, where conductivity is anticipated to be extremely high. In this context, the robust $EC_a$ estimation approach, proposed by [57], may enhance the reliability of $EC_a$ estimation over superconductive areas. Alternatively, ERT (e.g., [44]) or TDR (e.g., [45]) measurements carried out along the same transects can be used to calibrate $EC_a$ data to obtain more representative $EC_a$ values.

### 3.2. Soil Salinity Obtained from EMI vs. ERT

Figure 5 shows the two regional calibrations that were developed to predict $EC_e$, one using the σ obtained from the inversion of EMI data and the other using σ, obtained from the inversion of ERT, and their prediction results. Both models had a strong $R^2$, with EMI (0.86) being higher than ERT (0.75). The leave-one-out cross-validation resulted in acceptable and comparable RMSE and ME results. The obtained RMSE of 3.96 dS m$^{-1}$ for EMI, and of 4.72 dS m$^{-1}$ for ERT, were low in the measured $EC_e$ range (0.75–37.1 dS m$^{-1}$, Table 1) and comparable between them. ME of 3.22 dS m$^{-1}$ for EMI, and 2.32 dS m$^{-1}$ for ERT, mean a comparable underestimation of the predicted data for both techniques.

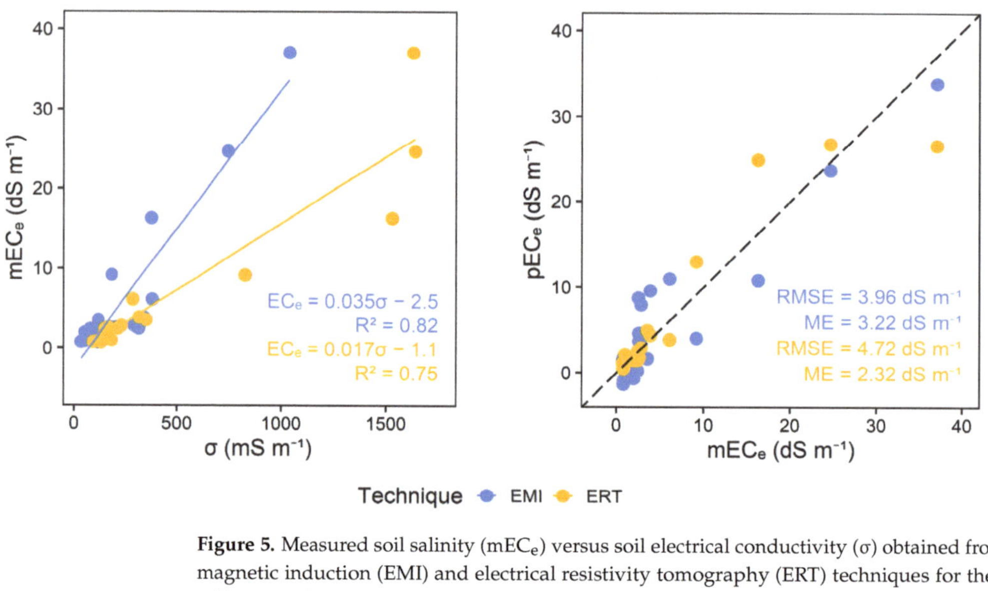


**Figure 5.** Measured soil salinity ($mEC_e$) versus soil electrical conductivity (σ) obtained from electromagnetic induction (EMI) and electrical resistivity tomography (ERT) techniques for the regional calibrations (**left**), and predicted soil salinity ($pEC_e$) obtained from the leave-one-out-cross-validation of the calibrations (**right**). Plots include calibrations and their coefficient of determination ($R^2$) (**left**), and their root mean square error (RMSE) and mean error (ME).

Figure 6 shows the plot of the differences between measured $EC_e$ and predicted $EC_e$, for each technique, against the measured $EC_e$. The proximity of the points to the 0-horizontal line indicates that there is good agreement between measured and predicted $EC_e$. Points above the line indicate underestimation, while points below the line indicate overestimation. In the non-saline classification interval of soil salinity, both techniques showed good agreement between measured and predicted $EC_e$. In the slightly saline classification interval, EMI provided mostly overestimated predictions but also some underestimated predictions, whereas ERT showed good agreement between measured and predicted $EC_e$. In the moderately saline classification interval, EMI provided one overestimated prediction, while ERT underestimated that same $EC_e$ measurement. In the very saline classification interval, EMI provided one underestimated prediction, while ERT overestimated that same $EC_e$ measurement. In the highly saline classification interval, the same happened as in the previous interval, except for one $EC_e$ measurement, which was underestimated by both techniques, but still classified as highly saline, for both techniques. Spearman's RCC calculated for EMI (0.24) indicated that there was an underestimation tendency as the magnitude of $EC_e$ (soil salinity) increased. In the case of ERT, Spearman's RCC of −0.12 indicated a slight overestimation tendency as the soil salinity grew.

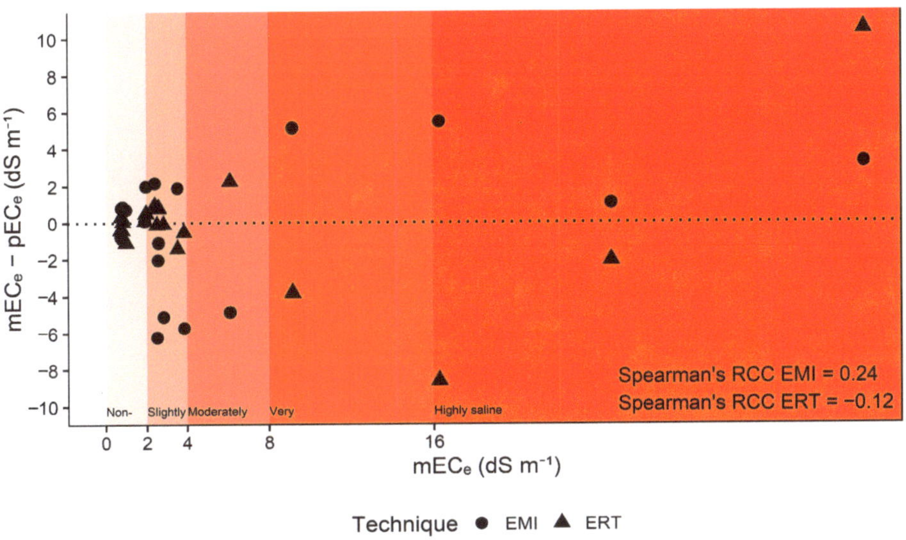


**Figure 6.** Agreement of measured soil salinity (mECe) and predicted soil salinity (pECe) obtained using electromagnetic induction (EMI) and electrical resistivity tomography (ERT) techniques, with Spearman's rank correlation coefficient (RCC).

A thorough examination of the calibrations reveals that, despite the EMI model significantly underestimating $\sigma$ in the superconductive zone at location 4, resulting in a distinct linear regression slope between $EC_e$ and $\sigma$ (EMI vs. ERT), this issue did not adversely affect the overall predictive capability of the regional calibration when compared to the results obtained from ERT. This is attributed to the fact that, although EMI underestimated $\sigma$ at location 4, the pattern of $\sigma$ distribution and its variations with depth aligned with those obtained from the ERT model.

To provide a better insight into the prediction ability of both methodologies in different locations, Figure 7 depicts the 2D vertical maps of soil salinity classification obtained from the conversion of $\sigma$ obtained by EMI and ERT for locations 1 to 4, using the corresponding calibration. The filled-in circles in the maps represent the position and classification of the soil samples (measured $EC_e$).

At location 1 there is total agreement between the predicted classification and the actual classification obtained from the samples, for both techniques. At location 2 soil salinity is overestimated by EMI at intermediate (0.9–1.2 m) and lower subsoil (1.2–1.5 m), while it is also overestimated by ERT but only in lower subsoil (1.2–1.5 m). At location 3, soil salinity is underestimated by EMI at the subsurface (0.3–0.9 m) and overestimated at intermediate (0.9–1.2 m) and lower subsoil (1.2–1.5 m), while it is underestimated by ERT at the subsurface (0.3–0.9 m) and upper subsoil (0.6–0.9 m). At location 4, soil salinity is underestimated by EMI from topsoil to upper subsoil (0–0.9 m), while there is total agreement between the predicted classification and the actual classification by ERT.

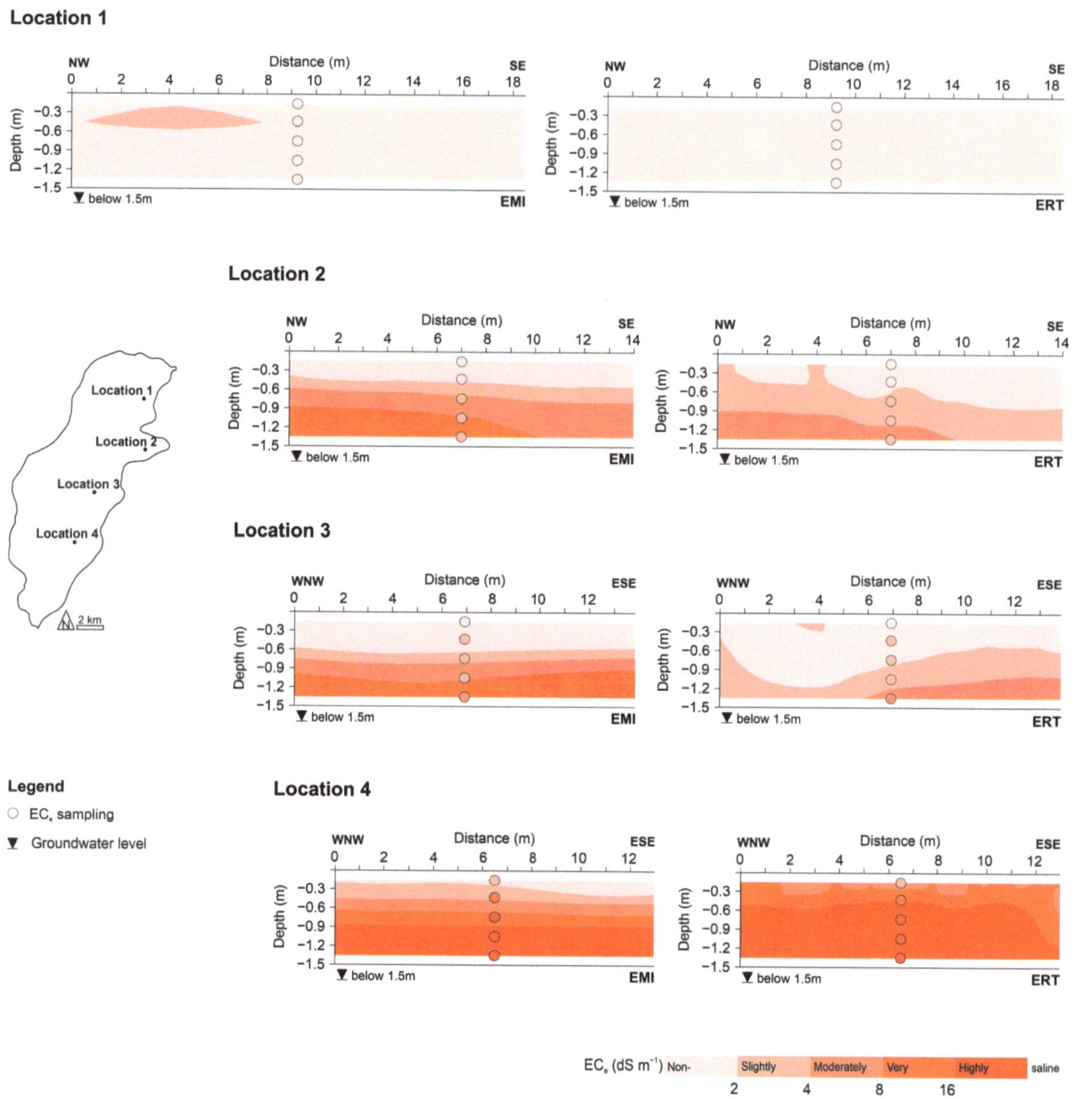


**Figure 7.** Two-dimensional vertical soil salinity classification maps obtained by electromagnetic induction (EMI, **left**) and electrical resistivity tomography (ERT, **right**) techniques for locations 1 to 4. The filled circles represent the position and classification of the soil samples (measured $EC_e$).

As anticipated, the salinity maps obtained from EMI and ERT are acceptably comparable, although EMI generally underestimated $\sigma$. Both techniques displayed similar levels of underestimations and overestimations, indicating a comparable level of prediction accuracy. The underestimation or overestimation of soil salinity based on $\sigma$ is not only related to the geophysical approach but also influenced by the variability of other soil properties along the transects, such as soil texture, moisture content, salinity type, and temperature. For instance, in our previous study in the same study area, we observed that at location 2 the presence of slightly higher clay content in the subsoil, combined with the lower range of soil salinity (compared to locations 3 and 4), made it challenging to estimate $EC_e$ from EMI data and regional calibration [38]. Also, in the same study area, a relatively larger variability of moisture content and soil temperature in the topsoil at all locations was a major factor limiting $EC_e$ prediction in the topsoil [4]. Addressing such variability in soil properties may require location-specific calibration to consider soil texture and moisture variations, as discussed in [38]. Additionally, developing multiple regression models can account for the influence of other parameters on $\sigma$ and soil salinity prediction. Nevertheless, since all measurements were conducted simultaneously at each location, we expect that comparable effects of other soil properties on both ERT and EMI techniques will not significantly impact the comparison between these two methodologies in assessing soil salinity, as they should have a similar effect on both.

## 4. Conclusions

In this study, EMI and ERT surveys and soil sampling were carried out at four locations with different salinity levels across the study area of Lezíria de Vila Franca, in dry season conditions, to analyse the agreement between the two techniques in estimating soil electrical conductivity and compare their ability in predicting soil salinity. While ERT may not offer a precise subsurface conductivity distribution of the subsoil (as no indirect method can achieve that), it stands out as one of the most reliable techniques for imaging the subsurface conductivity distribution. Conversely, EMI measurements are highly sensitive to various factors, including ground coupling, thermal drifts, and EM noise. Hence, it is sensible to consider an ERT inversion as a reference model that the EMI inversion should strive to approximate.

Based on the obtained results in this study, there was a reasonable agreement between the EMI and ERT models in three locations, where $\sigma$ ranged from 50 to 500 mS m$^{-1}$. In contrast, at location 4, where $\sigma$ surpassed 1000 mS m$^{-1}$, EMI notably underestimated $\sigma$ in comparison to ERT. However, EMI models could still predict the increasing trend of $\sigma$ well with depth. This suggests that the obtained EMI model may substantially underestimate $\sigma$ in an extremely saline area given the very high level of soil conductivity, which exhibits a non-monotonic relationship between the quadrature component of the EMI signal. Under this condition, the $\sigma$ values inferred from EMI modelling cannot be used alone to assess the soil salinity level without a location-specific regression or by applying a more robust approach to obtain a more representative $EC_a$ value. Further case studies across different soil types and salinity levels will offer more insights into the circumstances under which EMI performs optimally.

The regional calibrations based on both EMI and ERT demonstrated similar predictive capabilities. Despite the EMI model significantly underestimating $\sigma$ in the superconductive zone at location 4, leading to a distinct regression linear slope between $EC_e$ and $\sigma$, this issue did not markedly affect the overall predictive performance of the regional calibration when compared to the results from ERT. This is because despite the underestimation of $\sigma$ by EMI at location 4, the distribution pattern and depth-related variations in $\sigma$ similarly mirrored those obtained from the ERT model, resulting in comparable prediction abilities.

Our case study was limited to four plots with distinct soil salinity levels, but with the same soil type. In addition, the number of soil samples was relatively limited, with one borehole for each location. Additional case studies across areas with different soil types and high conductivities are necessary to further evaluate the precision of EMI in soil salinity assessments. Specifically, EMI vs. ERT studies across sites with inverted soil salinity gradient in depth, where there is a superconductive zone over a more resistive zone, are particularly needed to assess the prediction ability of EMI in contrasting conditions, compared to this study. Larger numbers of boreholes and soil samples can also enhance the soil salinity prediction ability and the evaluation of the EMI prediction ability.

Lastly, we advise caution to EMI practitioners when working with superconductive soils. While it is not feasible to establish a definitive limit based solely on a single experiment using specific EMI equipment and across a study area with a similar soil type, our experiment indicated a significant underestimation $\sigma$ in ranges above approximately 500 mS m$^{-1}$. It is important to note that this finding may vary in different experiments across diverse soil and salinity types, as well as when using different EMI sensors. However, this does not diminish the concern regarding the challenge of using EMI in superconductive soil.

**Author Contributions:** Conceptualization, M.C.P. and M.F.; Data curation, N.L.C., M.F. and M.C.P.; Formal analysis, M.C.P. and M.F.; Funding acquisition, A.M.P., M.C.G. and F.M.S.; Investigation, M.C.P., N.L.C., A.M.P., M.C.G. and M.F.; Methodology, M.C.P., A.M.P., F.M.S. and M.F.; Project administration, M.C.G., F.M.S. and M.F.; Resources, N.L.C., A.M.P. and M.C.G.; Software, F.M.S.; Validation, M.C.P. and M.F.; Visualization, M.C.P.; Writing—original draft, M.C.P. and M.F.; Writing—review and editing, M.C.P., N.L.C., A.M.P., M.C.G., F.M.S. and M.F. All authors have read and agreed to the published version of the manuscript.

**Funding:** This work was funded by the Portuguese research agency, Fundação para a Ciência e a Tecnologia (FCT), in the scope of project SALTFREE—ARIMNET2/0004/2015 SALTFREE and ARIMNET2/0005/2015 SALTFREE. This work was also supported by the European Joint Programme Cofund on Agricultural Soil Management (EJP SOIL grant number 862695), funded by the European Union's Horizon H2020 research and innovation project, and was carried out in the framework of the STEROPES of EJP-SOIL.

**Data Availability Statement:** The data presented in this study are available on request from the corresponding author. The data are not publicly available due to having been acquired at privately owned farms.

**Acknowledgments:** The authors express their appreciation to the Associação de Beneficiários da Lezíria Grande de Vila Franca de Xira for granting access to study sites and for their ongoing support. Special thanks are also extended to Manuel Fernandes and Fernando Pires from INIAV for their field assistance. This work was funded by the Portuguese Fundação para a Ciência e a Tecnologia (FCT) I.P./MCTES through national funds (PIDDAC)—UIDB/50019/2020 (https://doi.org/10.54499/UIDB/50019/2020 accessed on 21 February 2024), UIDP/50019/2020 (https://doi.org/10.54499/UIDP/50019/2020 accessed on 21 February 2024) and LA/P/0068/2020 (https://doi.org/10.54499/LA/P/0068/2020 accessed on 21 February 2024).

**Conflicts of Interest:** The authors declare no conflict of interests.

## References

1. Corwin, D.L.; Scudiero, E. Chapter One—Review of Soil Salinity Assessment for Agriculture across Multiple Scales Using Proximal and/or Remote Sensors. In *Advances in Agronomy*; Sparks, D.L., Ed.; Academic Press: Cambridge, MA, USA, 2019; Volume 158, pp. 1–130.
2. Corwin, D.L.; Lesch, S.M. Apparent Soil Electrical Conductivity Measurements in Agriculture. *Comput. Electron. Agric.* **2005**, *46*, 11–43. [CrossRef]
3. Rhoades, J.D.; Corwin, D.L.; Lesch, S.M. Geospatial Measurements of Soil Electrical Conductivity to Assess Soil Salinity and Diffuse Salt Loading from Irrigation. In *Assessment of Non-Point Source Pollution in the Vadose Zone*; American Geophysical Union (AGU): Washington, DC, USA, 1999; pp. 197–215. ISBN 9781118664698.
4. Paz, M.C.; Farzamian, M.; Paz, A.M.; Castanheira, N.L.; Gonçalves, M.C.; Santos, F.M. Assessing Soil Salinity Dynamics Using Time-Lapse Electromagnetic Conductivity Imaging. *SOIL* **2020**, *6*, 499–511. [CrossRef]
5. Nguyen, V.H.; Germer, J.; Duong, V.N.; Asch, F. Soil Resistivity Measurements to Evaluate Subsoil Salinity in Rice Production Systems in the Vietnam Mekong Delta. *Near Surf. Geophys.* **2023**, *21*, 288–299. [CrossRef]
6. Innocenti, A.; Pazzi, V.; Napoli, M.; Fanti, R.; Orlandini, S. Application of Electrical Resistivity Tomography (ERT) to Study to Soil Water and Salt Movement under Drip Irrigation in a Saline Soil Cultivated with Melon. In Proceedings of the EGU General Assembly Conference Abstracts, Vienna, Austria, 3–8 April 2022; pp. EGU22–4469.
7. Paz, A.M.; Castanheira, N.; Farzamian, M.; Paz, M.C.; Gonçalves, M.C.; Monteiro Santos, F.A.; Triantafilis, J. Prediction of Soil Salinity and Sodicity Using Electromagnetic Conductivity Imaging. *Geoderma* **2020**, *361*, 114086. [CrossRef]
8. da Silva, L.D.C.M.; Peixoto, D.S.; Azevedo, R.P.; Avanzi, J.C.; Junior, M.D.S.D.; Vanella, D.; Consoli, S.; Acuña-Guzman, S.F.; Borghi, E.; de Resende, Á.V.; et al. Assessment of Soil Water Content Variability Using Electrical Resistivity Imaging in an Oxisol under Conservation Cropping Systems. *Geoderma Reg.* **2023**, *33*, e00624. [CrossRef]
9. Beff, L.; G ¨ Unther, T.; Vandoorne, B.; Couvreur, V.; Javaux, M. Three-Dimensional Monitoring of Soil Water Content in a Maize Field Using Electrical Resistivity Tomography. *Hydrol. Earth Syst. Sci* **2013**, *17*, 595–609. [CrossRef]
10. Guan, Y.; Grote, K.; Schott, J.; Leverett, K. Prediction of Soil Water Content and Electrical Conductivity Using Random Forest Methods with UAV Multispectral and Ground-Coupled Geophysical Data. *Remote Sens.* **2022**, *14*, 1023. [CrossRef]

11. Ratshiedana, P.E.; Abd Elbasit, M.A.M.; Adam, E.; Chirima, J.G.; Liu, G.; Economon, E.B. Determination of Soil Electrical Conductivity and Moisture on Different Soil Layers Using Electromagnetic Techniques in Irrigated Arid Environments in South Africa. *Water* **2023**, *15*, 1911. [CrossRef]
12. de Jong, S.M.; Heijenk, R.A.; Nijland, W.; van der Meijde, M. Monitoring Soil Moisture Dynamics Using Electrical Resistivity Tomography under Homogeneous Field Conditions. *Sensors* **2020**, *20*, 5313. [CrossRef]
13. Acosta, J.A.; Gabarrón, M.; Martínez-Segura, M.; Martínez-Martínez, S.; Faz, Á.; Pérez-Pastor, A.; Gómez-López, M.D.; Zornoza, R. Soil Water Content Prediction Using Electrical Resistivity Tomography (ERT) in Mediterranean Tree Orchard Soils. *Sensors* **2022**, *22*, 1365. [CrossRef]
14. Shanahan, P.W.; Binley, A.; Whalley, W.R.; Watts, C.W. The Use of Electromagnetic Induction to Monitor Changes in Soil Moisture Profiles beneath Different Wheat Genotypes. *Soil Sci. Soc. Am. J.* **2015**, *79*, 459–466. [CrossRef]
15. Whalley, W.R.; Binley, A.; Watts, C.W.; Shanahan, P.; Dodd, I.C.; Ober, E.S.; Ashton, R.W.; Webster, C.P.; White, R.P.; Hawkesford, M.J. Methods to Estimate Changes in Soil Water for Phenotyping Root Activity in the Field. *Plant Soil* **2017**, *415*, 407–422. [CrossRef]
16. Zhao, X.; Wang, J.; Zhao, D.; Li, N.; Zare, E.; Triantafilis, J. Digital Regolith Mapping of Clay across the Ashley Irrigation Area Using Electromagnetic Induction Data and Inversion Modelling. *Geoderma* **2019**, *346*, 18–29. [CrossRef]
17. Triantafilis, J.; Lesch, S.M. Mapping Clay Content Variation Using Electromagnetic Induction Techniques. *Comput. Electron. Agric.* **2005**, *46*, 203–237. [CrossRef]
18. Huang, J.; Lark, R.M.; Robinson, D.A.; Lebron, I.; Keith, A.M.; Rawlins, B.; Tye, A.; Kuras, O.; Raines, M.; Triantafilis, J. Scope to Predict Soil Properties at Within-Field Scale from Small Samples Using Proximally Sensed $\gamma$-Ray Spectrometer and EM Induction Data. *Geoderma* **2014**, *232–234*, 69–80. [CrossRef]
19. Zare, E.; Li, N.; Khongnawang, T.; Farzamian, M.; Triantafilis, J. Identifying Potential Leakage Zones in an Irrigation Supply Channel by Mapping Soil Properties Using Electromagnetic Induction, Inversion Modelling and a Support Vector Machine. *Soil Syst.* **2020**, *4*, 25. [CrossRef]
20. Triantafilis, J.; Lesch, S.M.; La Lau, K.; Buchanan, S.M. Field Level Digital Soil Mapping of Cation Exchange Capacity Using Electromagnetic Induction and a Hierarchical Spatial Regression Model. *Soil Res.* **2009**, *47*, 651–663. [CrossRef]
21. Koganti, T.; Narjary, B.; Zare, E.; Pathan, A.L.; Huang, J.; Triantafilis, J. Quantitative Mapping of Soil Salinity Using the DUALEM-21S Instrument and EM Inversion Software. *Land Degrad. Dev.* **2018**, *29*, 1768–1781. [CrossRef]
22. Zhao, D.; Li, N.; Zare, E.; Wang, J.; Triantafilis, J. Mapping Cation Exchange Capacity Using a Quasi-3d Joint Inversion of EM38 and EM31 Data. *Soil Tillage Res.* **2020**, *200*, 104618. [CrossRef]
23. Zhao, X.; Wang, J.; Zhao, D.; Sefton, M.; Triantafilis, J. Mapping Cation Exchange Capacity (CEC) Across Sugarcane Fields with Different Comparisons by Using DUALEM Data. *J. Environ. Eng. Geophys.* **2023**, *27*, 191–205. [CrossRef]
24. Huang, J.; Pedrera-Parrilla, A.; Vanderlinden, K.; Taguas, E.V.; Gómez, J.A.; Triantafilis, J. Potential to Map Depth-Specific Soil Organic Matter Content across an Olive Grove Using Quasi-2d and Quasi-3d Inversion of DUALEM-21 Data. *CATENA* **2017**, *152*, 207–217. [CrossRef]
25. Jupp, D.L.B.; Vozoff, K. Stable Iterative Methods for the Inversion of Geophysical Data. *Geophys. J. R. Astron. Soc.* **1975**, *42*, 957–976. [CrossRef]
26. Monteiro Santos, F.A. 1-D Laterally Constrained Inversion of EM34 Profiling Data. *J. Appl. Geophys.* **2004**, *56*, 123–134. [CrossRef]
27. Moghadas, D. Probabilistic Inversion of Multiconfiguration Electromagnetic Induction Data Using Dimensionality Reduction Technique: A Numerical Study. *Vadose Zone J.* **2019**, *18*, 180183. [CrossRef]
28. Narciso, J.; Bobe, C.; Azevedo, L.; Van De Vijver, E. A Comparison between Kalman Ensemble Generator and Geostatistical Frequency-Domain Electromagnetic Inversion: The Impacts on near-Surface Characterization. *Geophysics* **2022**, *87*, E335–E346. [CrossRef]
29. EMTOMO. *Manual for EM4Soil: A Program for 1-D Laterally Constrained Inversion of EM Data*; EMTOMO: Lisbon, Portugal, 2018.
30. McLachlan, P.; Blanchy, G.; Binley, A. EMagPy: Open-Source Standalone Software for Processing, Forward Modeling and Inversion of Electromagnetic Induction Data. *Comput. Geosci.* **2021**, *146*, 104561.EM. [CrossRef]
31. Loke, M.H. Rapid 2D Resistivity Forward Modeling Using the Finite Difference and Finite Element Methods. *RES2DMOD Ver* **2002**, *3*, 1996–2002.
32. Rücker, C.; Günther, T.; Wagner, F.M. PyGIMLi: An Open-Source Library for Modelling and Inversion in Geophysics. *Comput. Geosci.* **2017**, *109*, 106–123. [CrossRef]
33. Blanchy, G.; Saneiyan, S.; Boyd, J.; McLachlan, P.; Binley, A. ResIPy, an Intuitive Open Source Software for Complex Geoelectrical Inversion/Modeling. *Comput. Geosci.* **2020**, *137*, 104423. [CrossRef]
34. Araújo, O.S.; Picotti, S.; Francese, R.G.; Bocchia, F.; Santos, F.M.; Giorgi, M.; Tessarollo, A. Frequency Domain Electromagnetic Calibration for Improved Detection of Sand Intrusions in River Embankments. *Lead. Edge* **2023**, *42*, 615–624. [CrossRef]
35. FAO. Global Map of Salt-Affected Soils. Available online: https://www.fao.org/soils-portal/data-hub/soil-maps-and-databases/global-map-of-salt-affected-soils/ar/ (accessed on 21 February 2024).
36. Stavi, I.; Thevs, N.; Priori, S. Soil Salinity and Sodicity in Drylands: A Review of Causes, Effects, Monitoring, and Restoration Measures. *Front. Environ. Sci.* **2021**, *9*, 330. [CrossRef]

37. Paz, A.M.; Amezketa, E.; Canfora, L.; Castanheira, N.; Falsone, G.; Gonçalves, M.C.; Gould, I.; Hristov, B.; Mastrorilli, M.; Ramos, T.; et al. Salt-Affected Soils: Field-Scale Strategies for Prevention, Mitigation, and Adaptation to Salt Accumulation. *Ital. J. Agron.* **2023**, *18*, 2166. [CrossRef]
38. Farzamian, M.; Paz, M.C.; Paz, A.M.; Castanheira, N.L.; Gonçalves, M.C.; Monteiro Santos, F.A.; Triantafilis, J. Mapping Soil Salinity Using Electromagnetic Conductivity Imaging—A Comparison of Regional and Location-Specific Calibrations. *L. Degrad. Dev.* **2019**, *30*, 1393–1406. [CrossRef]
39. Khongnawang, T.; Zare, E.; Srihabun, P.; Khunthong, I.; Triantafilis, J. Digital Soil Mapping of Soil Salinity Using EM38 and Quasi-3d Modelling Software (EM4Soil). *Soil Use Manag.* **2022**, *38*, 277–291. [CrossRef]
40. Lavoué, F.; van der Krak, J.; Rings, J.; André, F.; Moghadas, D.; Huisman, J.A.; Lambot, S.; Weiherrnüller, L.; Vanderborght, J.; Vereecken, H. Electromagnetic Induction Calibration Using Apparent Electrical Conductivity Modelling Based on Electrical Resistivity Tomography. *Near Surf. Geophys.* **2010**, *8*, 553–561. [CrossRef]
41. Minsley, B.J.; Smith, B.D.; Hammack, R.; Sams, J.I.; Veloski, G. Calibration and Filtering Strategies for Frequency Domain Electromagnetic Data. *J. Appl. Geophys.* **2012**, *80*, 56–66. [CrossRef]
42. Moghadas, D.; Jadoon, K.Z.; McCabe, M.F. Spatiotemporal Monitoring of Soil Water Content Profiles in an Irrigated Field Using Probabilistic Inversion of Time-Lapse EMI Data. *Adv. Water Resour.* **2017**, *110*, 238–248. [CrossRef]
43. von Hebel, C.; Rudolph, S.; Mester, A.; Huisman, J.A.; Kumbhar, P.; Vereecken, H.; van der Kruk, J. Three-Dimensional Imaging of Subsurface Structural Patterns Using Quantitative Large-Scale Multiconfiguration Electromagnetic Induction Data. *Water Resour. Res.* **2014**, *50*, 2732–2748. [CrossRef]
44. Von Hebel, C.; Van Der Kruk, J.; Huisman, J.A.; Mester, A.; Altdorff, D.; Endres, A.L.; Zimmermann, E.; Garré, S.; Vereecken, H. Calibration, Conversion, and Quantitative Multi-Layer Inversion of Multi-Coil Rigid-Boom Electromagnetic Induction Data. *Sensors* **2019**, *19*, 4753. [CrossRef] [PubMed]
45. Dragonetti, G.; Comegna, A.; Ajeel, A.; Deidda, G.P.; Lamaddalena, N.; Rodriguez, G.; Vignoli, G.; Coppola, A. Calibrating Electromagnetic Induction Conductivities with Time-Domain Reflectometry Measurements. *Hydrol. Earth Syst. Sci.* **2018**, *22*, 1509–1523. [CrossRef]
46. Dragonetti, G.; Farzamian, M.; Basile, A.; Monteiro Santos, F.; Coppola, A. In Situ Estimation of Soil Hydraulic and Hydrodispersive Properties by Inversion of Electromagnetic Induction Measurements and Soil Hydrological Modeling. *Hydrol. Earth Syst. Sci.* **2022**, *26*, 5119–5136. [CrossRef]
47. Fischer, G.; Nachtergaele, F.O.; Prieler, S.; Teixeira, E.; Toth, G.; van Velthuizen, H.; Verelst, L.; Wiberg, D. Global Agro-Ecological Zones (GAEZ v3.0)-Model Documentation 2012. Available online: https://www.gaez.iiasa.ac.at/docs/GAEZ_Model_Documentation.pdf (accessed on 21 February 2024).
48. Daliakopoulos, I.N.; Tsanis, I.K.; Koutroulis, A.; Kourgialas, N.N.; Varouchakis, A.E.; Karatzas, G.P.; Ritsema, C.J. The Threat of Soil Salinity: A European Scale Review. *Sci. Total Environ.* **2016**, *573*, 727–739. [CrossRef]
49. Barrett-Lennard, E.; Bennett, S.; Colmer, T. Standardising Terminology for Describing the Level of Salinity in Soils in Australia. In Proceedings of the 2nd International Salinity Forum. Salinity, Water and Society: Global Issues, Local Action, Adelaide, Australia, 30 March–3 April 2008; Future Farm Industries CRC: Perth, WA, Australia; University of Western Australia: Crawley, WA, Australia, 2008.
50. Kaufman, A.; Keller, G.V. Frequency and Transient Sounding Methods in Geochemistry and Geophysics, Vol. 16 A, A. Kaufman and G. V. Keller, Elsevier, Amsterdam, 1983 686 pp. £85.55/$144.75. *Geophys. J. Int.* **1984**, *77*, 935–937. [CrossRef]
51. deGroot-Hedlin, C.; Constable, S. Occam's Inversion to Generate Smooth, Two-dimensional Models from Magnetotelluric Data. *Geophysics* **1990**, *55*, 1613–1624. [CrossRef]
52. Allison, L.E.; Bernstein, L.; Bower, C.A.; Brown, J.W.; Fireman, M.; Hatcher, J.T.; Hayward, H.E.; Pearson, G.A.; Reeve, R.C.; Richards, L.A.; et al. *Diagnosis and Improvement of Saline Alkali Soils, Agricultural Handbook*; Richards, L.A., Ed.; United States Department of Agriculture: Washington, DC, USA, 1954.
53. R Core Team. *R: A Language and Environment for Statistical Computing*; R Foundation for Statistical Computing: Vienna, Austria, 2020.
54. Bland, J.M.; Altman, D.G. Measuring Agreement in Method Comparison Studies. *Stat. Methods Med. Res.* **1999**, *8*, 135–160. [CrossRef]
55. Guillemoteau, J.; Sailhac, P.; Boulanger, C.; Trules, J. Inversion of Ground Constant Offset Loop-Loop Electromagnetic Data for a Large Range of Induction Numbers. *Geophysics* **2015**, *80*, E11–E21. [CrossRef]
56. De Smedt, P.; Delefortrie, S.; Wyffels, F. Identifying and Removing Micro-Drift in Ground-Based Electromagnetic Induction Data. *J. Appl. Geophys.* **2016**, *131*, 14–22. [CrossRef]
57. Hanssens, D.; Delefortrie, S.; Bobe, C.; Hermans, T.; De Smedt, P. Improving the Reliability of Soil EC-Mapping: Robust Apparent Electrical Conductivity (REC$_a$) Estimation in Ground-Based Frequency Domain Electromagnetics. *Geoderma* **2019**, *337*, 1155–1163. [CrossRef]

*land*

*Article*

# Assessing the Impact of Brackish Water on Soil Salinization with Time-Lapse Inversion of Electromagnetic Induction Data

**Lorenzo De Carlo [1,*] and Mohammad Farzamian [2,3]**

1 Water Research Institute, National Research Council of Italy, 70132 Bari, Italy
2 Instituto Nacional de Investigação Agrária e Veterinária, 2780-157 Oeiras, Portugal; mohammad.farzamian@iniav.pt
3 Centro de Estudos Geográficos, Laboratório Associado TERRA, Instituto de Geografia e Ordenamento do Território, Universidade de Lisboa, 1600-276 Lisboa, Portugal
* Correspondence: lorenzo.decarlo@cnr.it

**Abstract:** Over the last decade, electromagnetic induction (EMI) measurements have been increasingly used for investigating soil salinization caused by the use of brackish or saline water as an irrigation source. EMI measurements proved to be a powerful tool for providing spatial information on the investigated soil because of the correlation between the output geophysical parameter, i.e., the electrical conductivity, to soil moisture and salinity. In addition, their non-invasive nature and their capability to collect a high amount of data over broad areas and in a relatively short time makes these measurements attractive for monitoring flow and transport dynamics, which are otherwise undetectable with conventional measurements. In an experimental field, EMI measurements were collected during the growth season of tomatoes and irrigated with three different irrigation strategies. Time-lapse data were collected over three months in order to visualize changes in electrical conductivity associated with soil salinity. A rigorous time-lapse inversion procedure was set for modeling the soil salinization induced by brackish irrigation water. A clear soil response in terms of an increase in electrical conductivity (EC) in the upper soil layer confirmed the reliability of the geophysical tool to predict soil salinization trends.

**Keywords:** soil salinization; apparent electrical conductivity; electromagnetic induction measurements; time-lapse inversion

**Citation:** De Carlo, L.; Farzamian, M. Assessing the Impact of Brackish Water on Soil Salinization with Time-Lapse Inversion of Electromagnetic Induction Data. *Land* **2024**, *13*, 961. https://doi.org/10.3390/land13070961

Academic Editor: Amrakh I. Mamedov

Received: 13 June 2024
Revised: 25 June 2024
Accepted: 27 June 2024
Published: 30 June 2024



## 1. Introduction

Soil salinization has become one of the major environmental and socioeconomic issues globally, and this is expected to be exacerbated further by projected climatic change. Salinity is one of the main soil threats that reduces soil fertility and affect crop production because it can decrease plant growth and water quality, resulting in lower crop yields and degraded stock water supplies [1–4]. Soil salinization increases when the overexploitation of groundwater in coastal areas leads to pumping from a brackish or saline irrigation source due to saltwater intrusion. In addition, treated wastewater, increasingly used in water-scarce environments to tackle climate change, can have a significant salt concentration, depending on the treatment strategy. In such conditions, it is crucial to develop soil monitoring systems that are able to capture spatial and temporal dynamics with a high degree of accuracy.

Sensors typically used for agricultural purposes are installed in a few sparse points at no more than two or three previously defined depths. Likewise, it is unrealistic to collect over time a high number of soil samples for deriving the electrical conductivity of a saturated soil extract (ECe), which is the most useful and reliable measure of soil salinity. In the last decade, apparent electrical conductivity, defined as ECa, has been increasingly used for investigating soil properties. ECa is a measure of the bulk electrical conductivity of the soil and is influenced by various factors, such as soil porosity, the concentration of dissolved electrolytes, texture, the quantity and composition of colloids, organic matter, and water content in the soil [5]. ECa is an apparent value because it has the equivalent electrical conductivity of a homogeneous half-space whose measured response depends not only on the mentioned soil properties but also on instrumental configurations.

Geospatial ECa measurements are very fast and easy to collect because they do not require ground contact and, for this reason, they allow a range of spatial coverage from a few meters to several hectares at different depths, depending on the target to be investigated and the electromagnetic sensor to be used. In addition, collecting ECa data over time along the same transects allows temporal variations in the ECa to be correlated with soil properties that change over time. Over time, the ECa parameter has been widely used as a proxy for investigating soil salinity, as recent studies confirm. Corwin et al. [6] improved guidelines to broaden the scope of the application of ECa-directed soil sampling to map field-scale salinity on orchards under drip irrigation. Vanderlinden et al. [7] highlighted the potential of ECa to perform salinity monitoring at the field or farm scale. Emmanuel et al. [8] validated the potential of extending the electromagnetic induction (EMI) to characterize wetland soil properties, such as salinity, improving sampling plans, and interpolating soil property estimates to unsampled regions. Scudiero et al. [9] performed the ECa survey to identify locations that can be repeatedly sampled to infer the frequency distribution of soil salinity. Rodrıguez-Perez et al. [10] evaluated the usefulness of apparent electrical conductivity (ECa) data to identify variations in soil chemical and physical properties and moisture content. Lesch et al. [11] documented the spatial salinity mapping using EM survey data. Herrero and Hudnall [12] proved the value of electromagnetic induction (EMI) to map the salinity of the rootable layer.

Although ECa is widely used in agronomic applications, its precise indication remains somewhat unclear. What does ECa mean? Is it really representative of the soil properties? Can ECa be effectively related to the saturated past extract of ECe?

ECa is a depth-weighted parameter giving limited information about the variation in the conductivity with depth. In fact, ECa does not provide a rigorous correlation between the Earth's electrical conductivity (EC) structure and measured responses, which are affected by several factors, such as coil distance and orientation, sensitivity, and data error.

In the last few years, an inversion procedure has been implemented in order to produce a reliable soil EC model. The use of inversion codes has grown rapidly as the need for an effective EC distribution in the subsoil has become crucial. Several inversion methods [13–15] and software tools [16,17] have been created to estimate the distribution of soil EC from measured ECa data. These codes solve the complex equations of Maxwell's electromagnetism to generate forward models used to minimize an objective function and derive models from apparent raw data.

Jadoon et al. [18] identified a quantitative distribution pattern of soil salinity by the joint inversion of multicomponent EMI measurements. Paz et al. [19] highlighted time-lapse EMI as a reliable tool for evaluating the risk of soil salinization and supporting the evaluation and adoption of proper agricultural management strategies. Farzamian et al. [20] inverted ECa data to model the salinity of an oasis in Tunisia, which may affect agricultural productivity and the sustainability of crop production. Dakak et al. [21] produced soil salinity maps at various depths through EMI inversion. Shaukat et al. [22] used EMI inversion as a robust and effective method for the risk assessment of new shrub plantations. A benefit of the inversion technique lies in its capacity to predict the depth-wise distribution of EC. This facilitates linking laboratory soil samples obtained from different depths at specific sites with the corresponding EC values at those depths. As emphasized by [20], this method permits the incorporation of all soil samples into a single calibration/validation procedure, leading to a more accurate calibration at any chosen depth. Furthermore, employing a larger number of soil samples for calibration and validation enhances the reliability of the outcomes. Many of these references show a high correlation between ECa or inverted EC against ECe, confirming their potentiality in agricultural applications.

In the proposed case study, a time-lapse inversion approach was tested on ECa data collected on an experimental field in the South of Italy where different irrigation strategies were applied during the growing season of the tomato. The primary objective of time-lapse inversion is to accurately detect variations in conductivity at specific locations across different time intervals. While independent data inversions can be conducted individually, providing insight into changes in modeling results over time through the subtraction of pixel-by-pixel values from a reference dataset, it is important to recognize that temporal changes in conductivity values may not exclusively reflect actual changes in subsurface conductivity depending on the data noise level and inversion artifacts. This is particularly true without considering the reference model and prior information [23].

Raw ECa maps, as well as those detailed on the experimental setup and soil properties, have already been published [24]. In this paper, we focused on the inversion of ECa data, both as a single snapshot and time-lapse imaging. For the proposed case study, a 2D reference transect was extracted from three different plots subjected to different irrigation strategies (irrigation with freshwater and agro-industrial treated wastewater).

The aim of the paper was (a) to highlight the capability of the inversion tool to produce a detailed 2D EC soil distribution; (b) to image the spatio-temporal EC evolution distribution over an irrigation season; and (c) to assess the impact of brackish water on soil salinization in the short term.

## 2. Materials and Methods

### 2.1. Basics of ECa Parameter

ECa is a sensor-based indirect measurement that is strictly affected by some physical and chemical properties, such as soil salinity, soil moisture, clay content, and cation exchange capacity (CEC). Archie [25] and Rhoades et al. [26] developed a theoretical basis for the relationship between ECa and soil properties, such as the soil water content, the electrical conductivity of the soil water, soil bulk density, and the electrical conductivity of the soil particles.

According to these premises, over the last few decades, ECa has been widely used as a soil quality indicator [27–29].

Over time, several devices have been manufactured both in the time and frequency domain for measuring ECa according to the electromagnetic induction (EMI) theory. Since only frequency domain measurements were used in this work, a brief description of this method was reported. An EMI sensor is made of two coils, including a transmitter and a receiver.

A time-varying current ($I_P$) circulating in the loop coil T, named the transmitter coil, generates a time-varying magnetic field, which is in phase with the current and with the same rate of change according to Ampere's law. This field, referred to as the primary magnetic field ($H_P$), induces electric currents in a conductive body, thus generating secondary EM fields. A receiver coil records a signal that is the sum of the primary and secondary fields (Figure 1).

The measured resulting field has an imaginary part of the signal, also called out-of-phase or quadrature (Q), and a real part of the signal, the in-phase (Ph) component. Under simplified conditions, typically defined as a low induction number (LIN), the EMI sensors directly provide the subsurface apparent electrical conductivity (ECa) through Equation (1) [30]:

$$\mathrm{EC_a} = \frac{2}{\pi \mathrm{f} \mathrm{s}^2 \mu_0}\left(\frac{\mathrm{H_s}}{H_p}\right)_{Qu} \tag{1}$$

where f is the frequency (Hz), s is the coil separation (m), $\mu_0$ is the magnetic permeability of free space ($4\pi \times 10^{-7}$ H/m), and ($H_S/H_P$) $Q_u$ is the $Q_u$ component of the secondary $H_S$ to the primary $H_P$ magnetic field coupling ratio.

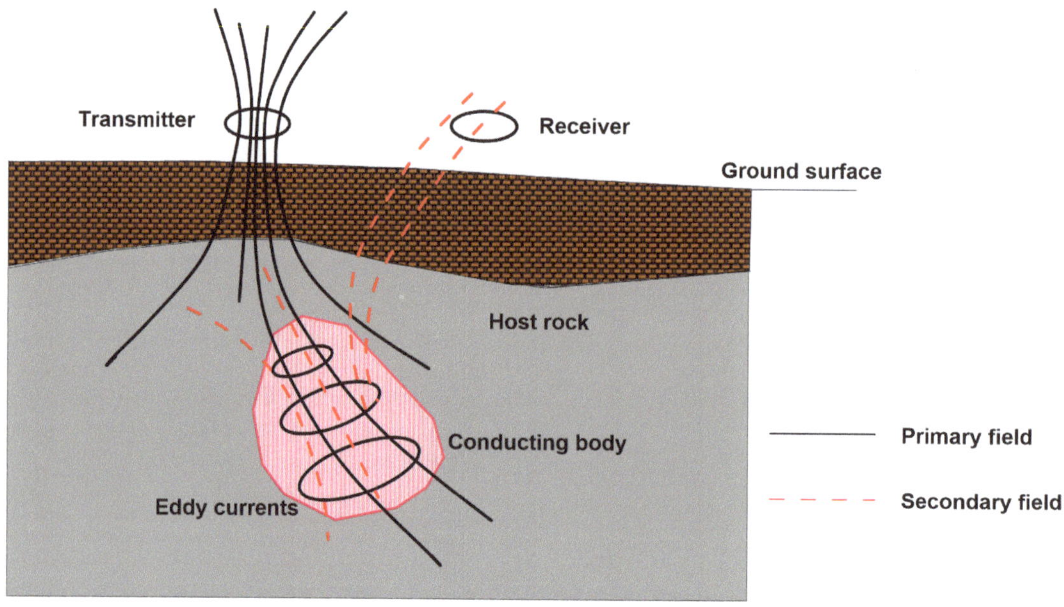


**Figure 1.** Basics of ECa measurements.

Conversely, the real part or the in-phase component of the measured signal is mainly affected by the magnetic permeability of the subsoil.

The ground response depends not only on the soil's electrical conductivity but also on instrumental factors, such as coil distance, coil orientation, and frequency. In particular, the coil distance and orientation affect the depth of penetration of the electromagnetic signal. Placing the coil in a vertical position (VCP coil configuration), the topsoil layers are investigated while rotating 90° of the coils along the main axis (HCP coil configuration), and the signal investigates deeper layers.

The instrumental configuration affects the soil response, as the analysis of the cumulative sensitivity (CS) function shows. This function, defined as the ratio between the variation in the output and the variation in the input [31], quantifies how much the complex electromagnetic response recorded by the device is affected by a variation in the conductivity and/or permeability of a particular point (area or section) of the subsurface.

Figure 2a,b plots the CS distribution against depth as a function of coil and orientation distances for the CMD-Mini-Explorer (GF Instrument s.r.o, Brno), which is the EMI sensor used for collecting field data, made of a cylindrical tube that is 1.3 m long, with a 30 kHz transmitter coil and three receiver coils with 0.32 m, 0.71 m, and 1.18 m offsets, respectively. As clearly observed, the sensitivity changes significantly for three different coil distances and coil orientations.

According to [30], given a specific configuration and under LIN conditions, the effective penetration depth of an EMI sensor corresponds to CS = 0.3. Therefore, ECa-derived depths are purely indicative because they depend not only on soil properties but also on instrumental factors.

On the basis of the aforementioned considerations, the assumption that ECa cannot be used to provide quantitative information about the soil properties confirms the need to invert raw data in order to obtain a reliable EC distribution in the subsoil.

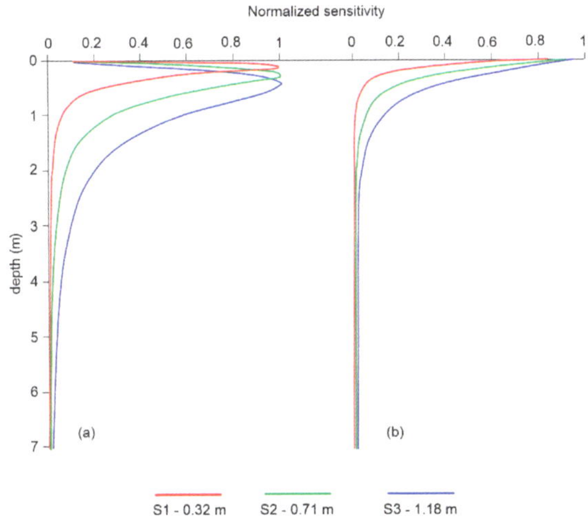


**Figure 2.** Normalized cumulative sensitivity (CS) for the three Mini-Explorer sensors S1, S2, and S3: (**a**) VCP configuration and (**b**) HCP coil configuration.

### *2.2. Time-Lapse ECa Dataset*

The dataset used for time-lapse inversion was collected during the growing season of a tomato crop belonging to a farm located in the South of Italy. The experimental field was randomized with three different irrigation treatments: (a) plots A and B were irrigated with agro-industrial-treated wastewater with different levels of fertigation; (b) plot C was irrigated with fresh water and conventional fertigation. Pre-transplant, fertilizers were applied to the soil in all investigated plots by distributing 30 kg ha$^{-1}$ N and 35 kg ha$^{-1}$ P. Throughout the crop cycle, for plots B and C, 75 kg ha$^{-1}$ N, 40 kg ha$^{-1}$ P, and 72.5 kg ha$^{-1}$ K were added through fertigation. Instead, 75 kg ha$^{-1}$ N and 40 kg ha$^{-1}$ P were added through fertigation in plot A. The EC of the irrigation water was about 2000–2500 µS/cm for plots A and B and 500 µS/cm for plot C. Tomato plants were covered with an anti-hail net and grown in a net house structure. Shading nets protect the plants from sunlight, thus allowing no significant soil temperature changes throughout the irrigation season. Irrigation water was supplied every two days through an underground irrigation system.

ECa data were collected at five time points approximately in the same conditions, i.e., soon before each irrigation event, thus ensuring a homogeneous and equalized moisture distribution across the nine parcels. The initial measurement was taken before the start of the irrigation season (26th June), which served as a reference time for recording the conductivity changes. The other four datasets were collected with time intervals of about 2 weeks (10th July, 24th July, 6th August, and 31st August) during the irrigation season. The data were collected in continuous measurement mode by selecting a time of measurement equal to 1 s, meaning that the conductivity and in-phase values were measured as the average of the values measured during the selected measuring period. The measurements were collected by hand, keeping the device as close as possible to the ground, almost trailing it on the ground, in order to minimize the air layer below the sensor.

During the last EMI campaign, on 31st August, six soil samples were collected at three points, each belonging to a single plot at two different depths, 0–0.2 m and 0.2–0.4 m, from the ground surface in order to provide the ground truth for soil salinity. EC was measured with a multi-range Cryson-HI8734 electrical conductivity meter (Crison Instruments, S.A., Barcelona, Spain). Details of the sampling procedure and soil analysis are described in [24]. In this paper, only the ECe values were extracted in order to provide a correlation function with the inverted EC. In order to test the inversion procedure and compare the findings, nine different 2D EMI transects corresponding to the locations of soil sampling (see Figure 3) were extrapolated from the three plots.

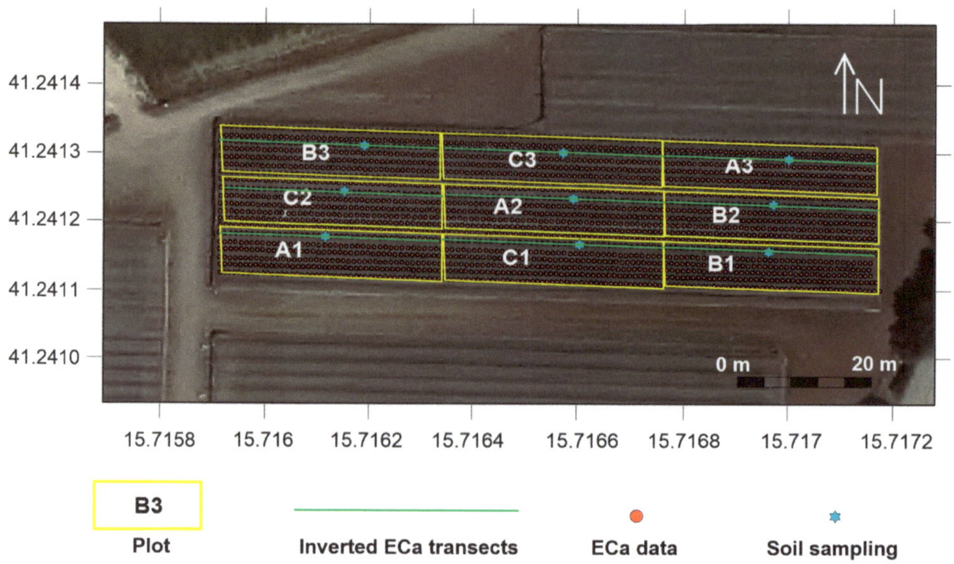


**Figure 3.** Distribution of ECa data collected in the experimental field. The soil sampling was carried out on 31st August during the last EMI campaign.

### 2.3. Time-Lapse Inversion Procedure

The time-lapse inversion procedure is aimed at estimating the electrical conductivity variations over time along the reference transects. The EM4Soil v4.5 software package [16], based on Occam's regularization [32], was used to invert the ECa data. The code uses a quasi-2D inversion, assuming that below each measured location, a 1-dimensional variation in calculated soil conductivity (EC, dS/m) is constrained by variations under neighboring locations [13].

In the time-lapse inversion procedure, the choice of some parameters is crucial for providing accurate inversion models: (a) the starting model for solving the inverse problem; (b) the inversion algorithm to be used; (c) the spatial damping factor ($\lambda$); and (d) the temporal damping factor ($\alpha$) [33].

Two inversion algorithms, S1 and S2 [34], provide a different level of constraint to the model parameters. In the S1 algorithm, the corrections to the model parameters at each iteration were calculated by solving the system of equations:

$$(J^TJ+\lambda C^T C)\ \delta_p = J^Tb \quad (2)$$

where $\delta_p$ is the vector comprising the corrections of the parameters (logarithm of conductivities, $p_j$) of an initial model; b is the vector containing the differences between the logarithm of the observed and calculated apparent conductivities. J is the Jacobian matrix, and $\lambda$ is the aforementioned damping factor.

Conversely, the S2 option has one more constraint (Equation (3)) and can produce smoother results than S1.

$$(J^TJ+\lambda C^T C)\ \delta p = J^Tb + \lambda C^T C\,(p - p_0) \quad (3)$$

where $p_0$ refers to a reference model.

The spatial smoothing or damping factor $\lambda$ [35] determines the amplitude of the parameter corrections in the space domain, which controls the balance between the data fit and model roughness. The selection of $\lambda$ is determined either through the "L curve" method [36] or through trial and error to determine the value that most accurately represents the expert expectations based on the study site and data fit. A smaller damping factor tends to refine more detailed model parameters, especially in areas with larger expected spatial variability [37]. The temporal damping factor $\alpha$ is a regularization factor that gives the weight for minimizing the temporal changes in the conductivity along the time [33]. As the $\alpha$ value increases, the resulting reference models from the inversion become more similar. A value of zero indicates that no temporal constraints are applied, resembling a traditional independent inversion.

A linear solution, based on the ECa cumulative response (CF) and a non-linear solution (FS), was used for forward calculations in order to convert depth-profile EC to ECa [30]. In our study case, both CF and FS modeling, as well as S1 and S2 algorithms, were tested with no substantial differences observed in the model output. The outputs obtained with CF modeling and the S2 algorithm are shown in the section Results section. A uniform starting model considering the average ECa value for each plot (ECa = 1.00 dS/m for plots A and B; ECa = 0.60 dS/m for plot C) was considered to solve the inversion problem, while $\lambda$ and $\alpha$ parameters were set to 0.07 and 0.05, respectively. These parameter sets were selected after conducting several tests and comparing the results in terms of inversion misfit.

In the pre-processing stage, a single dataset containing all the ECa readings obtained over space and time was defined as input for the EM4SOIL code.

## 3. Results

The inversion was performed for the nine selected EMI transects. As the findings of the inversion results were consistent with the plots subjected to the same irrigation strategies, for simplicity, we show a single time-lapse EMI inversion for plots A, B, and C.

The inversion results were visualized as static 2D EC images and normalized time-lapse EC differences. For a clear comparison of the soil response to different irrigation strategies, both results are scaled with the same color range.

Figure 4 shows the inverted EC for the transect corresponding to plot A, irrigated with brackish water, at five different time points.

At the reference time t1 (26th June), a two-layer model was identified as follows: (a) a low conductive top layer (EC < 0.60 dS/m at z < 0.5 m from ground surface); (b) a high conductive bottom layer (0.20 < EC < 0.60 dS/m). During the irrigation season, the top resistive layer was conductive, while the bottom conductive layer experienced a decrease in conductivity.

Variations in EC over time are clearly observed in Figure 5. After two weeks from the start of irrigation (Figure 5a), slight changes could be observed, with positive changes in the yellow area and negative changes in the green area. Over time, EC increasing in the topsoil layer and decreasing in the bottom layer became more prevalent, as observed in Figure 5b. This trend intensifies in Figure 5c, which corresponds to the EC distribution after 41 days from the start of irrigation and persists until the end of the irrigation season (Figure 5d).

To evaluate the accuracy of the inverted model, the observed vs. calculated fit was visualized for each coil configuration at every observation point. (Figure 6). The RMSE between calculated and observed data provides the error model. Generally, the datasets align along the fit line, although some misfits are evident. This trend is not surprising given the level of accuracy of the ECa data. In fact, the estimated RMSE of the geophysical model is 0.15 dS/m, which is the misfit between the observed data (field data) and the theoretical one (calculated data). In terms of percentage value, it is roughly equivalent to about 16%, expressed as the ratio between the RMSE in dS/m and the average value of the observations.

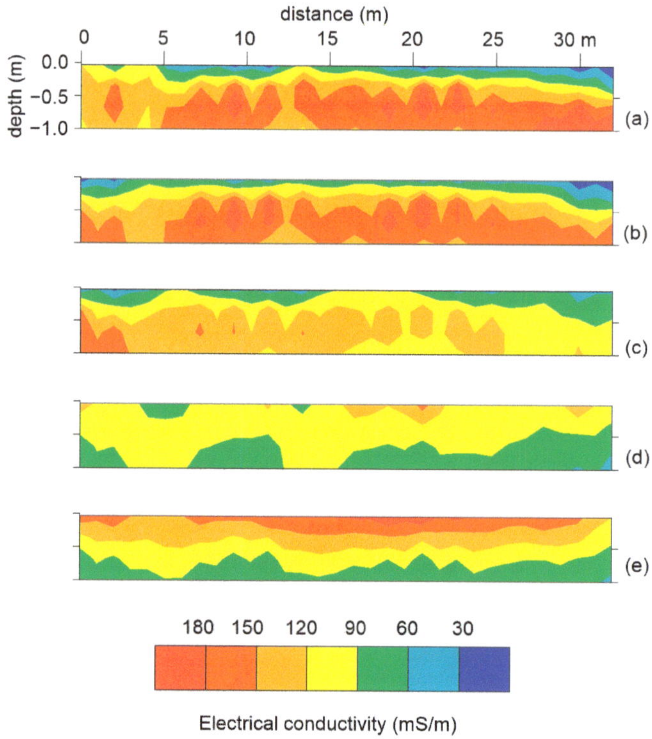


**Figure 4.** Inverted EC for the transect corresponding to plot A at five different time points: (**a**) 26th June; (**b**) 10th July; (**c**) 24th July; (**d**) 6th August; and (**e**) 31st August.

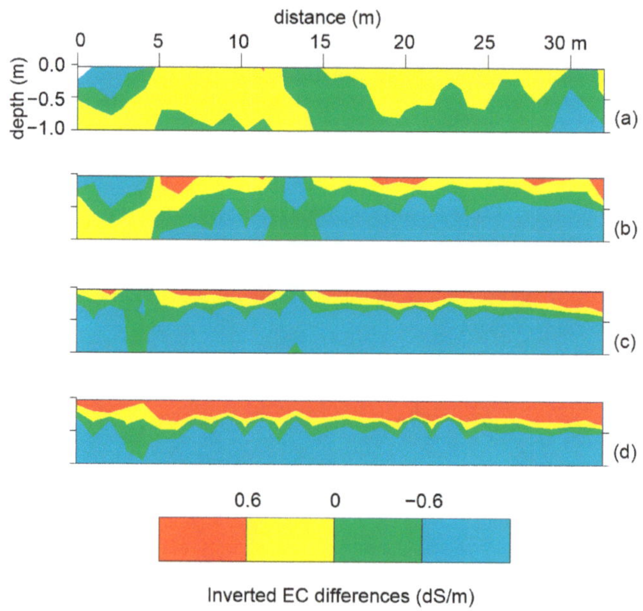


**Figure 5.** Inverted EC differences over time for plot A after (**a**) 14 days; (**b**) 28 days; (**c**) 41 days; and (**d**) 66 days.

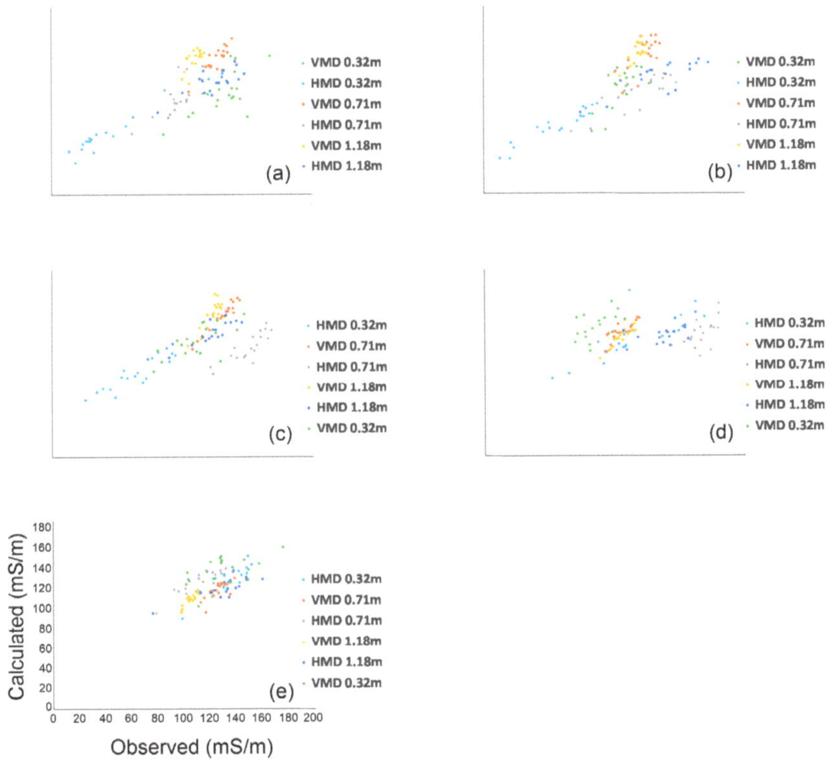


**Figure 6.** Observed vs. calculated data for each time point observation for plot A: (**a**) 26th June; (**b**) 10th July; (**c**) 24th July; (**d**) 6th August; and (**e**) 31st August.

Similar trends were recorded in the transect belonging to plot B. Figures 7 and 8 correspond to the static 2D images and normalized time-lapse EC differences, respectively. Small differences are visualized in the bottom layer, where a markedly lower decrease is recorded. In addition, the main increase in EC in the top layer occurs in Figure 7d, corresponding to the EC distribution after 66 days and, hence, later compared to plot A.

According to the same procedure used for plot A, the RMSE was estimated through the misfit between observed and calculated data, corresponding roughly to about 16%.

On the other hand, a different soil response was observed in plot C, irrigated with freshwater. Figures 9 and 10 show less marked variations in EC, both positive and negative. Given the poor salinity in the freshwater used for irrigation, the surficial increase can be attributed solely to the increase in moisture content in the soil during the irrigation season, while the decrease in EC observed in the cross-section can be attributed to the root water uptake.

The estimated RMSE for plot C is 9 dS/m, corresponding to about 17%.

Figure 11 shows the calibration function of inverted EC vs. EC, whose values are derived from [24]. To plot this graph, we extracted the inverted EMI values for all nine EMI transects. Only one EC point was removed from the dataset due to an unexplained drift from the general trend, probably as an error in soil sampling analysis.

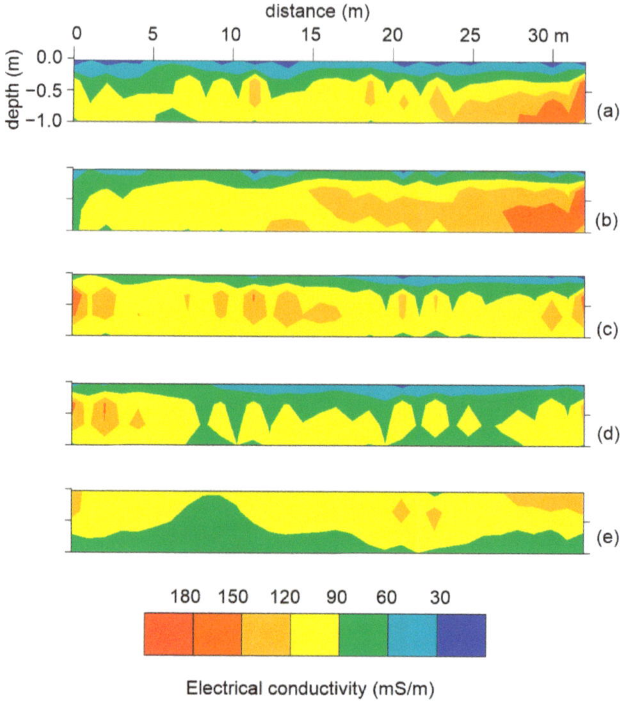


**Figure 7.** Inverted EC for the transect corresponding to plot B at five different time points: (**a**) 26th June; (**b**) 10th July; (**c**) 24th July; (**d**) 6th August; and (**e**) 31st August.

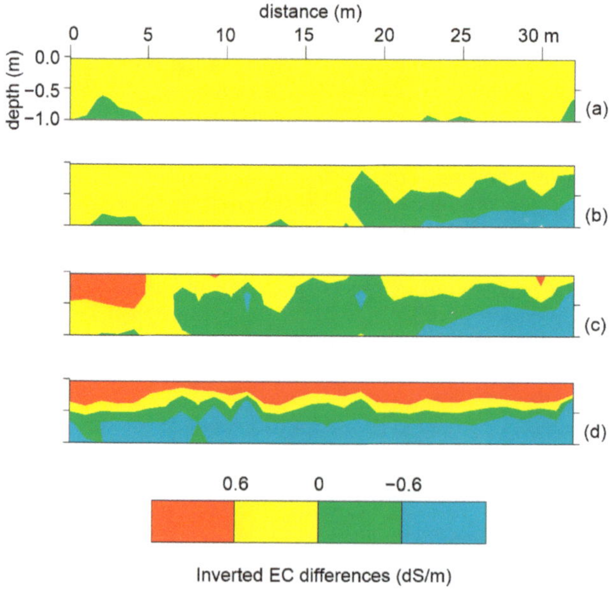


**Figure 8.** Inverted EC differences over time for plot B after (**a**) 14 days; (**b**) 28 days; (**c**) 41 days; and (**d**) 66 days.

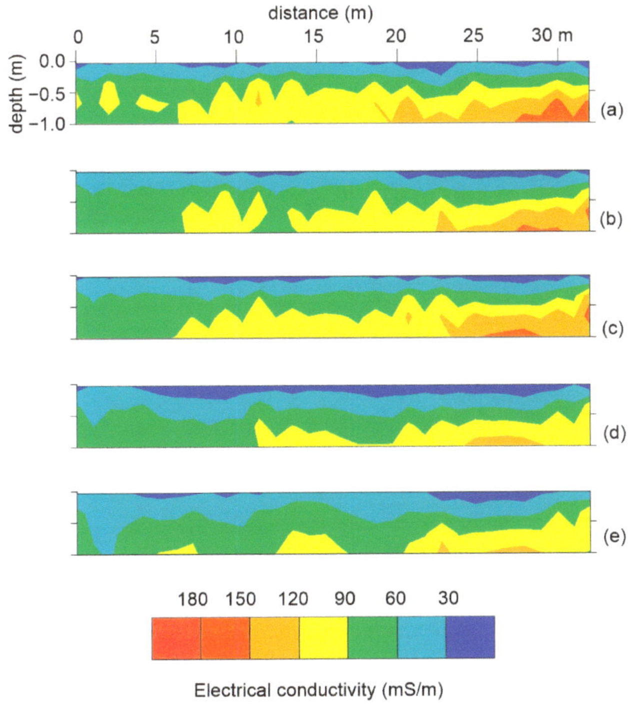


**Figure 9.** Inverted EC for transect belonging to plot C at five different time points: (**a**) 26th June; (**b**) 10th July; (**c**) 24th July; (**d**) 6th August; and (**e**) 31st August.

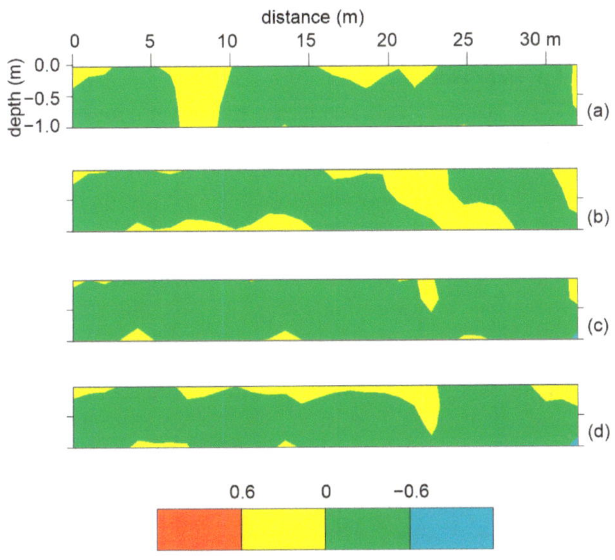


**Figure 10.** Inverted EC differences over time for plot C after (**a**) 14 days; (**b**) 28 days; (**c**) 41 days; and (**d**) 66 days.

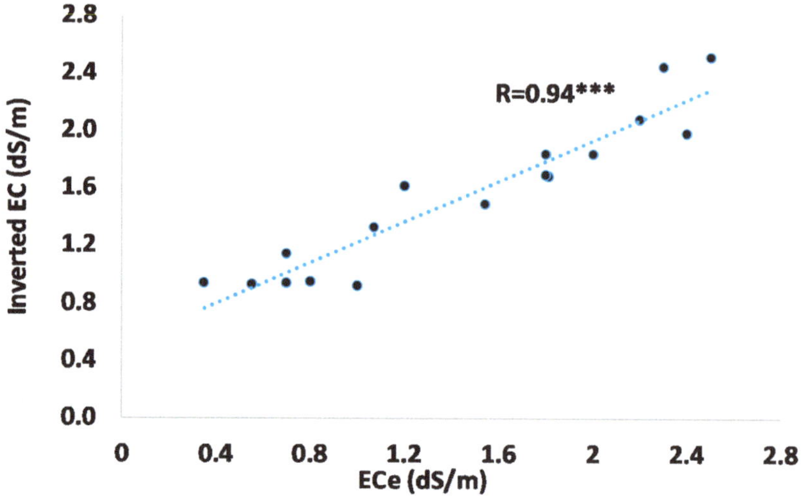


**Figure 11.** Inverted EC vs. ECe calibration function. Statistically significant at *p*-value significance level: *** 0.001 levels of significance.

Some differences between the two datasets can be observed. For low salinity values (less than 1.2 dS/m), EMI overestimates Ece, while for high salinity values, EMI slightly underestimates ECe. This is not surprising because, in soils with low salinity, other factors can increase the EMI signal (clay content, soil moisture, etc). On the other hand, overestimation could be due to data noise or inversion artifacts that could mainly affect the EC distribution for very shallow soil.

Apart from these differences, the correlation is clear, confirming the assumption that the inversion process improves the soil modeling.

## 4. Discussion

This study was conducted on tomato plants moderately tolerant to salinity [38,39]. Nevertheless, different levels of salt in the soil or in the irrigation water can induce changes in plant morphology and physiology and address severe consequences on crop yield [40–42]. In addition, the trend of accumulating salts during the irrigation season may have negative implications not only for soil health but also for groundwater quality. In fact, excess salt can be removed by winter rainfalls and pushed into the vadose zone until the aquifer is reached, causing the oversalinization of the water as a resource [43]. With these premises, the implementation of effective tools capable of monitoring the salinization dynamics in the soil plays a crucial role.

Two weeks after the start of the irrigation, slight changes in EC were observed (Figures 5a, 8a and 10a). The positive changes identified the soil-wetting effects caused by the increase in soil moisture. Conversely, decreases in EC observed below the top layer could be correlated to the root water uptake, which causes negative conductivity changes, as observed in previous works [44–47].

The changes in EC emphasize the soil–plant–water interaction during the irrigation season, as Figure 5, Figure 8, and Figure 10 highlight. In particular, at the end of the irrigation season, the inversion of the ECa data clearly distinguished the salt accumulation in the topsoil layer in plots irrigated with brackish water (plots A and B) and the root water uptake in the bottom layer of the three plots (Figure 12). For the higher EC differences observed in plots A and B compared with the observations in plot C, those irrigated with fresh water (plot C) had reduced root water uptake activity due to the high osmotic pressure that inhibits the water flow from the soil to the plant, as observed in [48–52]. In the context of saline or brackish groundwater management, this evidence should be taken into account in order to balance water requirement and consumption and, hence, to increase water saving and protect the soil. The lower the salt water required, the lower the accumulation of salts while preserving the crop yield.

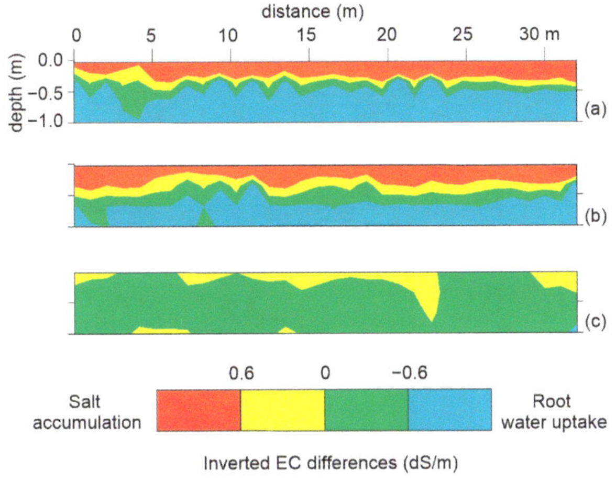


**Figure 12.** Changes in EC at the end of the irrigation season for (**a**) plot A; (**b**) plot B; and (**c**) plot C.

The model error recorded in the three plots reflects a certain level of noise in the raw ECa data. Such noise cannot be estimated with the CMD Mini-Explorer when the most commonly "continuous mode" of the measurement is used, i.e., the ECa data are collected when the instrument moves in the field. This is a critical aspect when the visualization of the results is based on ECa maps. Although ECa produces a rapid and fast visualization of the soil properties, it includes undetermined systematic or random errors. Conversely, inverting data allowed us to estimate a model error, remove spikes or noisy channels, and invert again the filtered datasets. In addition, as ECa is a depth-weighted parameter, it could not be correlated with the ECe or EC obtained through other sensors due to the different resolutions, depth of investigation, and sensitivity. In fact, according to the manufacturer's indications, the upper soil layer (0–0.50 m) is potentially investigated with three ECa measurements ($VCP_{0.32m}$, $VCP_{0.71m}$, and $HCP_{0.32m}$), leading to an ambiguous correlation function.

On the other hand, using the inverted EC data allowed us to plot a single EC vs. ECe calibration function because the inverted model provides an estimation of EC distribution at different depths. This is essential when a limited number of soil samples are available as "ground truth" because all inverted EC data can be used in a single calibration function, such as the proposed case study. It is worth mentioning that no temperature correction was applied either on the modeling results after the inversion procedure or over the raw data. For a more precise soil salinity assessment, such an option should be considered in future studies.

## 5. Conclusions

This study made use of ECa data collected and processed using time-lapse methods to assess the impact of brackish water used as an irrigation source during the tomato growth season. In three plots where different irrigation strategies were used, repeated ECa data were collected at five time points. This approach provided promising answers to the research questions highlighted in the aims of the study. These findings have highlighted how inverted EC allows us to accurately identify soil salinization when different irrigation strategies are used.

In fact, a clear soil response to different irrigations of water, brackish and freshwater, was observed.

Compared with the traditional raw ECa visualization, the processing of the data through time-lapse inversion allowed for a higher level of detail in the soil properties to be visualized. Although this study was conducted in the short term of a single irrigation season, a significant increase in EC in the upper layer could have strong implications in terms of the accumulation of salt. At the same time, the activity of water uptake from the roots was imaged, confirming the versatility of the geophysical tool in the agronomic investigation. The inversion of the ECa allowed for a single correlation function, EC vs. ECe, to be defined, although based on a few points, by comparing the EC data extracted in correspondence to the sampling points. Depending on the extension of the area to be investigated, a significant number of data points can strengthen the calibration function in order to accurately convert the geophysical outcome into hydrological properties of interest.

The capability of producing an accurate correlation function through the inverted model represents an added value with respect to the use of the ECa, which is a depth-weighted parameter and could address meaningless correlations with point scale measurements.

As electromagnetic data are increasingly widespread in the scientific landscape, it is strongly recommended that the inversion procedure be routinely used in the comprehension of soil properties and dynamics.

**Author Contributions:** Conceptualization, L.D.C. and M.F.; methodology, L.D.C. and M.F.; validation, M.F.; formal analysis, L.D.C.; investigation, L.D.C.; resources, L.D.C.; data curation, L.D.C.; writing—original draft preparation, L.D.C.; writing—review and editing, M.F.; visualization, M.F.; supervision, L.D.C. and M.F. All authors have read and agreed to the published version of the manuscript.

**Funding:** This research was co-funded by Regione Puglia as the project "Sistema innovativo di monitoraggio e trattamento delle acque reflue per il miglioramento della compatibilità ambientale ai fini di un'agricoltura sostenibile"—SMARTWATER—(No. 5ABY6PO) through the INNONETWORK CALL, 2017.

**Data Availability Statement:** Datasets are available on request from the authors.

**Acknowledgments:** The authors wish to thank Fiordelisi Farm for making the experimental plot facility available to carry out the field experimental activities. A special mention to Giovanni Berardi for his support in field activities and Domenico Bellifemine for his contribution in the technical aspects.

**Conflicts of Interest:** The authors declare no conflicts of interest.

## References


1. Singh, A. Soil salinity: A global threat to sustainable development. *Soil Use Manag.* **2022**, *38*, 39–67. [CrossRef]
2. Butcher, K.; Wick, A.F.; DeSutter, T.; Chatterjee, A.; Harmon, J. Soil Salinity: A Threat to Global Food Security. *Agron. J.* **2016**, *106*, 2189–2200. [CrossRef]
3. FAO. Saline Soils and Their Management. In *Salt-Affected Soils and Their Management*; Food and Agriculture Organization of the United Nations (FAO): Rome, Italy, 1988; Available online: https://www.fao.org/4/x5871e/x5871e00.htm (accessed on 20 October 2020).
4. Atta, K.; Mondal, S.; Gorai, S.; Singh, A.P.; Kumari, A.; Ghosh, T.; Roy, A.; Hembram, S.; Gaikwad, D.J.; Mondal, S.; et al. Impacts of salinity stress on crop plants: Improving salt tolerance through genetic and molecular dissection. *Front. Plant Sci.* **2023**, *14*, 1241736. [CrossRef] [PubMed]
5. Rhoades, J.D.; Raats, P.C.A.; Prather, R.J. Effects of liquid-phase electrical conductivity, water content, and surface conductivity on bulk soil electrical conductivity. *Soil. Sci. Soc. Am. J.* **1976**, *40*, 651–655. [CrossRef]
6. Corwin, D.L.; Scudiero, E.; Zaccaria, D. Modified ECa—ECe protocols for mapping soil salinity under micro-irrigation. *Agric. Water Manag.* **2022**, *269*, 107640. [CrossRef]
7. Vanderlinden, K.; Martínez, G.; Ramos, M.; Laguna, A.; Vanwalleghem, T.; Peña, A.; Carbonell, R.; Ordóñez, R.; Giráldez, J.V. Soil Salinity Patterns in an Olive Grove Irrigated with Reclaimed Table Olive Processing Wastewater. *Water* **2022**, *14*, 3049. [CrossRef]
8. Emmanuel, E.D.; Lenhart, C.F.; Weintraub, M.N.; Doro, K.O. Estimating Soil Properties Distribution at a Restored Wetland Using Electromagnetic Imaging and Limited Soil Core Samples. *Wetlands* **2023**, *43*, 39. [CrossRef]
9. Scudiero, E.; Skaggs, T.H.; Corwin, D.L. Simplifying field-scale assessment of spatiotemporal changes of soil salinity. *Sci. Total Environ.* **2017**, *587–588*, 273–281. [CrossRef] [PubMed]

10. Rodríguez-Pérez, J.R.; Plant, R.E.; Lambert, J.J.; Smart, D.R. Using apparent soil electrical conductivity (ECa) to characterize vineyard soils of high clay content. *Precis. Agric.* **2011**, *12*, 775–794. [CrossRef]
11. Lesch, S.M.; Corwin, D.L.; Robinson, D.A. Apparent soil electrical conductivity mapping as an agricultural management tool in arid zone soils. *Comput. Electron. Agric.* **2005**, *46*, 351–378. [CrossRef]
12. Herrero, J.; Hudnall, W.H. Measurement of soil salinity using electromagnetic induction in a paddy with a densic pan and shallow water table. *Paddy Water Environ.* **2014**, *12*, 263–274. [CrossRef]
13. Monteiro Santos, F.A. 1D laterally constrained inversion of EM34 profiling data. *J. Appl. Geophys.* **2004**, *56*, 123–134. [CrossRef]
14. Moghadas, D. Probabilistic Inversion of Multiconfiguration Electromagnetic Induction Data Using Dimensionality Reduction Technique: A Numerical Study. *Vadose Zone J.* **2019**, *18*, 180183. [CrossRef]
15. Narciso, J.; Bobe, C.; Azevedo, L.; Van De Vijver, E. A Comparison between Kalman Ensemble Generator and Geostatistical Frequency-Domain Electromagnetic Inversion: The Impacts on near-Surface Characterization. *Geophysics* **2022**, *87*, E335–E346. [CrossRef]
16. Triantafilis, J.; Terhune IV, C.H.; Monteiro Santos, F.A. An inversion approach to generate electromagnetic conductivity images from signal data. *Environ. Model. Softw.* **2013**, *43*, 88–95. [CrossRef]
17. McLachlan, P.; Blanchy, G.; Binley, A. EMagPy: Open-source standalone software for processing, forward modeling and inversion of electromagnetic induction data. *Comput. Geosci.* **2021**, *146*, 104561. [CrossRef]
18. Jadoon, K.Z.; Moghadas, D.; Jadoon, A.; Missimer, T.M.; Al-Mashharawi, S.K.; McCabe, M.F. Estimation of soil salinity in a drip irrigation system by using joint inversion of multicoil electromagnetic induction measurements. *Water Resour. Res.* **2015**, *51*, 3490–3504. [CrossRef]
19. Paz, M.C.; Farzamian, M.; Paz, A.M.; Castanheira, N.L.; Gonçalves, M.C.; Monteiro Santos, F. Assessing soil salinity dynamics using time-lapse electromagnetic conductivity imaging. *Soil* **2020**, *6*, 499–511. [CrossRef]
20. Farzamian, M.; Bouksila, F.; Paz, A.M.; Monteiro Santos, F.; Zemni, N.; Slama, F.; Slimane, A.B.; Selim, T.; Triantafilis, J. Landscape-scale mapping of soil salinity with multi-height electromagnetic induction and quasi-3d inversion in Saharan Oasis, Tunisia. *Agric. Water Manag.* **2023**, *284*, 108330. [CrossRef]
21. Dakak, H.; Huang, J.; Zouahri, A.; Douaik, A.; Triantafilis, J. Mapping soil salinity in 3-dimensions using an EM38 and EM4Soil inversion modelling at the reconnaissance scale in central Morocco. *Soil Use Manag.* **2017**, *33*, 553–567. [CrossRef]
22. Shaukat, H.; Flower, K.C.; Leopold, M. Soil Mapping Using Electromagnetic Induction to Assess the Suitability of Land for Growing Leptospermum nitens in Western Australia. *Front. Environ. Sci.* **2022**, *10*, 883533. [CrossRef]
23. Deiana, R.; Cassiani, G.; Kemna, A.; Villa, A.; Bruno, V.; Bagliani, A. An experiment of non-invasive characterization of the vadose zone via water injection and cross-hole time lapse geophysical monitoring. *Near Surf. Geophys.* **2007**, *5*, 183–194. [CrossRef]
24. De Carlo, L.; Vivaldi, G.A.; Caputo, M.C. Electromagnetic Induction Measurements for Investigating Soil Salinization Caused by Saline Reclaimed Water. *Atmosphere* **2022**, *13*, 73. [CrossRef]
25. Archie, G.E. The Electrical Resistivity Log as an Aid in Determining Some Reservoir Characteristics. *Trans. Am. Inst. Min.* **1942**, *146*, 54–62. [CrossRef]
26. Rhoades, J.D.; Manteghi, N.A.; Shrouse, P.J.; Alves, W.J. Soil electrical conductivity and soil salinity: New formulations and calibrations. *Soil. Sci. Soc. Am.* **1989**, *53*, 433–439. [CrossRef]
27. Corwin, D.L.; Lesch, S.M.; Oster, J.D.; Kaffka, S.R. Monitoring management-induced spatio–temporal changes in soil quality through soil sampling directed by apparent electrical conductivity. *Geoderma* **2006**, *131*, 369–387. [CrossRef]
28. Triantafilis, J.; Lesch, S.M. Mapping clay content variation using electromagnetic techniques. *Comp. Electron. Agric.* **2005**, *46*, 203–238. [CrossRef]
29. Atwell, M.A.; Wuddivira, M.N. Electromagnetic-induction and spatial analysis for assessing variability in soil properties as a function of land use in tropical savanna ecosystems. *SN Appl. Sci.* **2019**, *1*, 856. [CrossRef]
30. McNeill, J.D. Electromagnetic Terrain Conductivity Measurement at Low Induction Numbers. Mississauga, Canada. 1980. Geonics Limited. Available online: http://www.geonics.com/pdfs/technicalnotes/tn6.pdf (accessed on 1 November 2017).
31. Deidda, G.P.; Díaz de Alba, P.; Pes, F.; Rodriguez, G. Forward Electromagnetic Induction Modelling in a Multilayered Half-Space: An Open-Source Software Tool. *Remote Sens.* **2023**, *15*, 1772. [CrossRef]
32. Sasaki, Y. Two-dimensional joint-inversion of magnetotelluric and dipole–dipole resistivity data. *Geophysics* **1989**, *54*, 254–262. [CrossRef]
33. Farzamian, M.; Autovino, D.; Basile, A.; De Mascellis, R.; Dragonetti, G.; Monteiro Santos, F.; Binley, A.; Coppola, A. Assessing the dynamics of soil salinity with time-lapse inversion of electromagnetic data guided by hydrological modelling. *Hydrol. Earth Syst. Sci.* **2021**, *25*, 1509–1527. [CrossRef]
34. Sasaki, Y. Full 3-D inversion of electromagnetic data on PC. *J. Appl. Geophys.* **2001**, *46*, 45–54. [CrossRef]
35. Triantafilis, J.; Monteiro Santos, F.A. Electromagnetic conductivity imaging (EMCI) of soil using a DUALEM-421 and inversion modelling software (EM4Soil). *Geoderma* **2013**, *211–212*, 28–38. [CrossRef]
36. Hansen, P.C. *Rank-Deficient and Discrete Ill-Posed Problems: Numerical Aspects of Linear Inversion*; Society for Industrial and Applied Mathematics (SIAM): Philadelphia, PA, USA, 1998. [CrossRef]
37. Farzamian, M.; Ribeiro, J.A.; Khalil, M.A.; Monteiro Santos, F.A.; Kashkouli, M.F.; Bortolozo, C.A.; Mendonça, J.L. Application of Transient Electromagnetic and Audio-Magnetotelluric Methods for Imaging the Monte Real Aquifer in Portugal. *Pure Appl. Geophys.* **2019**, *176*, 719–735. [CrossRef]
38. Guo, M.; Wang, X.S.; Guo, H.D.; Bai, S.Y.; Khan, A.; Wang, X.M.; Gao, Y.M.; Li, J.S. Tomato salt tolerance mechanisms and their potential applications for fighting salinity: A review. *Front. Plant Sci.* **2022**, *13*, 949541. [CrossRef] [PubMed]

39. del Amor, F.M.; Martinez, V.; Cerdá, A. Salt Tolerance of Tomato Plants as Affected by Stage of Plant Development. *HortScience* **2001**, *36*, 1260–1263. [CrossRef]
40. Roșca, M.; Mihalache, G.; Stoleru, V. Tomato responses to salinity stress: From morphological traits to genetic changes. *Front. Plant Sci.* **2023**, *14*, 1118383. [CrossRef]
41. Al-Daej, M.I. Salt tolerance of some tomato (Solanum lycoversicum L.) cultivars for salinity under controlled conditions. *Am. J. Plant Physiol.* **2018**, *13*, 58–64. [CrossRef]
42. Zaki, H.E.M.; Yokoi, S. A comparative in vitro study of salt tolerance in cultivated tomato and related wild species. *Plant Biotechnol.* **2016**, *33*, 361–372. [CrossRef]
43. Greene, R.; Timms, W.; Rengasamy, P.; Arshad, M.; Cresswell, R. Soil and Aquifer Salinization: Toward an Integrated Approach for Salinity Management of Groundwater. In *Integrated Groundwater Management*; Jakeman, A.J., Barreteau, O., Hunt, R.J., Rinaudo, J.D., Ross, A., Eds.; Springer: Cham, Switzerland, 2016; pp. 377–412. [CrossRef]
44. Srayeddin, I.; Doussan, C. Estimation of the spatial variability of root water uptake of maize and sorghum at the field scale by electrical resistivity tomography. *Plant Soil* **2009**, *319*, 185–207. [CrossRef]
45. Ursino, N.; Cassiani, G.; Deiana, R.; Vignoli, G.; Boaga, J. Measuring and modelling water related soil–vegetation feedbacks in a fallow plot. *Hydrol. Earth Syst. Sci.* **2013**, *18*, 1105–1118. [CrossRef]
46. Cassiani, G.; Boaga, J.; Rossi, M.; Putti, M.; Fadda, G.; Majone, B.; Bellin, A. Soil–plant interaction monitoring: Small scale example of an apple orchard in Trentino, North-Eastern Italy. *Sci. Total Environ.* **2016**, *543*, 851–861. [CrossRef] [PubMed]
47. De Carlo, L.; Battilani, A.; Solimando, D.; Caputo, M.C. Application of time-lapse ERT to determine the impact of using brackish wastewater for maize irrigation. *J. Hydrol.* **2019**, *582*, 124465. [CrossRef]
48. Wang, L.; Ning, S.; Chen, X.; Li, Y.; Guo, W.; Ben-Gal, A. Modeling tomato root water uptake influenced by soil salinity under drip irrigation with an inverse method. *Agric. Water Manag.* **2021**, *255*, 106975. [CrossRef]
49. Chaali, N.; Comegna, A.; Dragonetti, G.; Todorovic, M.; Albrizio, R.; Hijazeen, D.; Lamaddalena, N.; Coppola, A. Monitoring and Modeling Root-uptake Salinity Reduction Factors of a Tomato Crop under Non-uniform Soil Salinity Distribution. *Procedia Environ. Sci.* **2013**, *19*, 643–653. [CrossRef]
50. Steudle, E. Water uptake by plant roots: An integration of views. *Plant Soil* **2000**, *226*, 45–56. [CrossRef]
51. Lu, Y.; Fricke, W. Salt Stress—Regulation of Root Water Uptake in a Whole-Plant and Diurnal Context. *Int. J. Mol. Sci.* **2023**, *24*, 8070. [CrossRef]
52. Mo, Z.; Li, L.; Ying, L.; Xiaolong, G. Effects of Sudden Drop in Salinity on Osmotic Pressure Regulation and Antioxidant Defense Mechanism of *Scapharca subcrenata*. *Front. Physiol.* **2020**, *11*, 884. [CrossRef]

*land*

MDPI

*Article*

# Indirect Prediction of Salt Affected Soil Indicator Properties through Habitat Types of a Natural Saline Grassland Using Unmanned Aerial Vehicle Imagery

**László Pásztor, Katalin Takács *, János Mészáros, Gábor Szatmári, Mátyás Árvai, Tibor Tóth, Gyöngyi Barna, Sándor Koós, Zsófia Adrienn Kovács, Péter László and Kitti Balog**

Institute for Soil Sciences, Centre for Agricultural Research, Herman Ottó Rd. 15., H-1022 Budapest, Hungary; pasztor.laszlo@atk.hu (L.P.); meszaros.janos@atk.hu (J.M.); szatmari.gabor@atk.hu (G.S.); arvai.matyas@atk.hu (M.Á.); toth.tibor@atk.hu (T.T.); barna.gyongyi@atk.hu (G.B.); koos.sandor@atk.hu (S.K.); kovacs.zsofia@atk.hu (Z.A.K.); laszlo.peter@atk.hu (P.L.); balog.kitti@atk.hu (K.B.)

* Correspondence: takacs.katalin@rissac.hu

**Abstract:** Salt meadows, protected within National Parks, cannot be directly surveyed, yet understanding their soil condition is crucial. Our study indirectly estimates soil parameters (Total Salt Content (TSC), Na, and pH) related to salinization/sodification/alkalinization using spectral indices and UAV survey-derived elevation model, focusing on continental lowland salt meadows. A vegetation map was created using 16 spectral indices and a Digital Elevation Model calculated from RGB orthophotos using photogrammetry. Field observations helped define habitat types based on the General National Habitat Classification System (Hungary), and quadrats with complete coverage of specific plant species were identified. Machine learning was employed on 84 training quadrats to develop a prediction algorithm for vegetation patterns. Five saline habitat types, representing variations in soil properties and topography, were identified. Spectral and topomorphometric indices derived from UAV were key to the spatial prediction of soil properties, employing random forest and co-kriging methods. TSC, Na, and pH data served as indicators of salt-affected soils (SAS), and thematic maps were generated for each indicator (57 samples). Overlapping with the vegetation map, the probability range of estimated SAS indicator values was determined. Consequently, a model-based estimation of soil pH, TSC, and Na conditions is provided for habitat types without disturbing protected areas.

**Keywords:** protected salt meadows; vegetation map; machine learning methods; UAV; spectral indices; SAS indicator prediction

**Citation:** Pásztor, L.; Takács, K.; Mészáros, J.; Szatmári, G.; Árvai, M.; Tóth, T.; Barna, G.; Koós, S.; Kovács, Z.A.; László, P.; et al. Indirect Prediction of Salt Affected Soil Indicator Properties through Habitat Types of a Natural Saline Grassland Using Unmanned Aerial Vehicle Imagery. *Land* **2023**, *12*, 1516. https://doi.org/10.3390/land12081516

Academic Editors: Tiago Brito Ramos, Maria da Conceição Gonçalves and Mohammad Farzamian

Received: 30 June 2023
Revised: 21 July 2023
Accepted: 25 July 2023
Published: 30 July 2023



## 1. Introduction

Natural semi-arid saline steppes and salt marshes, which are mostly protected as part of National Parks, are ecologically valuable ecosystems that play a crucial role in maintaining biodiversity and providing various ecosystem services [1–3]. Preserving these habitats is of high importance in the European Union ([4], Natura 2000 network of protected areas) [5]. However, due to their protected status, direct surveys such as disturbance of surface of these areas with excavations are often restricted, making it challenging to assess their soil condition. As a consequence of the current variable climatic conditions, the hydrological and soil formation conditions are also changing [6], thus understanding and monitoring the soil parameters related to salinization [7], sodification, and alkalinization is essential for effective management and conservation of these sensitive habitats.

Digital soil mapping (DSM, [8]) has emerged as a valuable tool providing spatial soil information across a wide range of soil-related applications [9], including precision agriculture, hydrology, environmental sciences, conservation biology, or spatial planning [10]. DSM approach offers an alternative to conventional mapping methods for the spatial assessment of soil properties such as soil salinity [11,12]. In DSM, a wide range of spatial data is collected, integrated, and analyzed using geostatistical methods and/or machine learning techniques for inferring the spatial variability of soil characteristics [13–16].

Remote sensing provides a wealth of information about land surface, with improving spatial, temporal, and spectral resolutions, which can be used in assessing the spatial variability of soil properties in different ways. (i) Bare surface soil characteristics can be directly obtained by remote sensing. Spectral reflectance features in the visible, near-infrared, and shortwave infrared spectrum can be used as a direct indicator for soil surface salinity [12,17]. Increased reflectance can be observed particularly in the blue band [18,19]. Surface soil salinity does not necessarily reflect actual salinity levels of the whole profile [20], but (ii) indirect information about subsurface salinity conditions can be gathered through vegetation type and plant growth as these are controlled by root zone soil properties [17]. (iii) Remote sensing based environmental covariates such as digital elevation models (DEM) and its derivatives (e.g., slope, aspect, and topographic wetness index), spectral band data, vegetation indices, land use, and land cover maps are proved to be useful in DSM for characterizing the most relevant environmental variables representing the soil forming factors [18,21].

Remote sensing is also a valuable, fast, and non-destructive tool to overcome the limitations of direct surveys for monitoring and assessing inaccessible areas such as wetlands [22,23] or protected areas [24]. In recent years, the application of aerial surveys using Unmanned Aerial Vehicles (UAVs) has shown great potential in the spatial assessment of a wide range of features in agriculture and soil science such as vegetation patterns [25,26], monitoring invasive plants [27], peat soil properties [28], soil erosion mapping [29], soil productivity [30], or soil water content mapping [31]. UAV-based data applications have shown their utility also in improving accuracy and providing more insights into soil salinity mapping [19,20,32,33].

In this study, we aim to indirectly estimate soil parameters, specifically Total Salt Content (TSC) [34], sodium (Na) concentration, and pH, by utilizing spectral indices calculated from RGB sensor based orthomosaic and DEM.

The database was derived from aerial surveys in Europe's largest continuous natural semi-arid steppe (in total 82,000 ha, of which 100 ha area was studied) in Hortobágy (Hortobágy National Park) which is a special part of the Great Hungarian Plain. According to our concept, distinct saline habitat types (vegetation patches) differentiated based on various vegetation colors, can be determined using spectral indices derived from orthophotos captured in the visible spectrum (RGB) [35]. As halophytic plant communities exhibit distinct elevation zones ranging from wet salt meadows to dry closed steppes [36,37], surface elevation plays a critical role [38]. Therefore, in our study, we utilized DEMs in conjunction with spectral indices [39] for vegetation mapping on the test area. This article describes the altitude-based distribution of saline habitats and estimates pH, TSC, and Na concentration value categories using a model constructed based on field-validated points for the distinct habitat types. The applied predictive model employing machine learning methods, namely random forest combined with kriging [40–42] was developed, to provide a reliable estimation of these salt affected soil indicator properties, namely pH, TSC, and Na concentration. Thus, indirect estimation of soil properties using remote sensing data and machine learning techniques has significant implications for the monitoring, management and conservation of these ecologically important habitats. The findings of this study will contribute to a better understanding of the soil condition in salt meadows, despite their protected status.

## 2. Materials and Methods

### 2.1. Study Site

The study site is a protected salt marsh, which is a part of the largest contiguous natural saline grassland area in Europe, located in the Eastern region of Hungary, in Hortobágy National Park (Hortobágy NP) (Figure 1). The sampling plot has a rectangular shape (corner coordinates of 47°22′10.67″ N 21°04′27.02″ E; 47°22′39.83″ N 21°04′28.14″ E; 47°22′39.08″ N 21°05′10.98″ E; 47°22′09.91″ N; 21°05′09.86″ E) fitting to a 1 km$^2$ grid in the Hungarian National Grid System (EOV/HD72—EPSG:23700) and belongs to the Central Tisza Region, microregion of Hortobágy according to [43]. The sample area is a plain lying between 85.9 and 87.5 m a.m.s.l. and covered with fine-grained sediments (clay, silt) [44]. Despite the relatively low relief, the area is rich in pedological and geomorphologic features [45]. The River Tisza has deposited silt into the loess-silt surface depressions, which had salinization/sodification process with various types of salts, including $NaHCO_3$, NaCl, $Na_2SO_4$, and $MgSO_4$, as documented by [46,47]. The area is characterized by deep saline, sodic soils, which make up a mosaic spatial structure with diverse saline soil complexes. Largest area is dominated by *Meadow solonetz* {Hungarian official nomenclature} soils (Solonetz according to [48]) with a clay loam texture utilized as saline pastures. Smaller areas are covered with *Steppe meadow solonetz* (Chernozems or Solonetz according to [48]) soil and *Solonetzic meadow* soils (Gleysol or Vertisol according to [48]) as well [43,49]. Groundwater level is between 2 and 4 m, chemically rich in sodium [46,47], which has contributed to the salinization of the soil in the sample area. Salt- and drought-tolerant plant communities live on these salt affected soils and the flora of the landscape is also extremely diverse. Even a decimeter difference in surface elevation also results in a different vegetation pattern, which is closely related to changes in soil salinity and moisture content [50,51]. The microregion of the study area has moderately warm, dry climate; the value of aridity index is 1.30–1.35. The annual precipitation average ranges from 510 to 550 mm with high temporal and spatial variability. Total annual solar radiation ranges between 1900–1940 h, whereas long-term mean annual temperature is 10.0–10.2 °C [43]. Although networked by many channels, the area has a negative water balance.

Summarizing the sedimentological, pedological, climatic and water-holding conditions of the sample area, it can be stated that this area presents a very diverse, mosaic picture, where the vegetation zones on saline soils reflect the spatial variation in soil salinity and moisture in correlation with the surface elevation.

### 2.2. Field Survey and Laboratory Analysis

In order to capture small scale spatial heterogeneity of the salt and textural pattern of the study site, both in situ field measurements and ex situ laboratory measurements were implemented during the research. In situ measurements included vegetation survey. Aerial assessment was conducted to collect imagery information by using UAV. These non-destructive methods can also be applied on soil surface of protected nature saline areas where invasive approaches are not permitted.

#### 2.2.1. Soil Sampling

For taking undisturbed soil samples for further laboratory analysis, 100 sampling points were designated on the 1 km$^2$ sample area, in a 100 × 100 m regular grid. Due to accessibility problems in the waterlogged parts of the site, and the difficulty of drilling through impenetrable hard near-surface layers on other parts of the site, 57 of the planned 100 points were drilled (Figure 1b), with motorized hand drill down to 1 m depth. Undisturbed soil sample columns were collected in plastic tubes having 10 cm in diameter for further analysis. Beyond, at 0–30 cm depth, composite soil samples were also taken in April 2020. In addition, at 3 characteristic locations of the sample area, soil profiles were excavated for soil sampling and soil classification in July 2020. Ex situ laboratory measurements were carried out, and numerous soil parameters were determined. In this study, total salt content (TSC), reaction (pH), and Na concentration of the soil (Na)—as parameters related to salinization/sodification/alkalinization—were investigated as predictors for thematic mapping.

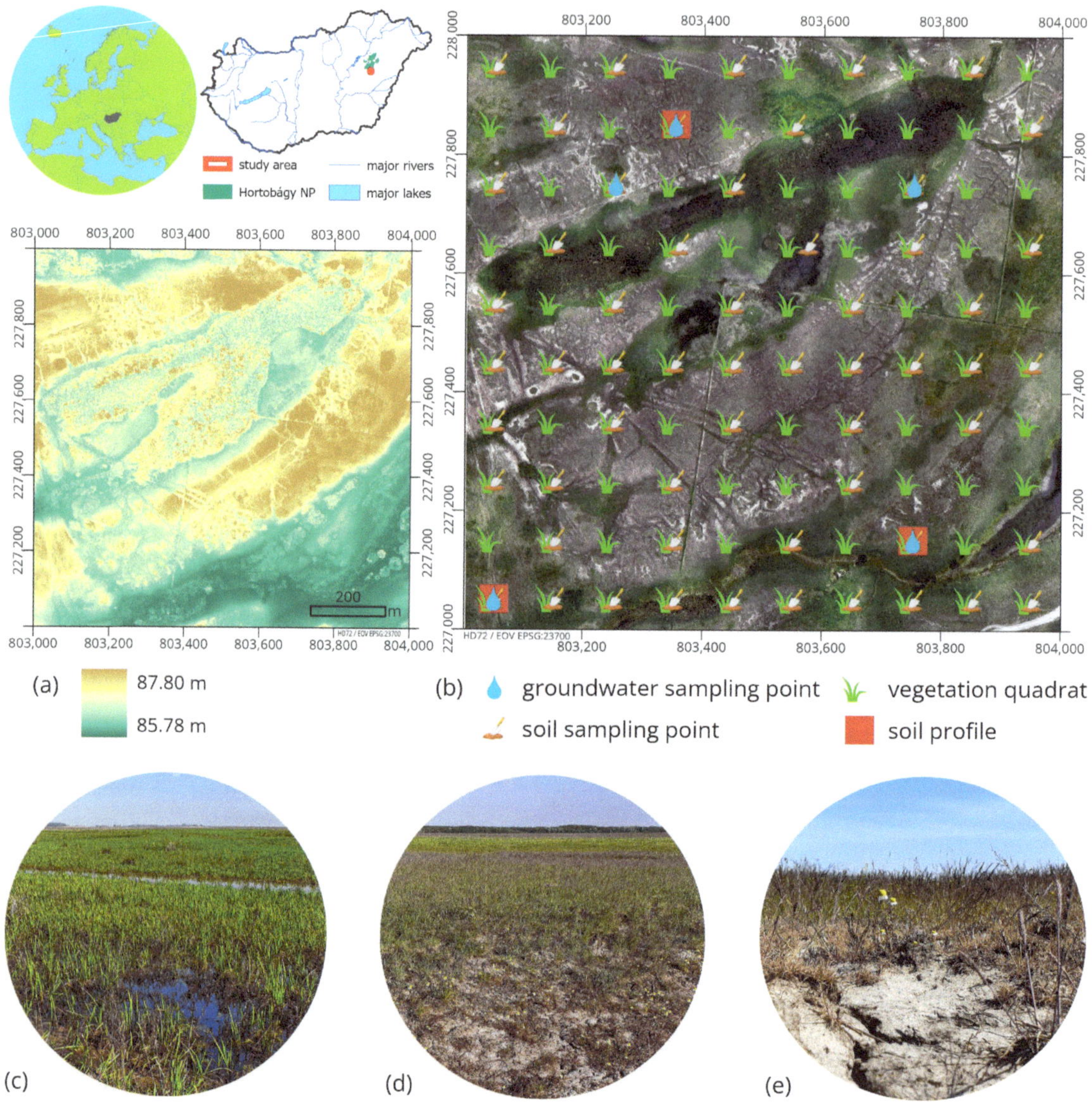


**Figure 1.** Location and characteristics of the study area. (**a**) Topography; (**b**) sampling sites on RGB orthomosaic background; (**c**) salt meadow; (**d**) transition of a bare spot (in the foreground) to Artemisia steppe (in the background); (**e**) bare spot with *Matricaria chamomilla* surrounded by *Puccinellia limosa* plants.

### 2.2.2. Vegetation Survey

The 1 km$^2$ sample area was divided into 1-hectare sections. Within each hectare, vegetation was assessed using the 10 × 10 m quadrat method. This resulted in 100 vegetation quadrats (Figure 1b), where the spatial percentage cover of each vegetation type within the quadrat was determined. The vegetation in each quadrat was then classified into habitat types according to the Hungarian General National Habitat Classification System (Á-NÉR; [52,53]), based on the occurring plant species ([52,53], Table 1).

**Table 1.** Habitat types present on the study area.

| Habitat Code | Description of Habitat Type | Note |
|---|---|---|
| B6 | Salt marshes | |
| F1b | Achillea steppes on *Meadow solonetz* | |
| F2 | Salt meadows | |
| F5 | Annual salt pioneer swards of steppes and lakes | *"padkásszik"* (microerosional mound) |
| F5bs [1] | Annual salt pioneer swards of steppes and lakes | *"vakszik"* (bare spot) |
| H5a | Closed steppes on loess | |
| U9 | Standing waters | |

[1] Added by the Authors, it is not part of the Á-NÉR system.

For habitat type F5, we have separated the *"vakszik"* (bare spot in vegetation of annual and perennial plants of usually small alkali mud surfaces) and the *"padkásszik"* (microerosional mound) types, because there is no surface height difference in the extent of the bare spot, while for the microerosional mound there can be a height difference of several cm, which can be an important difference for mapping. Therefore, the F5bs category was established solely for the bare spot in this study, and the genuine F5 was kept for the vegetation of microerosional mound habitat.

### 2.2.3. Proximal Soil Sensing

Aerial survey was conducted at the study site in April 2020, using UAV with a visible range (RGB) camera onboard. A 24 MP Fuji X-T20 camera was applied for the survey having an APS-C sensor; focal length of 14 mm; angle of view: 91°; automatic ISO speed; automated exposure time based on the sharpness, color saturation, and brightness of input images. Aerial survey was performed in a fully automatic flight mode with image overlap of 80% and sidelap of 60% for helping the proper photogrammetric processing. In terms of focal length and resolution, altitude of 120 m was found to be sufficient for recording and separating vegetation patches. Nine ground control points (GCP) were placed at the soil surface of the study site and were measured with a South Galaxy G1 type real-time kinematic correction GPS unit, having a nominal 1 cm horizontal and vertical accuracy. These GCPs were used to transform raw images into the Hungarian National coordinate system (EOV/HD72–EPSG:23700). During the photogrammetrical processing (orthorectification) of raw images, the RGB orthomosaic and DEM of the study area were generated. The whole workflow was performed in Agisoft Metashape Professional (Version 1.6.1). At the end, both datasets were transformed into the EOV/HD72 coordinate system and exported at 0.5 m spatial resolution.

As ancillary data, additional co-variable layers were calculated based on both datasets (RGB mosaic and DEM). With the help of the "uavRst" R package [54] 16 spectral indices were generated using the red, green, and blue bands of the mosaic. The different spectral indices can estimate different properties of vegetation and in some cases soil surface of salt affected soils, e.g., brightness index—bare spots, open water surfaces; redness index—biomass estimation and stand health. The combined use of the calculated spectral indices can help to complement multidimensional information and achieve more reliable results, thus increasing the accuracy of the vegetation map. Secondly, 20 topomorphometric layers were calculated using the SAGA GIS [55] Channels, Hydrology, and Morphometry libraries. The detailed description of all environmental co-variables can be found in Table 2.

**Table 2.** List of environmental co-variables used in vegetation mapping and spatial modelling of soil properties.

| | Environmental Co-Variable | Reference |
|---|---|---|
| Spectral indices | Red band (R) | |
| | Green band (G) | |
| | Blue band (B) | |
| | Visible Vegetation Index (VVI) | [56] |
| | Visible Atmospherically Resistant Index (VARI) | [57] |
| | Normalized Difference Turbidity Index (NDTI) | [58] |
| | Redness Index (RI) | [59] |
| | Soil Color Index (SCI) | [60] |
| | Brightness Index (BI) | [61] |
| | Spectral Slope Saturation Index (SI) | [61] |
| | Hue Index (HI) | [61] |
| | Triangular Greeness Index (TGI) | [62] |
| | Green Leaf Index (GLI) | [63] |
| | Normalized Green Red Difference Index (NGRDI) | [64] |
| | Green Leaf Area Index (GLAI) | [54] |
| | Overall Hue Index (HUE) | [65] |
| | Coloration Index (CI) | [65] |
| | Overall Saturation Index (SAT) | [65] |
| | Shape Index (SHP) | [65] |
| Topomorphometric indices | DEM | |
| | Slope | [66] |
| | Aspect | [66] |
| | Topographic Position Index (TPI) | [67] |
| | Terrain Ruggeddness Index (TRI) | [68] |
| | Surface roughness (SR) | [69] |
| | Flow direction (flowdir) | [70] |
| | Catchment area (carea) | [70] |
| | Modified catchment area (mcarea) | [71] |
| | General curvature (GC) | [70] |
| | Diurnal anisotropic heating (DAH) | [72] |
| | LS factor (LS) | [73] |
| | Mass Balance Index (MBI) | [74] |
| | Multi-resolution Ridge Top Flatness (MRRTF) | [75] |
| | Multi-resolution Valley Bottom Flatness (MRVBF) | [75] |
| | Real Surface Area (RSA) | [66] |
| | Stream Power Index (SPI) | [73] |
| | SAGA Wetness Index (SAGAWI) | [71] |
| | Vertical distance to channel network (vd2cn) | [66] |
| | Channel network base level (cnbl) | [66] |
| | Topographic Wetness Index (TWI) | [76] |

### 2.2.4. Laboratory Measurements

The collected groundwater samples were analyzed for pH and electrical conductivity (EC), furthermore, cation composition was measured according to the Hungarian standards (summarized in Table 3) in order to determine the sodium adsorption ratio (SAR).

pH and EC of groundwater were measured with Multi Line P4, WTW Multi 350i combined electrode and conductometer, respectively.

Reaction of groundwater was determined using direct potenciometry, following the Hungarian standard Nr. 1484-22:2009 (Table 3, Note 1).

**Table 3.** List of Hungarian standards of the measured groundwater and soil parameters.

| Hungarian Standard of the Measurement | Parameter | Unit | Instrument | Accuracy | Nr. of Data |
|---|---|---|---|---|---|
| MSZ 1484-22:2009 (Note 1) | pH of groundwater | - | Multi Line P4, WTW Multi 350i, Xylem Analytics Germany Sales GmbH & Co. KG, WTW, Weilheim, Germany | 0.004 | 5 |
| MSZ EN 27888:1998 (Note2) | Electrical conductivity of groundwater | dS/m | Multi Line P4, WTW Multi 350i, Xylem Analytics Germany Sales GmbH & Co. KG, WTW, Weilheim, Germany | 1 μS/cm | 5 |
| MSZ 1484-3:2006 (Note 3) | Calcium and Magnesium ion concentration of ground water | mg/L | Ultima-2 type ICP OES, Horiba Jobin Yvon SAS., Longjumeau, France | 0.5 μg/L | 5 |
| MSZ 1484-3:2006 (Note 3) | Sodium and Potassium ion concentration of groundwater | mg/L | Ultima-2 type ICP OES, Horiba Jobin Yvon SAS, Longjumeau, France | 0.5 (Mg), 1.0 (Na) μg/L | 5 |
| MSZ-08-0206-2:1978, 2.1 section (Note 4) | Reaction of the soil | pH | Radelkis OP-300, Radelkis Elektroanalitikai Műszergyártó Kft, Budapest, Hungary, digital pH measuring device, Sentron Europe B.V., Leek, The Netherlands | ±0.05 | 57 |
| MSZ-08-0206-2:1978, 2.4 section (Note 4) | Total salt content of soil | *w*/*w*% | Radelkis OK-102/1 conductometer, Radelkis Elektroanalitikai Műszergyártó Kft, Budapest, Hungary | 5–7.5 rel.% | 57 |
| MSZ 20135:1999, 5.1 (Note 5) | Na concentration of soil | mg/kg | iCAP 7400 ICP-OES Analyzer Thermo Scientific Duo View, Thermo Fisher Scientific (Praha) s.r.o., Praha, Czech republic | 4–7.5 rel.% | 57 |

The laboratory measurement data are the results of averaging three parallel measurements. Note 1: "Water quality. Part 22: Detemination of pH and pH in equilibrium state." MSZ 1484-22:2009. 2009. (in Hungarian); Note 2: "Water quality. Determination of electrical conductivity" MSZ EN 27888:1998 (ISO 7888:1985). 1998 (in Hungarian); Note 3: "Testing of waters, Part 3: Determination of dissolved, suspended and total metals in water by AAS and ICP-OES" MSZ 1484-3:2006. 2006 (in Hungarian); Note 4: "Evaluation of some chemical properties of the soil. Laboratory tests (pH value, phenolphtaleine alkalinity expressed in soda, all water soluble salts, hydrolite /y1 value/ and exchanging acidity /y2-value/" MSZ-08-0206-2:1978. 1978 (in Hungarian), Note 5: "Determination of the soluble nutrient element content of the soil" MSZ 20135:1999. 1999 (in Hungarian), The Hungarian standards are available via the following website: http://szabvanykonyvtar.mszt.hu/ (accessed on 29 June 2023).

Direct potenciometry relies on measuring the potential on the surface of an electrode submerged in an electrolyte solution. This potential is measured by calculating the voltage difference between the measuring electrode (glass electrode) and the reference electrode (e.g., Ag/AgCl). The glass electrode surface has a well-defined potential relative to the surrounding aqueous solution, which is linearly related to the pH of the solution. A combined electrode was used for pH measurement, integrating both the measuring and reference electrodes. This electrode consists of a double-walled glass tube containing a buffer solution inside and the reference electrode outside. Essentially, the electrode acts as a membrane that, upon contact with an aqueous solution, absorbs water and swells, establishing an ion exchange equilibrium with the solution's protons to be measured. The potential across the membrane is determined by the concentration ratio of $H^+$ ions on each side. The buffer solution inside the glass sphere ensures a constant concentration of $H^+$, allowing the electrode potential to depend solely on the external $H^+$ concentration, which is directly proportional to the pH of the solution.

Hungarian standard No. 27888:1998 (Table 3, Note 2) was used for determining water quality, specifically providing guidelines for measuring the electrical conductivity of water. The standard is fully aligned with the European standard EN 27888:1993. Electric conductivity measures the water's ability to conduct electricity, which is determined by the quantity and quality of water-soluble salts present. A higher conductivity value indicates a higher concentration of dissolved salts in the water. The conductometric measurement methodology is based on determining the resistance (electrical resistivity) of the solution between two electrodes, either flat or cylindrical, separated by a fixed distance. Electrical conductivity can be calculated as the reciprocal of electrical resistivity.

Hungarian standard No. 1484-3:2006 (Table 3, Note 3) specifies the possibility of measuring the dissolved Ca, Mg, Na, and K content of groundwater—among other elements—by inductively coupled plasma optical emission spectroscopy (ICP-OES). ICP-OES allowing the simultaneous detection and a sensitive, accurate quantification of 70–80 different elements.

The collected groundwater samples were filtered through a 0.45 μm pore mesh membrane filter. Then, it was sprayed inside the equipment, using argon as carrier gas. The components of the sample evaporate at a temperature of 6000 K inside the chamber, atomize, the resulting free atoms are excited. The excited atoms, as they transition to a lower energy level, emitting photons at wavelengths characteristic of the given element. We spectrally resolve the plasma light emission and measure the intensity of each element in a specific wavelength spectral line using detectors. The quantities of the investigated elements (Ca, Mg, Na, and K) are determined through calibration using a series of solutions with known element concentrations.

SAR is a water quality parameter used in soil science to express sodicity hazard of ground water by showing the relative activity of sodium ions in the exchange reactions with the soil relative to calcium and magnesium. SAR value was calculated according to Equation (1):

$$\mathrm{SAR} = \frac{\mathrm{Na}^+}{\sqrt{\frac{1}{2}\left(\mathrm{Ca}^{2+} + \mathrm{Mg}^{2+}\right)}} \tag{1}$$

where Na, Ca, and Mg are ion concentrations all displayed in meq/L.

SAR is widely used in the irrigation management of sodium-affected soils, and have to be assessed combined with EC, according to the interpretive guidelines of [77].

Soil chemical analysis was conducted on composite soil samples from the experimental site, from the depth of 0–30 cm. In the laboratory, pH, TSC, and Na concentration was measured according to the actual Hungarian standards (see Table 3).

pH was measured from 1:2.5 proportioned (6 g soil: 15 mL n KCl) suspension with Radelkis OP-300 digital pH-meter, using potenciometry method according to Hungarian standard No. 08-0206-2:1978, 2.1 section (Table 3, Note 4). The pH measurement is conducted using a procedure similar to that used for groundwater analysis, with the exception that the chemistry of a soil suspension is evaluated. To ensure accurate results, it is necessary to allow a minimum of 12 h for the ionic balance of the soil suspension to stabilize before the measurement is taken. It is important to note that the suspension should not be filtered prior to the pH assessment.

To determine the Total Salt Content (TSC), soil paste saturated with water up to its plasticity limit was used. The measurements were conducted using the Radelkis OK-102/1 conductometer, following the guidelines of Hungarian standard No. 08-0206-2:1978, 2.2.4 section (Table 3, Note 4). A known-capacity immersion electrode was carefully inserted into the soil paste. The resistance of the soil paste was measured, allowing the device to calculate the specific conductivity and the total salt content (%) based on the provided calibration table. The verification of electrode capacity was carried out using known concentration KCl solutions.

The determination of the soluble nutrient content of soil, including the measurement of sodium (Na) concentration available for plant uptake, is described in Hungarian standard No. 20135:1999 (Table 3, Note 5). In this method, the Na concentration in the soil was extracted using a solution of ammonium-lactate (AL) in a soil-to-solution ratio of 1:20. The extract was then filtered, and the Na concentration was determined using an ICP OES instrument, specifically the Thermo Scientific iCAP 7400 Duo View type.

### 2.3. *Methods*

#### 2.3.1. Vegetation Mapping

Based on the environmental co-variable dataset, the classification of the sample area was performed using the habitat type survey as training areas, where the same habitat type covered the whole quadrat. Unfortunately, there was no information about the actual spatial coverage of the different habitat types present inside the polygon for quadrats with mixed habitats, only their ratio, therefore they were removed from the dataset. Artificial quadrats of classes standing water (3) and bare spot (9) were added manually in the same spatial size as the habitat quadrats (10 × 10 m). In total 72 field and 12 additional quadrats were used as training areas.

Values of pixels falling under the quadrat polygons from all spectral and topographic layers were extracted (33,600 data points), later used as training and testing data (divided in 70/30 ratio in a balanced way regarding the number of pixels of each class present in the study area) for the classification. Class separability analysis was performed using the extracted data showing water as a completely separate category, bare spot as a slightly overlapping class with all other vegetation related categories. Latter ones have major overlaps to each other in general; however, they can be separated based on their average heights in the DEM and using the topomorphometric layers as well. The "ranger" machine learning R package (v0.15.1 [78]) was used to build up a random forest based classifier, the hyperparameters were optimized with the help of the "caret" package (v6.0-94 [79]) to find the optimal set, where min.node.size was held constant at a value of 1 and two other parameters (mtry and splitrule) were 40 and extratrees, respectively. Any other hyperparameter was defined on their default values. At the end, the raster dataset containing all mentioned environmental co-variables in layers was classified with the trained model.

For any further analysis, this map was considered as the vegetation and habitat type map of the study site.

#### 2.3.2. Spatial Modelling of Soil Properties

A hybrid machine learning and geostatistical approach was used for topsoil (0–30 cm) property estimations. In DSM random forest combined with kriging (RFK) is considered to be a new "workhorse" [13]. In this case, at first the spatial variation in soil properties was modelled with random forest (RF, [80]) based on the environmental covariates, which were generated from the DEM and RGB orthophoto (Table 2). RF is one of the most popular machine learning algorithms in DSM [81], which is a bagging type ensemble learning method [14]. The advantages of RF over other regression methods are (i) it does not require normally distributed soil data [13], (ii) it can fit complex non-linear relationships between soil data and auxiliary variables, and (iii) and the correlation between environmental covariates is not a limiting factor [40]. Then, at second part we used a geostatistical modelling method (ordinary kriging, OK) to spatially extend the derived residuals from RF model [82]. The outcome of the estimation is the sum of the RF model result and the kriged residuals.

2.3.3. Validation of Soil Property Estimations

To assess the performance of the spatial estimation of soil properties a 5-fold cross validation was applied. The following common validation measures were computed: (i) mean error (or bias, ME—Equation (2)) and (ii) root mean square error (RMSE—Equation (3))

$$ME = \frac{1}{n}\sum_{i=1}^{n}(P_i - O_i) \tag{2}$$

$$RMSE = \sqrt{\frac{1}{n}\sum_{i=1}^{n}(P_i - O_i)^2} \tag{3}$$

where n is the number of observations; $P_i$ and $O_i$ are the predicted and observed soil property for observation location, respectively.

The methodological steps of data processing are summarized in Figure 2.

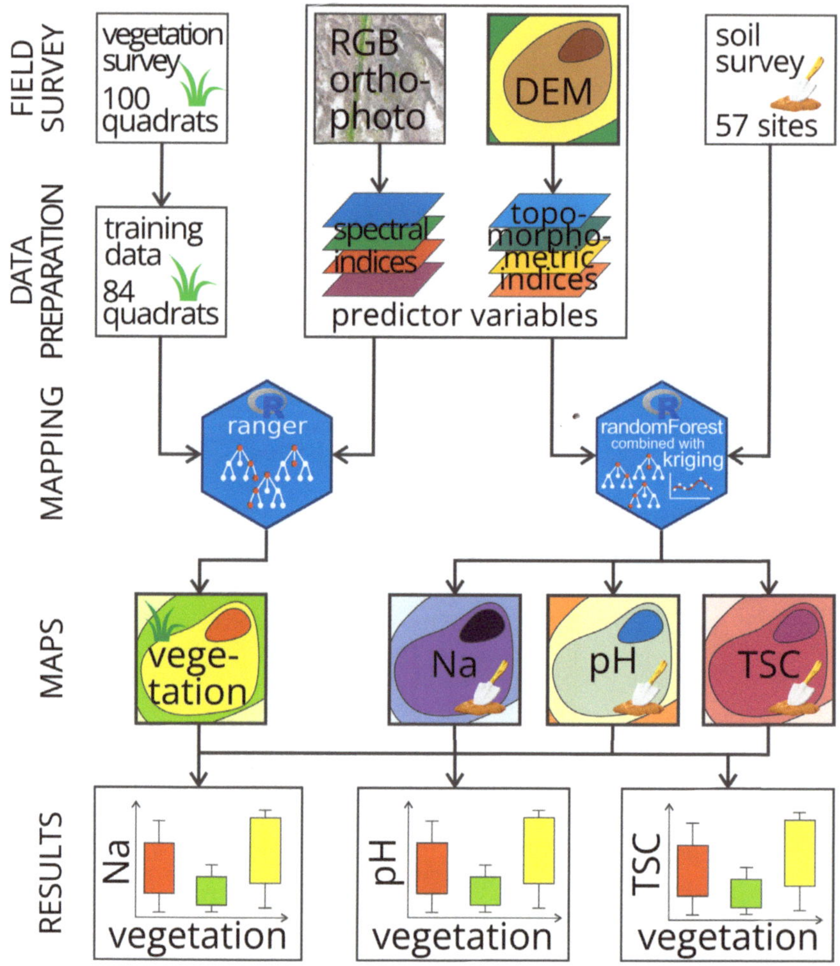


**Figure 2.** Methodological flowchart of the research work.

## 3. Results

### *3.1. Vegetation Map*

Figure 3 represents the final vegetation/habitat type map (Table S1). Generally, it resembles the DEM of the area: Achillea steppes (F1b) and closed steppes (H5a) can be found on the highest levels (eastern side and southwestern corner), salt marshes (B6) surrounded by salt meadows (F2) in their foregrounds dominates the lower levels (southern and center part). The intermediate, transitional zones are covered by the annual salt pioneer swards of steppes and lakes (F5) class where the surface is dissected by small canals (on the southern and northern side of the center area). The patches of bare spot can be found in these latter areas.

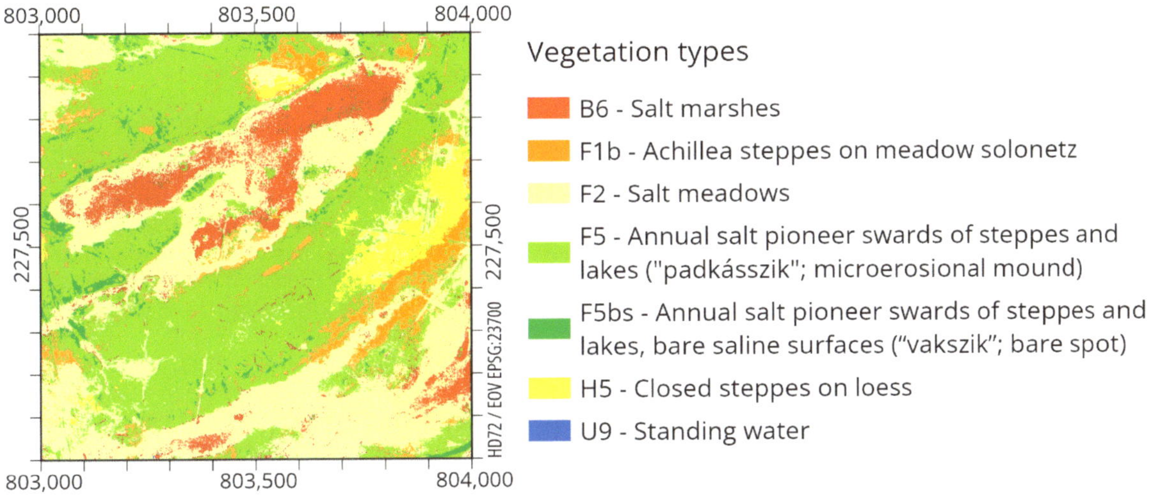


**Figure 3.** Classified habitat type map.

Analyzing the most important co-variables of the developed classifier model, it indicates the importance of morphometric variables (CNBL, DEM, MRRTF, and MRVBF; Table 2) in the top four position (Figure 4), followed by spectral variables of red, green, and blue bands and BI, VVI, and GLI. In summary, the morphometric variables can be useful to differentiate between habitats located on different altitude levels, RGB bands, and vegetation related spectral indices to separate various plant types and finally brightness index can highlight the bare spot because of their greyish-white surfaces. The bare spots are covered by Solonetz soils, where the Natric horizon is located either immediately on the surface or in close proximity. These spots remain bare due to the aggregate-dispersing effect caused by high sodicity, measured right at the surface. As Natric horizons exhibit the highest salt accumulation within the soil profile, their exposure leads to the highest salinity levels observed within the study site.

Furthermore, the map was tested against the test dataset also, with an average accuracy of 0.988 together with a Kappa value of 0.985 According to the detailed, by class accuracy metrics, the built-up classifier performed a very good classification in each habitat type class with balanced accuracy, precision, and recall values higher than 0.95.

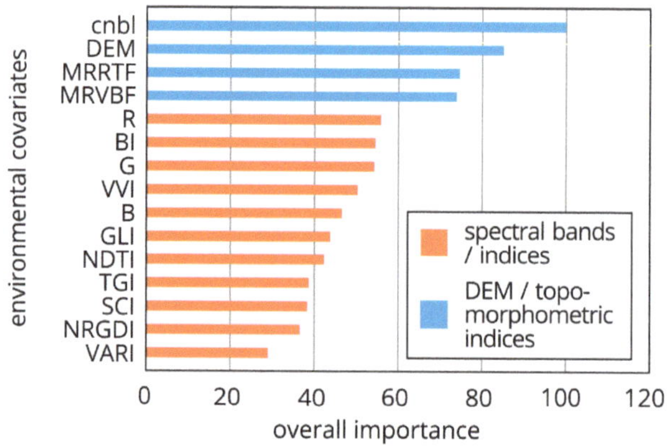


**Figure 4.** Variable importance of co-variables of the developed classifier model applied in vegetation mapping.

The standing water (U9) class had a perfect classification with balanced accuracy, precision, and recall values of 1.000 thankfully to its completely distinct characteristics and because it was over-represented in the training dataset compared to its spatial coverage on the study site (Table 4). The closed steppes on loess (H5a) had the same values, except balanced accuracy of 0.999. However, the H5a class is located strictly on the higher levels of altitude, making the class again a very distinct one. From the salt-related categories salt marshes (B6) were misclassified only with salt meadows (F2): 2 F2 data points as false positives to B6 salt marshes and 15 B6 test data as false negative to F2 salt meadows. While latter ones are generally located around salt marshes, both type of misclassifications could have happened on their mating edges. Because of the spectral similarity of the vegetation in the middle of spring, F2 points were also classified to F1b Achillea steppes (8) and F5 annual salt pioneer swards of steppes (5) habitat types; however, with 2341 true positive predictions, they still result in a balanced accuracy value of 0.995. Achillea steppes areas (F1b) are cross-misclassified with salt meadows (F2) and annual salt pioneer swards (F5) but the number of both false predictions are twice to the latter category than to F2 class. The reason behind this can be that F1b covered areas are "trapped" between these two habitats. However, despite of adjacent neighborhood with H5a closed steppes habitat in the eastern side of study site, there is no mixed classification with this class probably because of the completely different altitude levels what F1b (mean value: 86.86 m) and H5a (mean value: 87.16 m) habitats are occupied. Regarding the annual salt pioneer swards (F5) to bare spots (F5bs), latter is misclassified with only the F5 class in 36 test points, falsely predicted as F5 class and 11 F5 points falsely classified to F5bs class. The ratio of false predictions shows that F5bs misclassification to F5 category is three times more likely than in the other direction, resulting in the underestimation and underrepresentation of bare spots in the study area. The reason behind this can be that bright color of spots is very distinct compared to other classes (e.g., various kinds of green vegetation and dark water surfaces) but in topomorphometric properties, it is similar to F5 thus they are overlapping with each other in this domain.

The complete confusion matrix and detailed accuracy metrics of habitat type classes can be found in the Supplementary Materials (Tables S2 and S3).

**Table 4.** Area ratios of habitat categories in the study area according to the classified vegetation map.

| Habitat Code | Area ($m^2$) | Percent (%) |
|---|---|---|
| B6 | 82,825.25 | 8.28 |
| F1b | 102,861.25 | 10.29 |
| F2 | 314,694.75 | 31.47 |
| F5 | 439,985.75 | 44.00 |
| F5bs | 24,518.50 | 2.45 |
| H5a | 34,991.75 | 3.50 |
| U9 | 122.75 | 0.01 |

### *3.2. Thematic Soil Maps*

Figure 5 represents the results of soil property predictions for the study area. Soil properties were estimated independently, but the expected relationship between soil properties is clearly shown in the maps; areas with higher Na concentration also exhibit elevated TSC and increased (even alkaline) pH levels (Table S4).

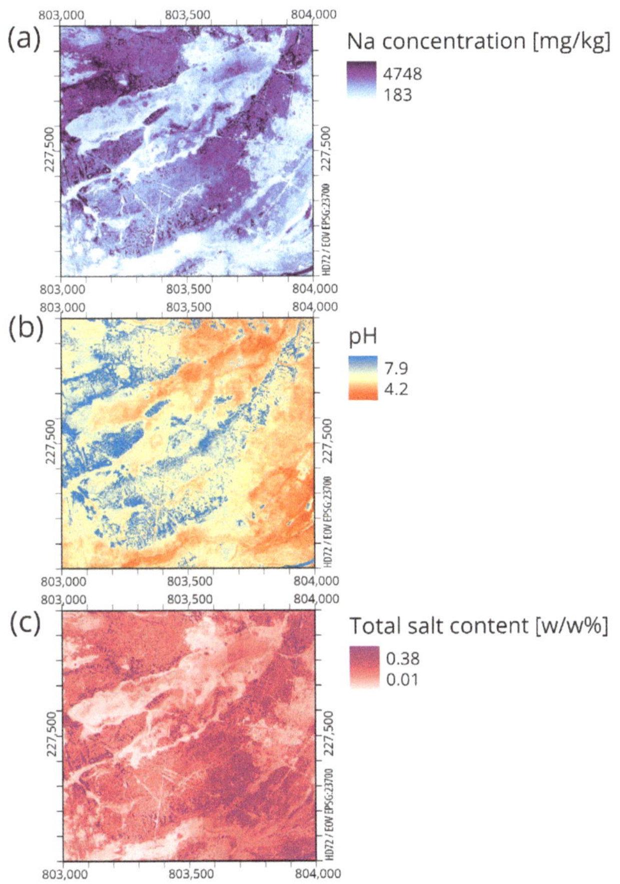


**Figure 5.** Soil property maps. (**a**) Na concentration; (**b**) pH; (**c**) total salt content.

The findings of variables importance of this study indicate that spectral indices found to be more informative than topomorphometric indices. Among the various spectral indices examined, including SHP, BI, TGI, GLI, VVI, RI, SI, B, CI, and SAT, these indices consistently ranked within the top 15 for every soil property under investigation (Figure 6).

The difference with the important variables of the classification in the previous section can be noted. While ranger is also random forest based algorithm [78], it can sort the covariates by importance summing the individual trees. Our examination showed that morphometric variables can separate the most distinct habitat types (e.g., water, salt marshes, and vegetation related habitats) at first on higher level of trees and on lower levels (on the leaves) vegetation related indices and spectral bands are more important to differentiate between various saline vegetation habitats. However, predicting a continuous numerical value is quite the opposite: in case of the three saline related parameters the fine spatial transitions in values indicated by changing which plant types (highlighted by vegetation indices) received the most importance, supplemented by topographic layers (DEM, CNBL, MRVBF, ASPECT, SPI, SAGAWI, SR, TWI, TRI, SL, and SLOPE) for the sudden changes (mostly found on areas with microerosional mounds).

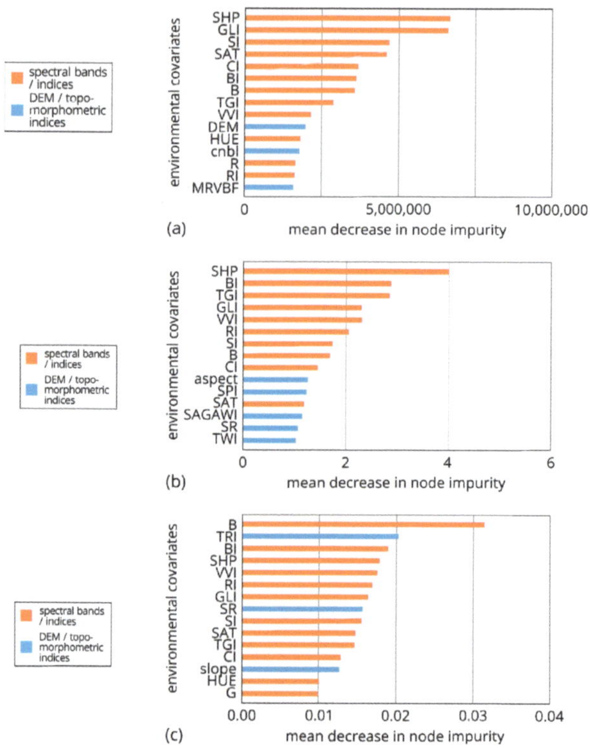


**Figure 6.** Variable importance. (**a**) Na; (**b**) pH; (**c**) TSC.

Validation results were assessed by a 5-fold cross validation and summarized in Table 5.

**Table 5.** Validation results of soil property estimations.

| Delete Column | Na | pH | TSC |
|---|---|---|---|
| ME | 19.10 | 0.03 | −0.00 |
| RMSE | 971.45 | 0.88 | 0.09 |

The mean Na concentration is 1959 mg/kg for the whole study area. The northwest part demonstrates a significant concentration of Na, but smaller patches of high Na content observed in the northeastern and southern parts as well. The average pH value for the study area is 5.76. Similar to Na content, more alkaline pH levels are predominantly observed in the northwest with mosaic-like distribution in the south and northeast. The mean TSC is 0.14 *w*/*w*%. In case of TSC, larger areas with higher-than-average TSC values can be found in the southern part, accompanied by smaller spots in the west and the northeast.

Overlapping the three thematic soil maps (Na, pH, and TSC) with the vegetation map, we obtained predicted statistics of soil parameters for each distinct habitat type (Figure 7).

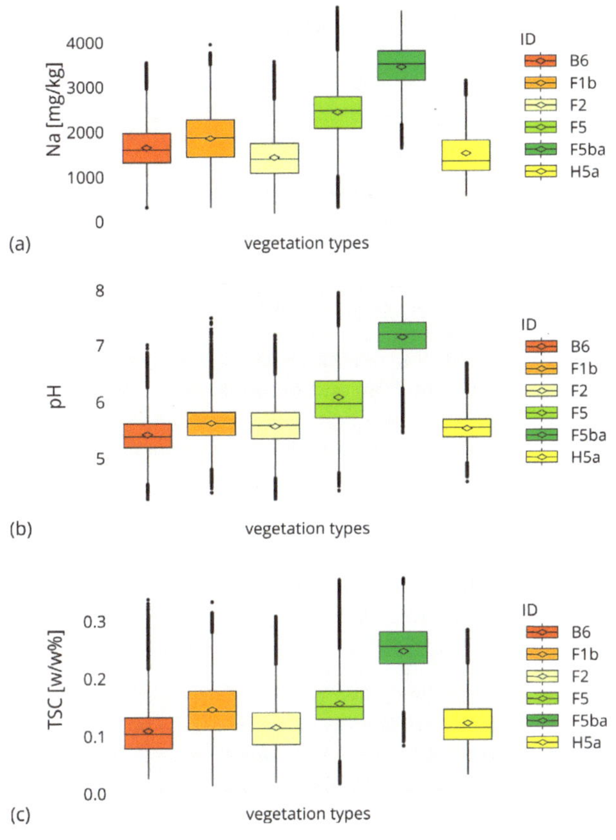


**Figure 7.** Boxplots of SAS parameters separated according to habitat types. (**a**) Na; (**b**) pH; (**c**) TSC. Legend: codes of Á-NÉR system: B6: salt marshes; F1b: Achillea steppes on *Meadow solonetz;* F2: salt meadow; F5: annual salt pioneer swards of steppes and lakes ("*padkásszik*" = microerosional mound); F5bs: annual salt pioneer swards of steppes and lakes, bare spot ("*vakszik*"); H5a: closed steppes on loess.

Salt marshes (B6), covering 8% of the area (Table 4), are intrazonal habitats generally characterized by strongly saline soils (Solonchaks) and vegetation coverage with saline water for a significant period of the growing season [52]. A significant portion of the water that provides moisture to the habitat may originate from the groundwater, which is also the source of the salts causing salinity. In our case, since the water in the area was partly derived from precipitation during the sampling period, the salts—that were previously present in the area in large quantities in the soil—were diluted and leached out. Therefore, the salt content is lower, categorizing it in the moderately saline range (0.075–0.13 *w*/*w*%). The relatively high sodium concentration indicates that a significant portion of the salts is sodium-based (1322–1976 mg/kg). These facts are consistent with the observation of [83], stating that compared with other salt lakes and marshes of the world, the alkaline lakes in Hungary are characterized by lower salt content but higher alkalinity. The formation and persistent existence of the habitat are linked to high groundwater levels and evaporative water management.

Achillea steppe in *Meadow solonetz* (F1b) (Endoprotocalcic Epistagnic Solonetz (Albic, Katoclayic) (Table S5) classified according to [48]) are steppe-like communities, which depend on adequate water supply and moderately saline soil conditions [52,53]. It occupies 10% of the total sampling area (Table 4) and characterized by species that tolerate long summer droughts and heavy textured soils. Since sodium-rich groundwater (Na 1170 mg/kg, SAR 26.5, EC 5.8 dS/m, Table S6) is present at shallow depths (3.3 m) and due to its mid-elevation position within the vegetation zone, this habitat type is the most exposed to salt accumulation in the groundwater fluctuation zone. A total of 1444–2264 mg/kg adsorbed sodium was detected in the soil (TSC: 0.11–0.18 *w/w*%), which tend to accumulate in higher amounts in the deeper layers, resulting in 5.4–5.8 pH in the topsoil. These habitats form a transition between meadows and loess grasslands in terms of water balance, dominated by generalist plant species.

The appearance of salt meadows (F2) (31.5% coverage, Table 5) requires adequate water supply and moderately saline soil. They are tall grasslands that are temporarily covered with water during the initial stages of the growing season (Figure 1c). They develop on wet areas and variously *Saline meadow* or *Solonchak* soils [52,53], in this case it was formed on Endoprotocalcic Protostagnic Solonetz (Albic, Epiloamic, Katoclayic) classified according to [48]. These habitats are widespread on saline soils throughout the Great Hungarian Plain and in many other countries in Europe as well. The soil of salt meadows is often less calcareous, with the upper 5–10 cm layer having higher organic matter content, resulting in 5.3–5.7 pH. While they are typically found on *Solonetz* soils, the characteristics of both *Solonetz* and *Solonchak* soils often coexist, resulting in transitional phenomena. Therefore, compared to other parts of the sampling area, sodium concentration is moderate (1086–1747 mg/kg). Ground water level of the soil of this habitat is 3.6 m depth, having 9.58 SAR value and 1.4 dS/m EC carrying moderate sodification effect (based on [77]). Salt meadows are situated between salt marshes and Festuca pseudovina-saline steppes, often alternating with salt pans in a fully developed zonation [52].

The areas of microerosional mound (F5) surrounding the bare spots are the vegetation zone with the largest extent (~44%, Table 3) in the sample area. The highs and lows show a regular repeating pattern of several plant communities that make up a mosaic. The soil chemistry is more variable than in other habitat types, with a pH ranging from 5.69 to 6.34. In terms of salt content, it can be clearly distinguished from bare spots, which contain on average 0.1% more salt (TSC: 0.13–0.18 *w/w*%). The Na content of the soil is 2067–2764 mg/kg, which is lower than that of the bare spot, but higher than that of the soils in other surrounding habitats. Annual salt pioneer swards of steppes and lakes evolved on Endoprotocalcic Solonetz (Albic, Katoclayic, Humic) soil classified according to [48], having a groundwater depth of 4.39 m, and SAR value of groundwater 11, which represents sodic hazard in the deeper soil layers.

Based on the boxplots shown in Figure 7, the bare spot (F5bs) exhibits the highest pH value (6.91–7.39), total salt content (0.22–0.28 *w/w*%), and sodium concentration (3100–3800 mg/kg) inside the study area, followed by the microerosional mound (F5), indicating a correlation between the soils of these interconnected habitats. These findings are consistent with the descriptions by [52,53]. Bare spots on the salt steppe vegetation cover the smallest percentage of the overall sample area (2.45%, Table 3). However, this feature is the most distinctive characteristic of the habitat, typically found in drier sections of the saline slope and saline zone at mid-elevation positions. The presence of bare spots is closely associated with areas with evaporative water management, contributing to their persistence. The habitat type found on bare spots is primarily shaped by the intensive effects of trampling, including herd paths and herding routes. They form small patches with low species diversity, in each and every instance displaying bare soil surfaces, making them reliably estimable using spectral indices [84–86]. Bare spot serves as distinctive elements of the saline vegetation zonation, holding significant landscape importance characterized by strong patches and mosaic-like patterns. Due to bare spot forming small patches inside the landscape, spatial resolution of our vegetation map was 0.5 m in order to accurately represent this habitat type alongside other, more extensive saline habitats.

The closed steppes on loess (H5a), representing 3.5% of the sampling area (Table 5), occur on loess or other soft bedrock-derived soils [52], in our case, on elevated ridges protruding from the saline steppes. As shown in Figure 8, the highest elevations within the study area are occupied by closed steppes on loess, based on the DEM created for the sampling area. Originally, these habitats are characteristic of areas with higher organic matter content and prone to erosion. Therefore, the sodium content (1128–1793 mg/kg) and salt content (0.09–0.14 *w*/*w*%) are moderate, while the pH (5.34–5.66) is lower than the average value typical for the area.

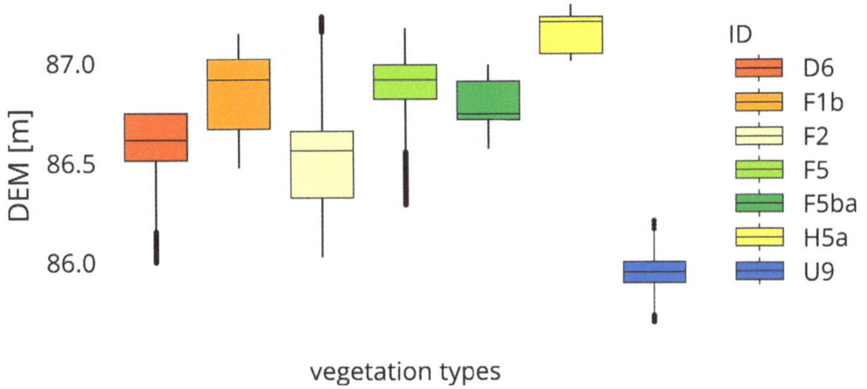


**Figure 8.** The arrangement of habitats by surface elevation (maBl) based on DEM. Legend: codes of Á-NÉR system: B6: salt marshes; F1b: Achillea steppes on *Meadow solonetz*; F2: salt meadow; F5: annual salt pioneer swards of steppes and lakes (*"padkásszik"* = microerosional mound); F5bs: annual salt pioneer swards of steppes and lakes, bare spot (*"vakszik"*); H5a: closed steppes on loess; U9: standing waters.

We conducted statistical analysis using boxplots to examine the relationship between thematic maps of salt affected soil indicators (pH, Na, and TSC) and habitat map patterns. Through our model-based estimation, we achieved successful indirect estimation of these SAS indicators for each specific habitat type, establishing characteristic thresholds for the soil parameters (Table 6, Figure S1).

**Table 6.** Summary table of the vegetation map-based SAS indicator estimation of the soils of the different Á-NÉR habitats.

| Habitat Types | Á-NÉR Codes | TSC (*w*/*w*%) | | Na (mg/kg) | | pH | |
|---|---|---|---|---|---|---|---|
| | | Threshold | | | | | |
| | | Low | High | Low | High | Low | High |
| Salt marshes | B6 | 0.08 | 0.13 | 1322.74 | 1976.77 | 5.18 | 5.60 |
| Achillea steppes on *Meadow solonetz* | F1b | 0.11 | 0.18 | 1444.19 | 2264.35 | 5.40 | 5.80 |
| Salt meadow | F2 | 0.08 | 0.14 | 1085.91 | 1747.34 | 5.32 | 5.79 |
| Annual salt pioneer swards of steppes and lakes (microerosional mound) | F5 | 0.13 | 0.18 | 2067.34 | 2763.57 | 5.69 | 6.34 |
| Annual salt pioneer swards of steppes and lakes (bare spot) | F5bs | 0.22 | 0.28 | 3126.08 | 3776.23 | 6.91 | 7.39 |
| Closed steppes on loess | H5a | 0.09 | 0.14 | 1128.18 | 1793.46 | 5.35 | 5.66 |
| Standing waters | U9 | 0.09 | 0.15 | 861.18 | 1639.39 | 4.86 | 5.45 |

## 4. Discussion

Several studies have been carried out on remote sensing of salt affected soils and its applicability, e.g., for estimating plant cover [87], for monitoring [88,89], combining with salt movement modelling [90], or testing salinity indices [91]. The applied method, UAV-based multispectral imagery is widespread, but is mostly used to solve problems in precision agriculture [92], e.g., estimate vegetation quality [93], nitrogen content of crops [94,95], or monitoring crop diseases [96]. Our approach involves the creation of a SAS vegetation map using 16 different spectral vegetation indices (VVI, VARI, NDTI, RI, SCI, BI, SI, TGI, GLI, NGRDI, GLAI, HUE, CI, HI, SHP, and SAT) and the calculation of a DEM from RGB orthophoto mosaics as a result of aerial survey using a UAV. Through field observations, we identified distinct habitat types based on the General National Habitat Classification System of Hungary [52], with quadrats representing specific plant species providing crucial data for our predictive model development. By integrating topomorphometric and spectral indices and applying random forest [97] and co-kriging methods [98], we estimated soil properties and generated thematic maps of salt-affected indicators (pH, TSC, and Na), validated using 57 soil samples from the field. Boxplots were generated in order to estimate the pH, TSC and Na concentration range in the soil under different vegetation patterns.

In the Hortobágy microregion, as well as in numerous other European salt steppes [99], the natural process of soil formation exhibits diverse patterns [100]. In certain instances, there is a gradual accumulation of salts, leading to progressive salinization, while in others, leaching and desalinization occur [5,101]. These processes not only manifest in spatial variations but, as described by [102,103], they also display temporal dynamics within specific areas. The alternation of leaching and salt accumulation is influenced by environmental conditions (e.g., hydrological conditions and texture [104] and shapes the characteristics of the developing saline soils [105].) Our research, as indicated by the boxplots at Figure 6, also demonstrates substantial variability and dispersion in soil salinity, sodium content, and pH within specific habitats. For instance, in bare spot (F5bs), there is a prevailing accumulation of salts in the topsoil, resulting in higher salinity and pH, and also sodium content due to Na-salts [106,107]. Conversely, in elevated and more exposed areas, e.g., closed steppes on loess (H5a) (Figure 8), as well as in *Solonetz* soils, the salinity, sodium content, and pH of the topsoil are all lower [108,109]. In these cases, the influence of humus content is also apparent.

The zonation of saline habitats, as determined by the increase in surface elevation, follows the sequence of salt meadow (F2) < salt marshes (B6) < annual salt pioneer swards of steppes and lakes, bare spots (F5bs) < annual salt pioneer swards of steppes and lakes (F5) = Achillea steppes on *Meadow solonetz* (F1b) < closed steppes on loess (H5a) (see Figure 8). This order of zonation aligns with the findings of [110,111]. The highest salt content, sodium content, and pH, considering the zonation of the area, are observed in the soils of vegetation belts located in the intermediate positions [112,113].

The novelty of our work is to employ a cost-effective and straightforward approach utilizing multispectral RGB imaging to produce a highly accurate (98.8%) vegetation map of the salt steppe.

Habitat types in the Hungarian General National Habitat Classification System (Á-NÉR [52]) are associated with specific soil types. Our research introduces a novel aspect by offering threshold values for salinity, sodicity, and alkalinity indicators (Na, TSC, and pH) corresponding to the saline habitat types in Á-NÉR. To estimate these values, we utilized a model combining the random forest and kriging (RFK) methods.

The study area we investigated is situated within Europe's largest continuous natural semi-arid steppe, which represents extensive Eurasian steppes with similar characteristics. Our modelling method can form the basis for the proximal, non-invasive surveying of protected saline areas and the model estimation of salinity indicators.

## 5. Conclusions

- The analysis of the classifier model's ("ranger" machine learning) most important co-variables in case of preparing vegetation map, highlights the significance of morphometric variables (CNBL, DEM, MRRTF, and MRVBF) in the top four positions, followed by spectral variables (red, green, blue bands, BI, VVI, and GLI). Morphometric variables differentiate habitats based on altitude levels, while RGB bands and vegetation-related spectral indices separate different plant types. The BI is particularly useful in identifying bare spots with greyish-white surfaces. The applied geostatistical model demonstrated high accuracy (0.9889) and a Kappa value of 0.9857 when tested against the dataset. The classification performance for each habitat type was excellent, with balanced accuracy, precision, and recall values exceeding 0.95.
- Correlation analysis of thematic maps of SAS indicators (pH, Na, and TSC) and habitat map patterns was carried out applying boxplots. Our model-based estimation was successful to indirectly estimate these SAS indicators for every distinct habitat type, defining characteristic thresholds for the soil parameters.
- For UAV-based RGB orthophotos, it was found that spectral indices (SHP, BI, TGI, GLI, VVI, RI, SI, B, CI, and SAT) provided more comprehensive information compared to topomorphometric indices when considering the importance of the variables in estimating all SAS parameters.

**Supplementary Materials:** The following supporting information can be downloaded at: https://www.mdpi.com/article/10.3390/land12081516/s1, Table S1: Summary table of the quadrat vegetation survey, forming the base data for the vegetation map. Table S2: Confusion matrix of test dataset with reference (columns) and predicted (rows) comparison. Table S3: Detailed accuracy metrics by class. Table S4: Laboratory measured soil parameters applied in thematic mapping. Table S5: Site Soil Investigation Reports of the 3 soil Profiles. Table S6: Groundwater chemistry data. Figure S1: 3D scatterplot of the represented habitat types in the light of the three mapped soil salinity parameters.

**Author Contributions:** L.P.: conceptualization, validation, formal analysis, resources, writing—review and editing, supervision, and funding acquisition. K.T.: conceptualization, visualization, software, formal analysis, methodology, and writing—review and editing. J.M.: field measurement and observation, methodology, software, formal analysis, validation, data curation, resources, writing—original draft preparation, and writing—review and editing. G.S.: software, methodology, validation, formal analysis, writing—original draft preparation, writing—review and editing, and visualization. M.Á.: field measurement and observation, methodology, validation, data curation, and visualization. T.T.: conceptualization, validation, formal analysis, field observation and mapping, investigation, supervision, and funding acquisition. G.B.: investigation, methodology, field measurement and observation, and validation. P.L.: field measurement and observation and methodology. Z.A.K.: data curation and investigation. S.K.: field measurement and observation and methodology. K.B.: methodology, validation, writing—original draft preparation, writing—review and editing, and project administration. All authors have read and agreed to the published version of the manuscript.

**Funding:** This research was funded by the National Research, Development and Innovation Office (NKFIH), grant number K-124290.

**Data Availability Statement:** The data presented in this study are available in Supplementary Materials.

**Acknowledgments:** We would like to thank the Hortobágy National Park for allowing us to carry out the investigation on its Natura 2000 site. Special thanks to Tibor József Novák for his help in the field works. The work of Gábor Szatmári was supported by the János Bolyai Research Scholarship of the Hungarian Academy of Sciences.

**Conflicts of Interest:** The authors declare no conflict of interest.

## References


1. Montoroi, J.-P. Soil Salinization and Management of Salty Soils. In *Soils as a Key Component of the Critical Zone 5: Degradation and Rehabilitation. Geosciences Series. Soils Set*; Valenton, C., Ed.; ISTE, Wiley: London, UK, 2018; pp. 97–126.
2. Zhao, M.; Wang, C.; He, Y.; He, T. Soil Characteristics and Response Thresholds of Salt Meadow on Lake Beaches of the Ordos Platform. *Front. Environ. Sci.* **2022**, *10*, 1050757. [CrossRef]
3. Lubińska-Mielińska, S.; Kącki, Z.; Kamiński, D.; Pétillon, J.; Evers, C.; Piernik, A. Vegetation of Temperate Inland Salt-Marshes Reflects Local Environmental Conditions. *Sci. Total Environ.* **2023**, *856*, 159015. [CrossRef]
4. IMEUH. *Interpretation Manual of European Union Habitats*; European Comission: Brussels, Belgium, 2007.
5. Ladányi, Z.; Balog, K.; Tóth, T.; Barna, G. Longer-Term Monitoring of a Degrading Sodic Lake: Landscape Level Impacts of Hydrological Regime Changes and Restoration Interventions (SE Hungary). *Arid. Land Res. Manag.* **2023**, *37*, 389–407. [CrossRef]
6. Eswar, D.; Karuppusamy, R.; Chellamuthu, S. Drivers of Soil Salinity and Their Correlation with Climate Change. *Curr. Opin. Environ. Sustain.* **2021**, *50*, 310–318. [CrossRef]
7. Schofield, R.V.; Kirkby, M.J. Application of Salinization Indicators and Initial Development of Potential Global Soil Salinization Scenario under Climatic Change. *Glob. Biogeochem. Cycles* **2003**, *17*, 4-1–4-13. [CrossRef]
8. McBratney, A.B.; Mendonça Santos, M.L.; Minasny, B. On Digital Soil Mapping. *Geoderma* **2003**, *117*, 3–52. [CrossRef]
9. Minasny, B.; McBratney, A.B. Digital Soil Mapping: A Brief History and Some Lessons. *Geoderma* **2016**, *264*, 301–311. [CrossRef]
10. Szatmári, G.; Bakacsi, Z.; Laborczi, A.; Petrik, O.; Pataki, R.; Tóth, T.; Pásztor, L. Elaborating Hungarian Segment of the Global Map of Salt-Affected Soils (GSSmap): National Contribution to an International Initiative. *Remote Sens.* **2020**, *12*, 4073. [CrossRef]
11. Gorji, T.; Tanik, A.; Sertel, E. Soil Salinity Prediction, Monitoring and Mapping Using Modern Technologies. *Procedia Earth Planet. Sci.* **2015**, *15*, 507–512. [CrossRef]
12. Suleymanov, A.; Gabbasova, I.; Komissarov, M.; Suleymanov, R.; Garipov, T.; Tuktarova, I.; Belan, L. Random Forest Modeling of Soil Properties in Saline Semi-Arid Areas. *Agriculture* **2023**, *13*, 976. [CrossRef]
13. Keskin, H.; Grunwald, S. Regression Kriging as a Workhorse in the Digital Soil Mapper's Toolbox. *Geoderma* **2018**, *326*, 22–41. [CrossRef]
14. Szabó, B.; Szatmári, G.; Takács, K.; Laborczi, A.; Makó, A.; Rajkai, K.; Pásztor, L. Mapping Soil Hydraulic Properties Using Random-Forest-Based Pedotransfer Functions and Geostatistics. *Hydrol. Earth Syst. Sci.* **2019**, *23*, 2615–2635. [CrossRef]
15. Nabiollahi, K.; Taghizadeh-Mehrjardi, R.; Shahabi, A.; Heung, B.; Amirian-Chakan, A.; Davari, M.; Scholten, T. Assessing Agricultural Salt-Affected Land Using Digital Soil Mapping and Hybridized Random Forests. *Geoderma* **2021**, *385*, 114858. [CrossRef]
16. Heuvelink, G.B.M.; Webster, R. Spatial Statistics and Soil Mapping: A Blossoming Partnership under Pressure. *Spat. Stat.* **2022**, *50*, 100639. [CrossRef]
17. Metternicht, G.I.; Zinck, J.A. Remote Sensing of Soil Salinity: Potentials and Constraints. *Remote Sens. Environ.* **2003**, *85*, 1–20. [CrossRef]
18. Mulder, V.L.; de Bruin, S.; Schaepman, M.E.; Mayr, T.R. The Use of Remote Sensing in Soil and Terrain Mapping—A Review. *Geoderma* **2011**, *162*, 1–19. [CrossRef]
19. Sahbeni, G.; Ngabire, M.; Musyimi, P.K.; Székely, B. Challenges and Opportunities in Remote Sensing for Soil Salinization Mapping and Monitoring: A Review. *Remote Sens.* **2023**, *15*, 2540. [CrossRef]
20. Ivushkin, K.; Bartholomeus, H.; Bregt, A.K.; Pulatov, A.; Franceschini, M.H.D.; Kramer, H.; van Loo, E.N.; Jaramillo Roman, V.; Finkers, R. UAV Based Soil Salinity Assessment of Cropland. *Geoderma* **2019**, *338*, 502–512. [CrossRef]
21. Richer-de-Forges, A.C.; Chen, Q.; Baghdadi, N.; Chen, S.; Gomez, C.; Jacquemoud, S.; Martelet, G.; Mulder, V.L.; Urbina-Salazar, D.; Vaudour, E.; et al. Remote Sensing Data for Digital Soil Mapping in French Research—A Review. *Remote Sens.* **2023**, *15*, 3070. [CrossRef]
22. Dronova, I.; Kislik, C.; Dinh, Z.; Kelly, M. A Review of Unoccupied Aerial Vehicle Use in Wetland Applications: Emerging Opportunities in Approach, Technology, and Data. *Drones* **2021**, *5*, 45. [CrossRef]
23. Lyu, X.; Li, X.; Dang, D.; Dou, H.; Wang, K.; Lou, A. Unmanned Aerial Vehicle (UAV) Remote Sensing in Grassland Ecosystem Monitoring: A Systematic Review. *Remote Sens.* **2022**, *14*, 1096. [CrossRef]
24. Wang, Y.; Lu, Z.; Sheng, Y.; Zhou, Y. Remote Sensing Applications in Monitoring of Protected Areas. *Remote Sens.* **2020**, *12*, 1370. [CrossRef]
25. Zhang, J.; Okin, G.S.; Zhou, B.; Karl, J.W. UAV-Derived Imagery for Vegetation Structure Estimation in Rangelands: Validation and Application. *Ecosphere* **2021**, *12*, e03830. [CrossRef]
26. Zhang, H.; Feng, Y.; Guan, W.; Cao, X.; Li, Z.; Ding, J. Using Unmanned Aerial Vehicles to Quantify Spatial Patterns of Dominant Vegetation along an Elevation Gradient in the Typical Gobi Region in Xinjiang, Northwest China. *Glob. Ecol. Conserv.* **2021**, *27*, e01571. [CrossRef]
27. Lehmann, J.R.K.; Prinz, T.; Ziller, S.R.; Thiele, J.; Heringer, G.; Meira-Neto, J.A.A.; Buttschardt, T.K. Open-Source Processing and Analysis of Aerial Imagery Acquired with a Low-Cost Unmanned Aerial System to Support Invasive Plant Management. *Front. Environ. Sci.* **2017**, *5*, 44. [CrossRef]
28. Mustaffa, A.A.; Mukhtar, A.N.; Rasib, A.W.; Suhandri, H.F.; Bukari, S.M. Mapping of Peat Soil Physical Properties by Using Drone- Based Multispectral Vegetation Imagery. *IOP Conf. Ser. Earth Environ. Sci.* **2020**, *498*, 012021. [CrossRef]

29. Oh, S.; Chang, A.; Yang, Y.E.; Kim, H.S.; Lim, K.J.; Jung, J. Recent Advances in UAS Based Soil Erosion Mapping. *Mod. Concepts Dev. Agron.* **2020**, *7*. [CrossRef]
30. Takata, Y.; Yamada, H.; Kanuma, N.; Ise, Y.; Kanda, T. Digital Soil Mapping Using Drone Images and Machine Learning at the Sloping Vegetable Fields in Cool Highland in the Northern Kanto Region, Japan. *Soil Sci. Plant Nutr.* **2023**, *69*, 221–230. [CrossRef]
31. Bertalan, L.; Holb, I.; Pataki, A.; Szabó, G.; Kupásné Szalóki, A.; Szabó, S. UAV-Based Multispectral and Thermal Cameras to Predict Soil Water Content—A Machine Learning Approach. *Comput. Electron. Agric.* **2022**, *200*, 107262. [CrossRef]
32. Wei, G.; Li, Y.; Zhang, Z.; Chen, Y.; Chen, J.; Yao, Z.; Lao, C.; Chen, H. Estimation of Soil Salt Content by Combining UAV-Borne Multispectral Sensor and Machine Learning Algorithms. *PeerJ* **2020**, *2020*, e9087. [CrossRef] [PubMed]
33. Ma, Y.; Zhu, W.; Zhang, Z.; Chen, H.; Zhao, G.; Liu, P. Fusion Level of Satellite and UAV Image Data for Soil Salinity Inversion in the Coastal Area of the Yellow River Delta. *Int. J. Remote Sens.* **2022**, *43*, 7039–7063. [CrossRef]
34. Kahaer, Y.; Tashpolat, N. Estimating Salt Concentrations Based on Optimized Spectral Indices in Soils with Regional Heterogeneity. *J. Spectrosc.* **2019**, *15*, 2402749. [CrossRef]
35. Silver, M.; Tiwari, A.; Karnieli, A. Identifying Vegetation in Arid Regions Using Object-Based Image Analysis with RGB-Only Aerial Imagery. *Remote Sens.* **2019**, *11*, 2308. [CrossRef]
36. Last, W.M.; Ginn, F.M. Saline Systems of the Great Plains of Western Canada: An Overview of the Limnogeology and Paleolimnology. *Saline Syst.* **2005**, *1*, 10. [CrossRef] [PubMed]
37. Dítě, D.; Šuvada, R.; Tóth, T.; Dítě, Z. Inventory of the Halophytes in Inland Central Europe. *Preslia* **2023**, *95*, 215–240. [CrossRef]
38. Deák, B.; Valkó, O.; Alexander, C.; Mücke, W.; Kania, A.; Tamás, J.; Heilmeier, H. Fine-Scale Vertical Position as an Indicator of Vegetation in Alkali Grasslands—Case Study Based on Remotely Sensed Data. *Flora-Morphol. Distrib. Funct. Ecol. Plants* **2014**, *209*, 693–697. [CrossRef]
39. Themistocleous, K. DEM Modeling Using RGB-Based Vegetation Indices from UAV Images. In Proceedings of the Seventh International Conference on Remote Sensing and Geoinformation of the Environment (RSCy2019), Paphos, Cyprus, 18–21 March 2019; Themistocleous, K., Papadavid, G., Michaelides, S., Ambrosia, V., Hadjimitsis, D., Eds.; SPIE: Bellingham, WA, USA, 2019; Volume 11174, pp. 499–506.
40. Szatmári, G.; Pásztor, L. Comparison of Various Uncertainty Modelling Approaches Based on Geostatistics and Machine Learning Algorithms. *Geoderma* **2019**, *337*, 1329–1340. [CrossRef]
41. Wang, L.; Wu, W.; Liu, H. Bin Digital Mapping of Topsoil PH by Random Forest with Residual Kriging (RFRK) in a Hilly Region. *Soil Res.* **2019**, *57*, 387–396. [CrossRef]
42. Hengl, T.; Nussbaum, M.; Wright, M.N.; Heuvelink, G.B.M.; Gräler, B. Random Forest as a Generic Framework for Predictive Modeling of Spatial and Spatio-Temporal Variables. *PeerJ* **2018**, *2018*, e5518. [CrossRef]
43. Dövényi, Z. (Ed.) *Magyarország Kistájainak Katasztere [Inventory of Microregions in Hungary]*; MTA Földrajztudományi Intézet: Budapest, Hungary, 2010.
44. Novák, T.J.; Tóth, C.A. Development of Erosional Microforms and Soils on Semi-Natural and Anthropogenic Influenced Solonetzic Grasslands. *Geomorphology* **2016**, *254*, 121–129. [CrossRef]
45. Tóth, C.; Novák, T.; Rakonczai, J. Hortobágy Puszta: Microtopography of Alkali Flats. In *Landscapes and Landforms of Hungary*; Lóczy, D., Ed.; Springer International Publishing: Cham, Switzerland, 2015; pp. 237–246. ISBN 978-3-319-08997-3.
46. Tóth, T.; Szendrei, G. A Sókivirágzások Elterjedésének És Képződésének Összefüggése a Környezeti, Ezen Belül Talajtani Tényezőkkel [Relationship between Salt Efflorescences and Environmental Conditions with Special Emphasis on Edaphological Conditions]. *Topogr. Mineral. Hung.* **2006**, *IX*, 79–90.
47. Tóth, T.; Kuti, L. Összefüggés a Talaj Sótartalma És Egyes Földtani Tényezők Között a Hortobágyi "Nyírőlapos" Mintaterületen. I. Általános Földtani Jellemzés, a Felszín Alatti Rétegek Kalcittartalma És PH Értéke [Geological Factors Affecting the Salinization of the Nyírőlapos Sample Area (Hortobágy, Hungary). I. General Geological Characterization, Calcite Concentration and PH Values of Subsurface Layers]. *Agrokémia Talajt.* **1999**, *48*, 431–446.
48. IUSS Working Group WRB. *World Reference Base for Soil Resources 2014, Update 2015. World Soil Resources Reports 106*; FAO: Rome, Italy, 2015.
49. Michéli, E.; Fuchs, M.; Hegymegi, P.; Stefanovits, P. Classification of the Major Soils of Hungary and Their Correlation with the World Reference Base for Soil Resources (WRB). *Agrokémia Talajt.* **2006**, *55*, 19–28. [CrossRef]
50. Gallai, B.; Árvai, M.; Mészáros, J.; Barna, G.; Pásztor, L.; Szatmári, G.; Ulicsni, V.; Tóth, B.; Novák, T.; Tóth, T. A Talaj És Növényzet Összefüggése a Hortobágyi Ágota-Pusztán [The Relationship between Soil and Vegetation Ágota-Puszta, Hortobágy]. In Proceedings of the Talajtani Vándorgyűlés, Sárvár, Hungary, 25 September 2020.
51. Tóth, T.; Rajkai, K. Soil and Plant Correlations in a Solonetzic Grassland. *Soil Sci.* **1994**, *157*, 253–262. [CrossRef]
52. Bölöni, J.; Molnár, Z.; Kun, A. (Eds.) *Magyarország Élőhelyei. A Hazai Vegetációtípusok Leírása És Határozója. ÁNÉR 2011 [Habitats of Hungary]*; MTA Ökológiai és Botanikai Kutatóintézete: Vácrátót, Hungary, 2011.
53. Molnár, Z.; Biró, M.; Bölöni, J.; Horváth, F. Distribution of the (Semi-)Natural Habitats in Hungary I. Marshes and Grasslands. *Acta Bot. Hung.* **2009**, *50*, 59–105. [CrossRef]
54. Reudenbach, C.; Meyer, H. UavRst R package: Unmanned Aerial Vehicle Remote Sensing Tools 2022. Available online: https://gisma.github.io/uavRst/ (accessed on 28 June 2023).
55. Conrad, O.; Bechtel, B.; Bock, M.; Dietrich, H.; Fischer, E.; Gerlitz, L.; Wehberg, J.; Wichmann, V.; Böhner, J. System for Automated Geoscientific Analyses (SAGA) v. 2.1.4. *Geosci. Model Dev.* **2015**, *8*, 1991–2007. [CrossRef]

56. Mohammadi, H.; Samadzadegan, F. An Object Based Framework for Building Change Analysis Using 2D and 3D Information of High Resolution Satellite Images. *Adv. Space Res.* **2020**, *66*, 1386–1404. [CrossRef]
57. Gitelson, A.A.; Kaufman, Y.J.; Stark, R.; Rundquist, D. Novel Algorithms for Remote Estimation of Vegetation Fraction. *Remote Sens. Environ.* **2002**, *80*, 76–87. [CrossRef]
58. Lacaux, J.P.; Tourre, Y.M.; Vignolles, C.; Ndione, J.A.; Lafaye, M. Classification of Ponds from High-Spatial Resolution Remote Sensing: Application to Rift Valley Fever Epidemics in Senegal. *Remote Sens. Environ.* **2007**, *106*, 66–74. [CrossRef]
59. Madeira, J.; Bedidi, A.; Cervelle, B.; Pouget, M.; Flay, N. Visible Spectrometric Indices of Hematite (Hm) and Goethite (Gt) Content in Lateritic Soils: The Application of a Thematic Mapper (TM) Image for Soil-Mapping in Brasilia, Brazil. *Int. J. Remote Sens.* **1997**, *18*, 2835–2852. [CrossRef]
60. Novák, J.; Lukas, V.; Kren, J. Estimation of Soil Properties Based on Soil Colour Index. *Agric. Conspec. Sci.* **2018**, *83*, 71–76.
61. Mathieu, R.; Pouget, M.; Cervelle, B.; Escadafal, R. Relationships between Satellite-Based Radiometric Indices Simulated Using Laboratory Reflectance Data and Typic Soil Color of an Arid Environment. *Remote Sens. Environ.* **1998**, *66*, 17–28. [CrossRef]
62. Raymond Hunt, E.; Daughtry, C.S.T.; Eitel, J.U.H.; Long, D.S. Remote Sensing Leaf Chlorophyll Content Using a Visible Band Index. *Agron. J.* **2011**, *103*, 1090–1099. [CrossRef]
63. Louhaichi, M.; Borman, M.M.; Johnson, D.E. Spatially Located Platform and Aerial Photography for Documentation of Grazing Impacts on Wheat. *Geocarto Int.* **2001**, *16*, 65–70. [CrossRef]
64. Tucker, C.J. Red and Photographic Infrared Linear Combinations for Monitoring Vegetation. *Remote Sens. Environ.* **1979**, *8*, 127–150. [CrossRef]
65. Escadafal, R.; Belghit, A.; Ben-Moussa, A. Indices Spectraux Pour La Télédétection de La Dégradation Des Milieux Naturels En Tunisie Aride. In Proceedings of the Actes du 6eme Symposium International sur les Mesures Physiques et Signatures en Télédétection, Val d'Isère, France, 17–21 January 1994; pp. 253–259.
66. Olaya, V. A Gentle Introduction to SAGA GIS. Available online: http://sourceforge.net/saga-gis/ (accessed on 28 June 2023).
67. Wilson, J.; Gallant, J. Primary Topographic Attributes. In *Terrain Analysis: Principles and Applications*; Wilson, J., Gallant, J., Eds.; John Wiley & Sons, Inc.: Hoboken, NJ, USA, 2000; p. 520.
68. Riley, S.; DeGloria, S.; Elliot, R. A Terrain Ruggeddness Index That Quantifies Topographic Heterogeneity. *Int. J. Sci.* **1999**, *5*, 23–27.
69. Iwahashi, J.; Pike, R.J. Automated Classifications of Topography from DEMs by an Unsupervised Nested-Means Algorithm and a Three-Part Geometric Signature. *Geomorphology* **2007**, *86*, 409–440. [CrossRef]
70. Olaya, V. Chapter 6 Basic Land-Surface Parameters. *Dev. Soil Sci.* **2009**, *33*, 141–169. [CrossRef]
71. Böhner, J.; Selige, T. Spatial Prediction of Soil Attributes Using Terrain Analysis and Climate Regionalisation. In *SAGA—Analysis and Modelling Applications. Göttinger Geographische Abhandlungen 115*; Boehner, J., McCloy, K., Strobl, J., Eds.; Geographischen Instituts der Universität Göttingen: Göttingen, Germany, 2006; pp. 13–28.
72. Böhner, J.; Antonić, O. Chapter 8 Land-Surface Parameters Specific to Topo-Climatology. *Dev. Soil Sci.* **2009**, *33*, 195–226. [CrossRef]
73. Moore, I.D.; Grayson, R.B.; Ladson, A.R. Digital Terrain Modelling: A Review of Hydrological, Geomorphological, and Biological Applications. *Hydrol. Process.* **1991**, *5*, 3–30. [CrossRef]
74. Friedrich, K. Multivariate Distance Methods for Geomorphographic Relief Classification. In *Land Inforamtion Systems—Developments for Planning the Sustainable Use of Land Resources. European Soil Bureau—Research Report 4, EUR 17729 EN*; Heinecke, H., Eckelmann, W., Thomasson, A., Jones, J., Montanarella, L., Buckley, B., Eds.; Office for Oficial Publications of the European Communities: Ispra, Italy, 1998; pp. 259–266.
75. Gallant, J.C.; Dowling, T.I. A Multiresolution Index of Valley Bottom Flatness for Mapping Depositional Areas. *Water Resour. Res.* **2003**, *39*, 1347. [CrossRef]
76. Beven, K.J.; Kirkby, M.J. A Physically Based, Variable Contributing Area Model of Basin Hydrology. *Hydrol. Sci. Bull.* **1979**, *24*, 43–69. [CrossRef]
77. Ayers, R.S.; Westcot, D.W. *Water Quality for Agriculture. FAO Irrigation and Drainage Paper 29 Rev. 1*; FAO: Rome, Italy, 1985.
78. Wright, M.N.; Ziegler, A. Ranger: A Fast Implementation of Random Forests for High Dimensional Data in C++ and R. *J. Stat. Softw.* **2017**, *77*, 1–17. [CrossRef]
79. Kuhn, M. Building Predictive Models in R Using the Caret Package. *J. Stat. Softw.* **2008**, *28*, 1–26. [CrossRef]
80. Breiman, L. Random Forests. *Mach. Learn.* **2001**, *45*, 5–32. [CrossRef]
81. Hengl, T.; MacMillan, R.A. *Predictive Soil Mapping with R*; OpenGeoHub Foundation: Wageningen, The Netherlands, 2019.
82. Hengl, T. *A Practical Guide to Geostatistical Mapping of Environmental Variables. EUR 22904 EN*; Office for Official Publications of the European Communities: Luxembourg, 2007; ISBN 978-92-79-06904-8.
83. Šefferová Stanová, V.; Janák, M.; Ripka, J. *Management of Natura 2000 Habitats. 1530 *Pannonic Salt Steppes and Salt Marshes*; Technical Report; European Comission: Brussels, Belgium, 2008; Available online: https://ec.europa.eu/environment/nature/natura2000/management/habitats/pdf/1530_Pannonic_salt_steppes.pdf (accessed on 28 June 2023).
84. Elnaggar, A.A.; Noller, J.S. Application of Remote-Sensing Data and Decision-Tree Analysis to Mapping Salt-Affected Soils over Large Areas. *Remote Sens.* **2009**, *2*, 151–165. [CrossRef]
85. Zhang, Z.; Niu, B.; Li, X.; Kang, X.; Hu, Z. Estimation and Dynamic Analysis of Soil Salinity Based on UAV and Sentinel-2A Multispectral Imagery in the Coastal Area, China. *Land* **2022**, *11*, 2307. [CrossRef]

86. Zhao, W.; Zhou, C.; Zhou, C.; Ma, H.; Wang, Z. Soil Salinity Inversion Model of Oasis in Arid Area Based on UAV Multispectral Remote Sensing. *Remote Sens.* **2022**, *14*, 1804. [CrossRef]
87. Tóth, T.; Csillag, F.; Biehl, L.L.; Michéli, E. Characterization of Semivegetated Salt-Affected Soils by Means of Field Remote Sensing. *Remote Sens. Environ.* **1991**, *37*, 167–180. [CrossRef]
88. Mougenot, B.; Pouget, M. Remote Sensing of Salt Affected Soils. *Remote Sens. Rev.* **1993**, *7*, 241–259. [CrossRef]
89. Tóth, T.; Kertész, M.; Pásztor, L. New Approaches in Salinity/Sodicity Mapping in Hungary. *Agrokémia Talajt.* **1998**, *47*, 76–86.
90. Farifteh, J.; Farshad, A.; George, R.J. Assessing Salt-Affected Soils Using Remote Sensing, Solute Modelling, and Geophysics. *Geoderma* **2006**, *130*, 191–206. [CrossRef]
91. Jamali, A.A.; Montazeri Naeeni, M.A.; Zarei, G. Assessing the Expansion of Saline Lands through Vegetation and Wetland Loss Using Remote Sensing and GIS. *Remote Sens. Appl.* **2020**, *20*, 100428. [CrossRef]
92. Benos, L.; Tagarakis, A.C.; Dolias, G.; Berruto, R.; Kateris, D.; Bochtis, D. Machine Learning in Agriculture: A Comprehensive Updated Review. *Sensors* **2021**, *21*, 3758. [CrossRef]
93. García-Fernández, M.; Sanz-Ablanedo, E.; Rodríguez-Pérez, J.R. High-Resolution Drone-Acquired RGB Imagery to Estimate Spatial Grape Quality Variability. *Agronomy* **2021**, *11*, 655. [CrossRef]
94. Fu, Z.; Yu, S.; Zhang, J.; Xi, H.; Gao, Y.; Lu, R.; Zheng, H.; Zhu, Y.; Cao, W.; Liu, X. Combining UAV Multispectral Imagery and Ecological Factors to Estimate Leaf Nitrogen and Grain Protein Content of Wheat. *Eur. J. Agron.* **2022**, *132*, 126405. [CrossRef]
95. Li, Z.; Zhou, X.; Cheng, Q.; Fei, S.; Chen, Z. A Machine-Learning Model Based on the Fusion of Spectral and Textural Features from UAV Multi-Sensors to Analyse the Total Nitrogen Content in Winter Wheat. *Remote Sens.* **2023**, *15*, 2152. [CrossRef]
96. Yamamoto, S.; Nomoto, S.; Hashimoto, N.; Maki, M.; Hongo, C.; Shiraiwa, T.; Homma, K. Monitoring Spatial and Time-Series Variations in Red Crown Rot Damage of Soybean in Farmer Fields Based on UAV Remote Sensing. *Plant Prod. Sci.* **2023**, *26*, 36–47. [CrossRef]
97. Sahbeni, G. Prediction of Soil Salinity Using a Random Forest-Based Model between 2000 and 2016. A Case Study in the Great Hungarian Plain. In Proceedings of the Global Symposium on Salt-Affected Soils, online, 20–22 October 2021.
98. Hateffard, F.; Balog, K.; Tóth, T.; Mészáros, J.; Árvai, M.; Kovács, Z.A.; Szűcs-Vásárhelyi, N.; Koós, S.; László, P.; Novák, T.J.; et al. High-Resolution Mapping and Assessment of Salt-Affectedness on Arable Lands by the Combination of Ensemble Learning and Multivariate Geostatistics. *Agronomy* **2022**, *12*, 1858. [CrossRef]
99. Dajic-Stevanovic, Z.; Pecinar, I.; Kresovic, M.; Vrbnicanin, S.; Tomovic, L. Biodiversity, Utilization and Management of Grasslands of Salt Affected Soils in Serbia. *Community Ecol.* **2008**, *9*, 107–114. [CrossRef]
100. Szabolcs, I.; Rédly, M. State and Possibilities of Soil Salinization in Europe. *Agrokémia Talajt.* **1989**, *38*, 537–558.
101. Szabolcs, I.; Máté, F. A Hortobágyi Szikes Talajok Genetikájának Kérdéséhez. [On the Genesis of Alkaline Soils of Hortobágy]. *Agrokémia Talajt.* **1955**, *4*, 31–38.
102. Kovda, V.A. *Origin of Saline Soils and Their Regime I*; USSR Academy of Sciences: Moscow, Russia, 1946. (In Russian)
103. Kovda, V. *Origin of Saline Soils and Their Regime II*; USSR Academy of Sciences: Moscow, Russia, 1947. (In Russian)
104. Várallyay, G. Extreme Moisture Regime as the Main Limiting Factor of the Fertility of Salt Affected Soils. *Agrokémia Talajt.* **1981**, *30*, 73–96.
105. Wagenet, R.J.; Jury, W.A. Movement and Accumulation of Salts in Soils. In *Soil Salinity under Irrigation. Ecological Studies*; Shainberg, I., Shalhevet, J., Eds.; Springer: Berlin/Heidelberg, Germany, 1984; Volume 51, pp. 100–129.
106. Szabolcs, I.; Várallyay, G.; Darab, K. Soil and Hydrologic Surveys for the Prognosis and Monitoring of Salinity and Alkalinity. In Proceedings of the Prognosis of Salinity and Alkalinity. Report of an Expert Consultation, Rome, Italy, 3–6 June 1975; FAO Soil Bulletin 31. FAO: Rome, Italy, 1976; pp. 119–130.
107. Várallyay, G. Environmental Stresses Induced by Salinity/Alkalinity in the Carpathian Basin (Central Europe). *Agrokémia Talajt.* **2004**, *51*, 233–242. [CrossRef]
108. Dítě, Z.; Šuvada, R.; Tóth, T.; Jun, P.E.; Píš, V.; Dítě, D. Current Condition of Pannonic Salt Steppes at Their Distribution Limit: What Do Indicator Species Reveal about Habitat Quality? *Plants* **2021**, *10*, 530. [CrossRef]
109. Csontos, P.; Tamás, J.; Kovács, Z.; Schellenberger, J.; Penksza, K.; Szili-Kovács, T.; Kalapos, T. Vegetation Dynamics in a Loess Grassland: Plant Traits Indicate Stability Based on Species Presence, but Directional Change When Cover Is Considered. *Plants* **2022**, *11*, 763. [CrossRef] [PubMed]
110. Molnár, Z. Ősi És Másodlagos Eredetű Tiszántúli Szikes Puszták Zonációja [Zonation of Primary and Secondary Solonetz Alkaline Steppes in the Crisicum (Pannonicum)]. *Acta Biol. Debrecina. Suppl. Oecol. Hung.* **2010**, *22*, 181–189.
111. Kertész, M.; Tóth, T. Soil Survey Based on Sampling Scheme Adjusted to Local Heterogeneity. *Agrokémia Talajt.* **1994**, *43*, 113–132.
112. Steven, M.; Malthus, T.; Danson, F.; Jaggard, K.; Andrieu, B. Monitoring Response of Vegetation to Stress. In Proceedings of the Proceedings Remote Sensing Society Annual Conference, Dundee, UK, 15–17 September 1992; pp. 369–377.
113. Tóth, T. Salt-Affected Soils and Their Native Vegetation in Hungary. In *Sabkha Ecosystems: Volume III: Africa and Southern Europe*; Öztürk, M., Böer, B., Barth, H.-J., Clüsener-Godt, M., Khan, M.A., Breckle, S.-W., Eds.; Springer: Dordrecht, The Netherlands, 2011; pp. 113–132. ISBN 978-90-481-9673-9.

land

Article

# Application of Machine Learning Algorithms for Digital Mapping of Soil Salinity Levels and Assessing Their Spatial Transferability in Arid Regions

Magboul M. Sulieman [1], Fuat Kaya [2], Mohammed A. Elsheikh [1], Levent Başayiğit [2] and Rosa Francaviglia [3,*]

1 Department of Soil and Environment Sciences, Faculty of Agriculture, University of Khartoum, Shambat 13314, Sudan; magboul@uofk.edu or magboul.musa@gmail.com (M.M.S.); melsheik@uofk.edu or mohamedelsheik@gmail.com (M.A.E.)
2 Department of Soil Science and Plant Nutrition, Faculty of Agriculture, Isparta University of Applied Sciences, 32260 Isparta, Türkiye; fuatkaya@isparta.edu.tr (F.K.); leventbasayigit@isparta.edu.tr (L.B.)
3 Council for Agricultural Research and Economics, Research Centre for Agriculture and Environment, 00184 Rome, Italy
* Correspondence: r.francaviglia@gmail.com

**Abstract:** A comprehensive understanding of soil salinity distribution in arid regions is essential for making informed decisions regarding agricultural suitability, water resource management, and land use planning. A methodology was developed to identify soil salinity in Sudan by utilizing optical and radar-based satellite data as well as variables obtained from digital elevation models that are known to indicate variations in soil salinity. The methodology includes the transfer of models to areas where similar conditions prevail. A geographically coordinated database was established, incorporating a variety of environmental variables based on Google Earth Engine (GEE) and Electrical Conductivity (EC) measurements from the saturation extract of soil samples collected at three different depths (0–30, 30–60, and 60–90 cm). Thereafter, Multinomial Logistic Regression (MNLR) and Gradient Boosting Algorithm (GBM), were utilized to spatially classify the salinity levels in the region. To determine the applicability of the model trained at the reference site to the target area, a Multivariate Environmental Similarity Surface (MESS) analysis was conducted. The producer's accuracy, user's accuracy, and Tau index parameters were used to evaluate the model's accuracy, and spatial confusion indices were computed to assess uncertainty. At different soil depths, Tau index values for the reference area ranged from 0.38 to 0.77, whereas values for target area samples ranged from 0.66 to 0.88, decreasing as the depth increased. Clay normalized ratio (CLNR), Salinity Index 1, and SAR data were important variables in the modeling. It was found that the subsoils in the middle and northwest regions of both the reference and target areas had a higher salinity level compared to the topsoil. This study highlighted the effectiveness of model transfer as a means of identifying and evaluating the management of regions facing significant salinity-related challenges. This approach can be instrumental in identifying alternative areas suitable for agricultural activities at a regional level.

**Keywords:** dryland; digital soil mapping; environmental similarity; Google Earth Engine; remote sensing; SAR; Sentinel 2 MSI; salinization; transfer learning

**Citation:** Sulieman, M.M.; Kaya, F.; Elsheikh, M.A.; Başayiğit, L.; Francaviglia, R. Application of Machine Learning Algorithms for Digital Mapping of Soil Salinity Levels and Assessing Their Spatial Transferability in Arid Regions. *Land* **2023**, *12*, 1680. https://doi.org/10.3390/land12091680

Academic Editors: Maria da Conceição Gonçalves, Mohammad Farzamian and Tiago Brito Ramos

Received: 31 July 2023
Revised: 23 August 2023
Accepted: 25 August 2023
Published: 28 August 2023



## 1. Introduction

The majority of salt-affected soils globally are located in arid and semi-arid climate zones [1]. Saline soils can be formed naturally by the effects of soil formation factors, and their formation can be accelerated as a result of anthropogenic factors [2]. Specifically, soil salinity is a major soil constraint that threatens soil fertility, agricultural sustainability, and food security in arid and semi-arid regions [3–10]. The acceleration of the process of soil salinization constitutes a significant threat to crop production and can reduce agricultural productivity at regional, national, and even local scales [11].

Since 2015, when the Sustainable Development Goals (SDGs) [12] were announced, half of the time needed to achieve the 2030 SDGs has now passed. Food security and sustainability of agriculture, especially in rain-fed or irrigated areas in arid regions, are under significant pressure from soil constraints such as salinity [2,13–15]. To assess the impact of salinization on agriculture, especially in the mentioned regions, there is a need for useful spatial information on the salinity levels in the topsoil and effective root zone that can be integrated into decision-support processes. As is well known, to achieve SDG′-2 and SDG′-15, it is essential to spatially accurately identify the variations of soil constraints to allow for the best management of soils [16–18]. Rapid and reliable determination of the current levels of soil salinity, its edaphological suitability for crop cultivation, or the constraints it presents can help identify salinity management strategies to reduce the vulnerability of crops to salt content.

Over the last quarter century, the science of pedometry has made significant advances by combining remote sensing, geographic information systems, and advanced statistical and mathematical spatial modeling applications for soil mapping [19,20]. Of course, this branch of science has been supported by the increasing number of satellites and sensors from public and private initiatives, as well as the increasingly open-access global availability of Earth Observation (EO) data. Indeed, the increasing digital representation of the spatial distribution of soil formation factors has led to initiatives that can be integrated into policymaking and decision-support systems [21–23].

Machine learning algorithms (MLAs) have been effectively used in the spatial mapping of a soil constraint such as soil salinity with a pedometric approach [24,25]. While salinity indicators determined quantitatively in the laboratory can be modeled with regression-based ML algorithms [26] in continuous data types, discrete data classified according to certain criteria (salinity classes in our study) can be effectively spatially modeled with classification-based ML algorithms. Kaplan et al. [27] emphasized that the European Space Agency's (ESA) optical Sentinel 2 remote sensing data and MLAs can effectively map EC (dS/m) in hyper arid areas in continuous data types. ESA's new generation Sentinel 1 synthetic aperture radar (SAR) data, which has a higher capacity to penetrate the soil surface [5], is emphasized as an important data source in determining the salinity level of soils [28–32].

Traditional approaches to the determination of soil salinity levels, especially fieldwork, are costly and time-consuming. Nowadays, EO data have been robustly demonstrated to be essential tools for accurately estimating soil salinity in different parts of the world [33,34]. These developments have been widely used in studies, especially in vegetation, soil, and salinity indices, which are very different in their effectiveness while offering great potential for regions of the world where vegetation cover is reduced or seasonally absent [24,27,32,35]. Another important aspect concerns developments in processing algorithms such as MLAs [5]. Supervised learning algorithms make it possible to model the relationships and dependencies between the target prediction output [36] and input data/features to predict salinity constraint output values in new areas by learning from the data from areas where salinity threats exist.

The pedometric approach and digital soil mapping (DSM) have enabled regional [2], continental [33], and global [1] applications of soil salinity mapping at various spatial and temporal scales. However, most of the DSM research in the specific area of salinity threats focuses on modeling soil properties at a specific site. Kaya et al. [2] spatially mapped the threat of soil salinity in an area with complex land uses in the Mediterranean region using a random forest (RF) and support vector regression (SVR) algorithm. Guo et al. [37] presented an unsupervised approach to generating salinity management zones in coastal Central China. Konyushkova et al. [38] successfully utilized remote sensing data to improve assessment and decision support for sustainable management of soil and water resources in salt-affected croplands. Golestani et al. [39] systematically compared decision tree (DT), artificial neural network (ANN), RF, and SVR algorithms to spatially map salinity during the winter and summer seasons in Sirjan Playa, Iran. Kabiraj et al. [40] used the RF algorithm for spatial mapping of salinity classes in the Gulf of Mannar, India, and

Lekka et al. [41] effectively used the logistic regression algorithm to assess spatial patterns of soil salinity in agricultural fields in Lesvos Island, Greece.

The principle that similar soil-forming factors lead to similar soils has found an important place in the DSM on a global scale [42]. Regionally, indeed, areas with similar soil-forming factors develop similar soils over time [42]. In line with this assumption, there may be a possibility that a categorical or continuous soil model learned in one area may be transferable to a similar area. Of course, this possibility is based on the availability of digital data on existing soil formation factors in the area where the model was learned and in the transferred area. This application is organized in such a way that quantitative digital data are similarly measured for the target and the reference area. In the specific case of our study, this process is an opportunity to reduce the relatively high costs and time required to produce soil salinity maps in an arid region by focusing the transfer of models learned from a reference area to the target area.

Sudan is a country with agricultural areas, abundant water, two branches of the Nile River, and high agricultural potential [43]. Sudan, one of the largest countries in Africa, has over 80 million hectares of arable land, of which only 20 percent have been cultivated so far [44]. With direct diversion from rivers and groundwater, many industrial crops can be produced in Sudan [44]. However, it is necessary to manage the risk of soil salinization during the first 10–15 years of irrigated agricultural production in arid areas. In addition, the need to map existing saline areas and identify appropriate salinity management strategies is necessary to develop methods and approaches to identify, monitor, and assess the extent of salt-affected soils in Sudan, contributing to the development of strategies to help mitigate climate change impacts.

Transfer learning is the process of applying the model learned from a reference area to a target area [45]. The transfer learning approach has been demonstrated to be applicable in pedometrics, especially in studies on the prediction of soil properties by creating and using spectral reflectance libraries [45–47]. By integrating the transfer learning approach, relevant DSM studies were conducted, such as the parent material [48], organic carbon at the local scale [49], USDA Soil Taxonomy at the sub-group level [50], USDA Soil Taxonomy at the soil great group level [51], soil organic carbon in cropland soils [52], and soil particle fractions [53].

The spatial variability of soil salinity constraints is one of the most important causes of variability in crop production and is important information for spatial planning according to the sensitivity and tolerance level of the plant to be grown. Although there have been many field-based studies on the spatial prediction of salinity in drylands by integrating RS and ML [2,27,35,39], no studies on the transferability of the models have been carried out.

This study was the first to integrate "transfer learning" into mapping soil salinity levels in an arid region. Hence, we hypothesized that the utilization of transfer learning-based MLAs in conjunction with open-access EO data within this study can offer opportunities for mapping soil salinity within an arid region. The present research deals with the transferability of salinity class models derived from a reference area to a target area whose spatial similarity is quantified by a similarity index. In particular, the objectives of the study were: (i) to develop a classification model for the salinity of soils at three different depths in Eastern Sudan, (ii) to demonstrate the effectiveness of the Multivariate Environmental Similarity Surface (MESS) technique in applying the model learned from the reference site to the target site, and (iii) to evaluate the importance of the environmental variables used in the modeling within the soil scientist framework and to identify environmental variables that could be used in similar study areas.

## 2. Materials and Methods

Section 2.1 provided general information about the study area, and Section 2.2 provided detailed information about the soil sampling methodology and design. Section 2.3 presented information about the analyses performed on soil samples. Section 2.4 details the various environmental variables produced by the Google Earth Engine. Section 2.5 explains the modeling process and the transfer learning process. Section 2.6 details the importance of digital variables in modeling, accuracy, and uncertainty assessments of models.

### 2.1. Study Area

This study was conducted at the lower Atbara Nile, which extends about 270 km SE of Atbara town in River Nile State and nearly 288 km from Khartoum, the capital of Sudan. The study area is located between 16°44′ N and 16°55′ N Latitude and 34°50′ E and 35°2′ E Longitude and covers a total area of about 7600 ha (distributed as 4200 ha for the reference area and 3400 ha for the target area). The study area falls within the desert climatic zone of the country, with an average annual precipitation of 63.2 mm (mainly between July and August), an average annual temperature of 29.6 °C, and an average annual relative humidity of 28.3%. The soil is characterized by hyper-thermic and aridic soil temperature and moisture regimes, respectively. The soil is classified as Aridisols according to soil taxonomy [54,55].

### 2.2. Field Study and Sampling Strategy

A semi-detailed soil survey was used to perform this study using a scale of 1:45,000. We used a grid design to determine the targeted sample locations. The total auger locations for reference and target areas were 202 and 144 sites, respectively. We used a handheld GPS (Garmin Montana 680t) to determine the precise sites of the auger samples. Figure 1A shows the geographical location of the study area overlaid on the Sentinel 2 MSI natural color band combination map. Figure 1B presents the field distribution of auger samples on the DEM map. Soil samples were taken from a three-depth systematic sampling design [55] at 450 m intervals at both studied areas: 0–30 cm, 30–60 cm, and 60–90 cm, with approximately 0.5 to 1 kg of soil material gathered from each depth. The total number of samples collected was 1041 (608 from the reference area and 432 from the target area).

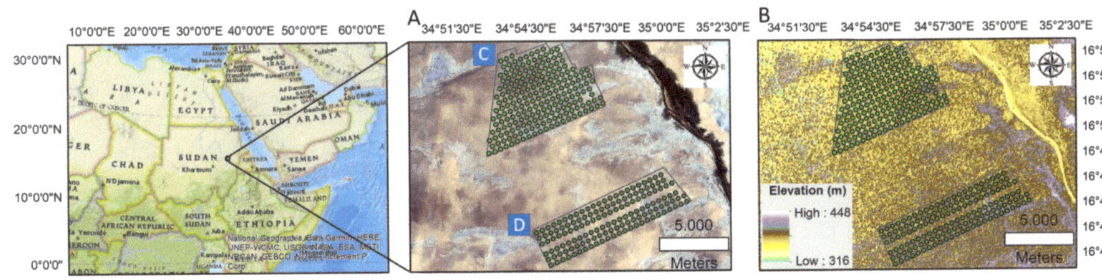


**Figure 1.** Spatial distribution of soil sampling points overlaid on Sentinel 2 MSI natural color band combination map (**A**) and digital elevation model (DEM) (**B**), including reference area (**C**) and target area (**D**). The polygons define the study areas, while the green dots define the soil sampling points.

### 2.3. Sample Analysis

Soil samples were air dried at ambient room temperature (≈25 °C), ground, and passed through a 2 mm sieve to isolate soil material from rock fragments. Electrical conductivity (ECe) as an indicator of salinity was determined in the extracts of the soil paste [56] using a digital EC meter (Jenway, 4510, UK). According to the FAO salinity classification [34,57] electrical conductivity data (dS m$^{-1}$) are classified into three classes: None (<2 dS m$^{-1}$), Moderate (between 2 and 4 dS m$^{-1}$), and Strong (> 4 dS m$^{-1}$).

### 2.4. Environmental Covariates via Google Earth Engine

To estimate salinity variations along the soil depth direction in the study area, relevant environmental covariates were selected due to their influence on salinity levels. Salinity, vegetation, and soil indices based on Sentinel 2 MSI [58], as well as horizontal transmit and vertical receive (HV) and horizontal transmit and horizontal receive (HH) polarization mode backscattering coefficient data from PALSAR-2 [59], along with derived digital elevation model derivatives, were generated using the Google Earth Engine (GEE) data catalog and platform [60]. All digital covariates were extracted from GEE to be aligned on a 10 × 10 m grid and subsequently utilized for mapping purposes.

### 2.4.1. Synthetic Aperture Radar Data

Since no available images could be obtained in the frames of the Sentinel 1 SAR satellite for the study area, Global PALSAR-2/PALSAR Yearly Mosaic [61,62], version 2 data were transferred from the GEE data catalog to the GEE code editor section [60], taking into account the years closest to the soil sampling dates. The PALSAR/PALSAR-2 mosaic was acquired at 25 m resolution [61]. This dataset is a seamless global SAR image created by mosaicking SAR images from PALSAR/PALSAR-2. In this study, 2018, 2019, and 2020 image collections in HH and HV polarization were then cropped according to the study area scope using the ".filterBounds" script in the GEE code editor [60]. Finally, using the ".mean" script, the mean of their collections was calculated for the study area to reduce data volume and for faster analysis. Polarization data can be obtained as 16 bit digital numbers (DN) and converted to backscatter coefficient values in decibel units (dB) using the following equation [61,63]:

$$\gamma^0 = 10\log_{10}(\mathrm{DN}^2) - 83 \quad (1)$$

where −83.0 is the calibration factor (dB) for the PALSAR-2 mosaics.

This equation was executed in ArcGIS 10.8—Arctoolbox—Spatial Analyst Tools—Map Algebra—Raster Calculator [64].

### 2.4.2. Multispectral Satellite Data

Sentinel-2 MSI: MultiSpectral Instrument, Level-2A product was called from the GEE catalog, and 180 images were taken from the catalog by running the ".filter('CLOUDY_PIXEL _PERCENTAGE < 5')" script within 1 year close to the soil sampling date. Using the study area shapefile and the ".filterBounds" script, the satellite image collection was clipped. Again, using the ".mean" script to reduce data volume and for faster analysis, mean synthesis images were calculated for Band 2, Band 3, Band 4, Band 8, Band 11, and Band 12 using all image collections among the respective dates. Salinity, vegetation, and soil indices in Table 1 were generated and used as environmental covariates.

### 2.4.3. Digital Elevation Model Data

NASADEM Merged DEM Global 1 arc second V001 data [65] was called from the GEE catalog and cut using the ".filterBounds" script according to the study area shapefile. In addition to the elevation data, the slope in degrees was used as an environmental covariate produced by the "ee.Terrain.slope" script.

**Table 1.** Environmental covariates are used for predicting soil salinity levels.

| Remote Sensing (RS) (Sentinel 2) OPTICAL-Based Covariates | Equations [27,32,35,58,66] |
|---|---|
| Band 2 | Blue (Central Wavelength: 490 nm) |
| Band 3 | Green (Central Wavelength: 560 nm) |
| Band 4 | Red (Central Wavelength: 665 nm) |
| Band 8 | NIR (Central Wavelength: 842 nm) |
| Band 11 | SWIR1 (Central Wavelength: 1610 nm) |
| Band 12 | SWIR1 (Central Wavelength: 2190 nm) |
| Normalized Difference Vegetation Index (NDVI) | $(NIR - Red\ / NIR + Red)$ |
| Carbonate Normalized Ratio (CNR) | $(Red - Green\ /\ Red + Green)$ |
| Clay Normalized Ratio (CLNR) | $(SWIR1 - SWIR2\ /\ SWIR1 + SWIR2)$ |
| Ferrous Normalized Ratio (FNR) | $(SWIR1 - NIR\ /\ SWIR1 + NIR)$ |
| Iron Normalized Ratio (INR) | $(Red - SWIR2\ /\ Red + SWIR2)$ |
| Normalized Difference Moisture Index (NDMI) | $(NIR - SWIR1/NIR + SWIR1)$ |
| Rock Outcrop Normalized Ratio (RONR) | $(SWIR1 - Green\ /\ SWIR1 + Green)$ |
| Green-Red vegetation index (GRVI) | $(Green - Red\ /Green + Red)$ |
| Saturation index (SatInd) | $(Red - Blue\ /\ Red + Blue)$ |
| Green Normalized Difference Vegetation Index (GNDVI) | $(NIR - Green/NIR + Green)$ |

**Table 1.** *Cont.*

| Remote Sensing (RS) (Sentinel 2) OPTICAL-Based Covariates | Equations [27,32,35,58,66] |
|---|---|
| Salinity Index 1 | $\sqrt{Blue \times Red}$ |
| Salinity Index 2 | $\sqrt{Green \times Red}$ |
| Salinity Index 3 | $(Blue - Red/Blue + Red)$ |
| Salinity Index 4 | $(Green \times Red)/(Blue)$ |
| Salinity Index 5 | $(Blue \times Red)/(Green)$ |
| Salinity Index 6 | $(NIR \times Red)/(Green)$ |
| Remote Sensing (RS) (PALSAR/PALSAR-2 mosaic) synthetic aperture RADAR-based covariates [59,61] | |
| AVG_HH_dB-polarization backscattering coefficient | For horizontal transmit and horizontal receive |
| AVG_HV_dB-polarization backscattering coefficient | For horizontal transmit and vertical receive |
| DEM-based primary covariates at NASA JPL [65] | |
| Elevation | m unit |
| Slope | Degree unit |

NIR: Near infrared, SWIR: Shortwave infrared.

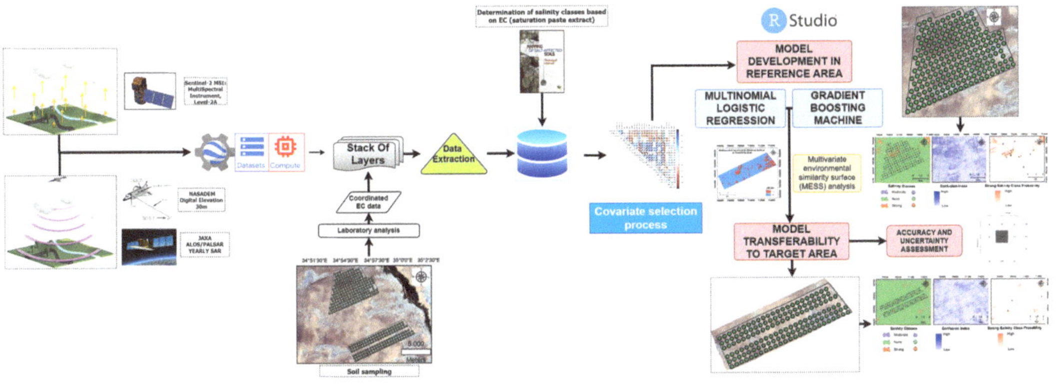


**Figure 2.** Flowchart of the methodology.

*2.5. Modelling Salinity Levels and Transferability of Models*

This study followed the DSM framework and involved several steps in the modeling process: (1) enabling and curating soil data; (2) obtaining environmental covariates from open sources; (3) extracting georeferenced sample points from the digital covariate data and preparing geodatabases [67]; (4) selecting environmental covariates through the use of "findCorrelation" functions to identify and eliminate highly correlated covariates; (5) performing classification-based modeling of salinity levels; and (6) transferring the models. The flowchart of the study is depicted in Figure 2.

The "findCorrelation" function in the "caret" package [68] was run to identify highly correlated covariates that could also compromise the performance of the model. Covariates with Spearman correlation coefficients above 0.8 were removed (Figure A1) [69,70].

To build a statistical model between environmental covariates and the predicted soil salinity classes, 2 different mathematically-based ML algorithms were systematically compared: Multinomial Logistic Regression [71,72] and Gradient Boosting Machine [73,74].

In the study, soil salinity classes are the outcome variables. In the process of data import in R core environment software [75], the categories of salinity classes were coded alphabetically as 1 (None), 2 (Moderate), and 3 (Strong). Specifically, two logit functions are needed in the three-outcome category model. The modeler can decide which outcome category to use as the reference, for which the class "1 (None)" was chosen in numerical order. Logit functions comparing the other 2 classes with the reference were created. All these processes were carried out with the "multinom" function in the "caret" package [68]. Due to the nature of the multinomial logistic regression algorithm, a pixel can belong to all three different soil salinity classes with given probabilities [76]. However, the salinity class with the highest probability is assigned to the pixel.

The Gradient Boosting Machine (GBM) is one of the most powerful MLAs for classification problems [77] involved in our study. Like tree-based learners in RF, GBM is an ensemble method based on decision trees [78]. However, unlike RF, this method generates trees serially, with each tree attempting to improve the prediction by correcting the errors of the previous one.

The hyperparameters of each ML algorithm were set using their respective packages "nnet" [72] and "gbm" [74] (Table 2). Using R Core Environment software (Version 4.2.1) [75] and RStudio IDE [79], soil salinity classes at 3 different depths in the reference study area and CLNR, FNR, NDVI, AVG_HH_dB, AVG_HV_dB, Elevation, Slope, and Salinity Index 1 were selected and estimated using these environmental covariates.

**Table 2.** Parameters for the machine learning algorithms used and final environmental covariates included for predicting soil salinity levels.

| Selected Covariates | Target Soil Variable | Algorithm | Tuning Hyperparameter |
|---|---|---|---|
| AVG_HH_dB, AVG_HV_dB, CLNR, Salinity index 1, FNR, NDVI, Slope, Elevation | 0–30 cm EC class | MNLR | decay = 0.0001 |
| | | GBM | shrinkage: 0.1, interaction.depth: 1, n.minobsinnode: 10, n.trees: 50 |
| | 30–60 cm EC class | MNLR | decay = 0.1 |
| | | GBM | shrinkage: 0.1, interaction.depth: 1, n.minobsinnode: 10, n.trees: 50 |
| | 60–90 cm EC class | MNLR | decay = 0.1 |
| | | GBM | shrinkage: 0.1, interaction.depth: 1, n.minobsinnode: 10, n.trees: 50 |

Abbreviations. GBM: Gradient Boosting Machine, MNLR: Multinomial Logistic Regression, AVG_HH_dB: for horizontal transmit and vertical receive, AVG_HV_dB: for horizontal transmit and horizontal receive, CLNR: Clay Normalized Ratio, FNR: Ferrous Normalized Ratio, NDVI: Normalized Difference Vegetation Index.

Descriptive statistical parameters were computed for the values of the eight chosen digital covariates within both the reference and target regions. Furthermore, Multivariate Environmental Similarity Surfaces were calculated [80] to compare the compatibility of the values of environmental variables in the dataset in the reference area with those in the target area to be transferred. This method can be used to measure the similarity between the selected covariates at the location of the training samples and the target area to be transferred [81,82]. Values lower than zero indicate prediction locations in both feature and geographic areas that are not explained by the training samples [82]. The MESS map for the target region was generated using the "mess" function in the "dismo" package [80].

### 2.6. *Importance of Used Covariates in Models, Accuracy, and Uncertainty Evaluations*

The relative importance levels of different digital environmental variables in the prediction models of salinity classes were calculated using the "varImp" function in the caret package [68].

In digital soil mapping, user accuracy (UA) and producer accuracy (PA) are used to validate the performance of different algorithms in both reference and target areas [66]. The "cvms" R package [83] was used to estimate the performance measures of the classification models through the confusion matrix, while the Tau index, whose performance on unbalanced datasets is emphasized by Rossiter et al. [84], was calculated using the "tauW" function in the "aqp" package [85]. When the value of the Tau index approaches 1, it indicates a strong indication of perfect agreement. In the study, both algorithms calculate probability values for each salinity class on a pixel basis, and for uncertainty evaluation, the confusion index (CI) is calculated, which spatially measures the confusion between the most probable salinity class and the second most probable class [86,87].

## 3. Results

Section 3.1 provides descriptive statistics on continuous and categorical soil salinity data. While Section 3.2 presents the performance measures of two different algorithms, Section 3.3 contains findings on the transfer of models to the target area. Section 3.4 includes maps of soil salinity classes and confusion index maps produced by two different algorithms at three depths. Section 3.5 contains information about the importance of environmental variables used in models.

### *3.1. Results of Measured Electrical Conductivity and Assessment of Salinity Classes*

Descriptive statistics and histograms of the reference area and target area sample sets taken from three different depths are shown in Figure A2. The distribution of reference area and target area samples according to salinity classes was relatively unbalanced (Figure 3a,b). Both in the reference region and in the target region, the number of observations of the "strong" salinity class increased with depth, while the "None" salinity class decreased (Figure 3a,b).

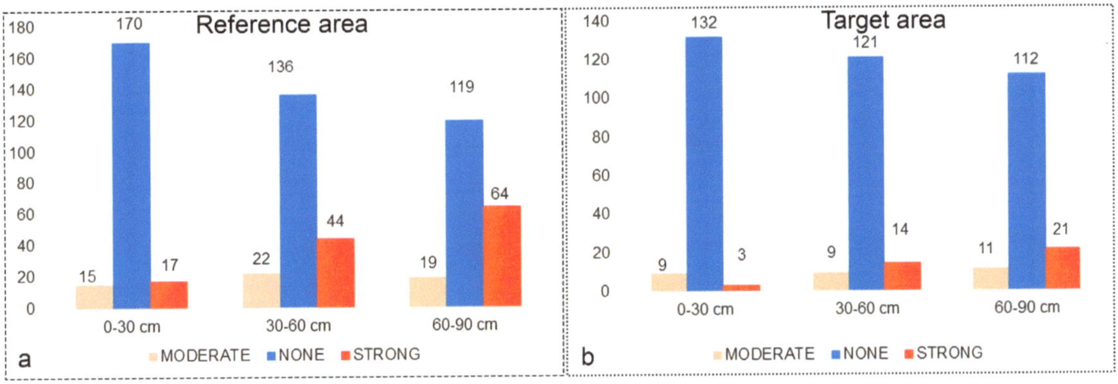


**Figure 3.** Distribution of observations by salinity classes in the reference (**a**) and target (**b**) area. The *y* axis is the number of samples.

### *3.2. Performance of the Different Classification Algorithms*

The validation statistics of soil salinity classes for each algorithm are presented in Table 3. The results of the confusion matrix from which the table was generated are presented in Figure A3. The highest Tau index values were obtained for both the reference area and the target area in the 0–30 cm samples, which can be considered surface samples (Table 3). The decrease in Tau index values followed a linear trend as the depth increased (Table 3). MNLR and GBM algorithms indeed provided very close performance measures when Tau index values were considered (Table 3).

In the surface samples (0–30 cm), the user's accuracy values for the "none" class were above 90% for the target area (Table 3). However, in both models, the remaining two classes failed to be predicted. After a careful examination of the confusion matrix (Figure A3), the models assigned the "strong" and "moderate" classes to the "none" classes.

Considering the distribution of the number of classes at different depths (Figure 3), this may be due to the fact that these models do not have enough observations to learn the classes. As a matter of fact, the user's accuracy values for the "strong" class could compute an increase in depth (30–60 cm and 60–90 cm). As the depth increased, the number of observations of the "strong" class also increased.

**Table 3.** Summary of machine learning algorithms performance criteria for reference area and transferred area.

| Depth (cm) | Soil Salinity Levels | Model | Reference Area | | | Target Area | | |
|---|---|---|---|---|---|---|---|---|
| | | | Producer's Accuracy | User's Accuracy | Tau Index | Producer's Accuracy | User's Accuracy | Tau Index |
| 0–30 | None | MNLR | 94 | 86 | GBM: 0.75<br>MNLR: 0.77 | 100 | 92 | GBM: 0.88<br>MNLR: 0.88 |
| | | GBM | 95 | 89 | | 100 | 93 | |
| | Moderate | MNLR | 0 | NaN * | | 0 | NaN | |
| | | GBM | 0 | NaN | | 0 | NaN | |
| | Strong | MNLR | 24 | 67 | | 0 | NaN | |
| | | GBM | 53 | 43 | | 34 | 50 | |
| 30–60 | None | MNLR | 98 | 75 | GBM: 0.61<br>MNLR: 0.61 | 96 | 85 | GBM: 0.72<br>MNLR: 0.72 |
| | | GBM | 97 | 76 | | 96 | 85 | |
| | Moderate | MNLR | 5 | 100 | | 0 | NaN | |
| | | GBM | 0 | NaN | | 0 | NaN | |
| | Strong | MNLR | 39 | 74 | | 8 | 17 | |
| | | GBM | 44 | 68 | | 8 | 17 | |
| 60–90 | None | MNLR | 90 | 67 | GBM: 0.38<br>MNLR: 0.47 | 98 | 79 | GBM: 0.66<br>MNLR: 0.66 |
| | | GBM | 100 | 59 | | 100 | 78 | |
| | Moderate | MNLR | 0 | NaN | | 0 | NaN | |
| | | GBM | 0 | NaN | | 0 | NaN | |
| | Strong | MNLR | 40 | 59 | | 10 | 34 | |
| | | GBM | 0 | NaN | | 0 | NaN | |

* NaN indicates unpredicted classes. GBM: Gradient Boosting Machine, MNLR: Multinomial Logistic Regression.

### 3.3. Transferability of Models according to Multivariate Environmental Similarity Surface

Since eight environmental variables were used in the modeling process, the selected variables in Table 2 were used for similarity analysis during the transfer of the models. Descriptive statistics of selected radar, optic-based, and terrain covariates at the sampling locations in the reference and target areas are shown in Table 4. The minimum values of the radar-based covariates are quite close for both areas (Table 4). A similar situation was found in the optical-based FNR, NDVI, and CLNR covariates (Table 4). In particular, the distributions of the land covariates are also basically similar across the regions, and the standard deviation values are quite close to each other (Table 4).

**Table 4.** Descriptive statistics for environmental variables for both reference and target areas.

| Covariate | Area | Minimum | Mean | Median | Maximum | Standard Deviation |
|---|---|---|---|---|---|---|
| AVG_HH_dB | Reference<br>Target | −30.07<br>−30.67 | −26.35<br>−25.47 | −26.63<br>−25.86 | −18.31<br>−12.27 | 1.97<br>3.02 |
| AVG_HV_dB | Reference<br>Target | −39.77<br>−38.86 | −36.71<br>−36.22 | −36.92<br>−36.27 | −25.33<br>−31.34 | 1.41<br>1.35 |
| CLNR | Reference<br>Target | 0.005<br>0.010 | 0.015<br>0.018 | 0.016<br>0.019 | 0.023<br>0.025 | 0.004<br>0.003 |
| Salinity index 1 | Reference<br>Target | 2555.61<br>2557.05 | 2809.86<br>2874.53 | 2823.33<br>2887.32 | 3252.18<br>3237.19 | 113.83<br>102.14 |
| FNR | Reference<br>Target | 0.037<br>0.030 | 0.061<br>0.054 | 0.060<br>0.054 | 0.085<br>0.073 | 0.006<br>0.007 |

**Table 4.** *Cont.*

| Covariate | Area | Minimum | Mean | Median | Maximum | Standard Deviation |
|---|---|---|---|---|---|---|
| NDVI | Reference | 0.034 | 0.042 | 0.041 | 0.060 | 0.004 |
| | Target | 0.033 | 0.041 | 0.040 | 0.057 | 0.005 |
| Slope | Reference | 0.00 | 4.62 | 4.017 | 24.62 | 2.97 |
| | Target | 0.00 | 4.45 | 4.016 | 12.52 | 2.53 |
| Elevation | Reference | 365.0 | 380.15 | 380.0 | 395.0 | 3.71 |
| | Target | 375.0 | 384.24 | 384.0 | 398.0 | 3.74 |

According to the environmental variable values of the observations in the reference area and the MESS results of the target region (Figure 4), the models can be effectively transferred for regions with values above 0. However, MESS values below 0 in the southeast of the target area are associated with the accumulation of wind-borne materials. Again, the partial excavation of the surface soil in the central part of the study area proves that this area does not have similar environmental variable values (smaller than 0) to the reference region.

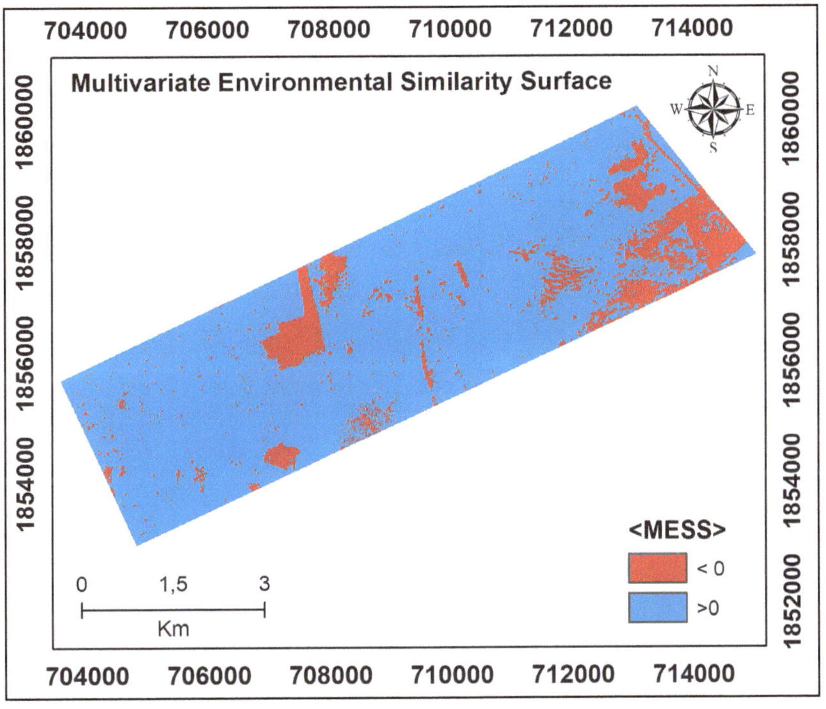


**Figure 4.** Multivariate Environmental Similarity Surfaces (MESS) for the target area, calculated with the selected covariates according to the reference area.

### *3.4. Spatial Prediction of Soil Salinity Levels in Reference and Target Areas*

Digital maps of the salinity classes of the 0–30 cm samples are presented in Figure 5 for the reference and target areas by applying MNLR and GBM. The "strong" class was mapped with high probability by the models in the reference area (Figure 5i,j). In the surface samples, MNLR and GBM models produced maps with similar salinity class patterns for both reference and target areas (Figure 5i,j).

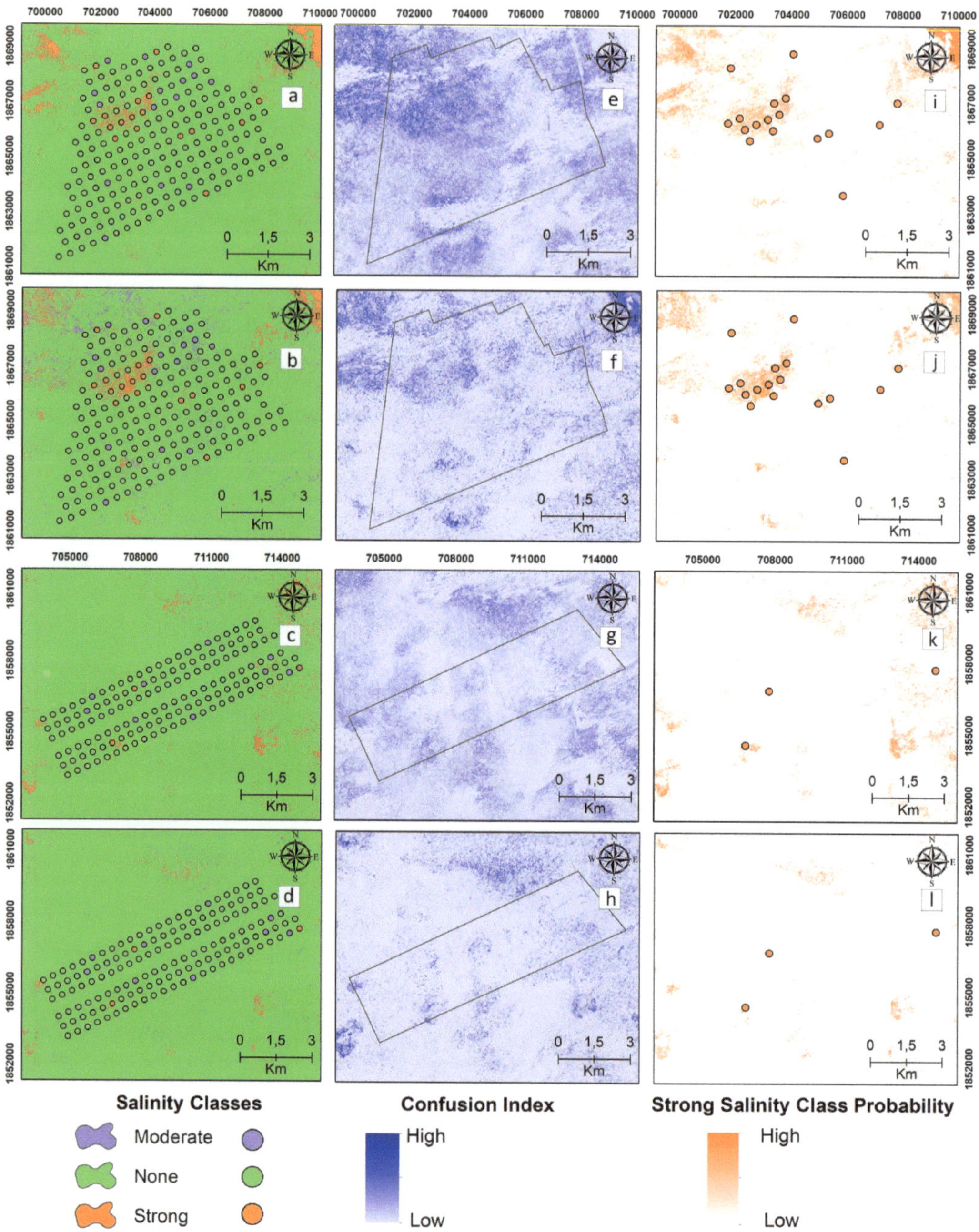


**Figure 5.** Digital maps of salinity classes (0–30 cm). Generated by applying the MNLR (**a**) and GBM (**b**) models for the reference area as well as the MNLR (**c**) and GBM (**d**) models for the target area. For the reference area, confusion index maps for MNLR (**e**) and GBM (**f**) as well as MNLR (**g**) and GBM (**h**) for the target area. Probability map of the "Strong" salinity class in the reference area obtained by applying the MNLR model (**i**) and the GBM model (**j**) as well as the MNLR model (**k**) and the GBM model (**l**) for the target area.

Digital maps of the salinity classes of the 30–60 cm samples are presented in Figure 6 for the reference and target areas by applying MNLR and GBM. Unlike the 0–30 cm maps, the presence of the "strong" salinity class increased in the northwest of the reference area (Figure 6a,b). The "strong" class was mapped with a higher probability by the GBM model in the reference area (Figure 6j). In the 30–60 cm samples, the MNLR and GBM models do not seem to be effective in spatially predicting the "strong" salinity class for the target area. The CI values, which show the difference between the probability values of the most probable and 2nd most probable classes, are higher at 30–60 cm compared to the surface samples (0–30 cm) (Figure 6e–h).

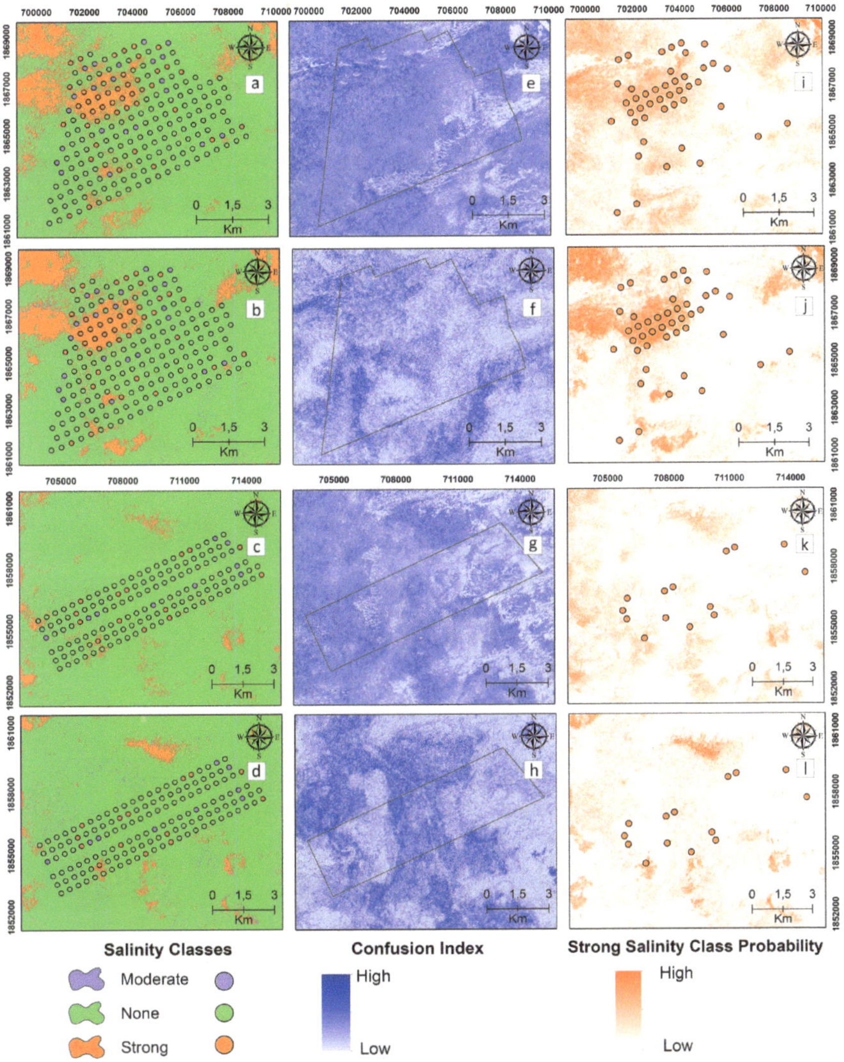


**Figure 6.** Digital maps of salinity classes (30–60 cm). Generated by applying the MNLR (**a**) and GBM (**b**) models for the reference area as well as the MNLR (**c**) and GBM (**d**) models for the target area. For the reference area, confusion index maps for MNLR (**e**) and GBM (**f**) as well as MNLR (**g**) and GBM (**h**) for the target area. Probability map of the "Strong" salinity class in the reference area obtained by applying the MNLR model (**i**) and the GBM model (**j**) as well as the MNLR model (**k**) and the GBM model (**l**) for the target area.

Digital maps of the salinity classes of the 60–90 cm samples are presented in Figure 7 for the reference and target areas by applying MNLR and GBM. Unlike the previous two depth maps, the presence of the "strong" salinity class increased northwest of the reference area at 60–90 cm, where the deepest sampling occurred (Figure 7a,b). In the 60–90 cm samples, the MNLR and GBM models were not effective in spatially predicting the "strong" salinity class for the target area. The CI values, which show the difference between the probability values of the most probable and 2nd most probable classes, are higher at 60–90 cm compared to the two depth maps (Figure 7e–h).

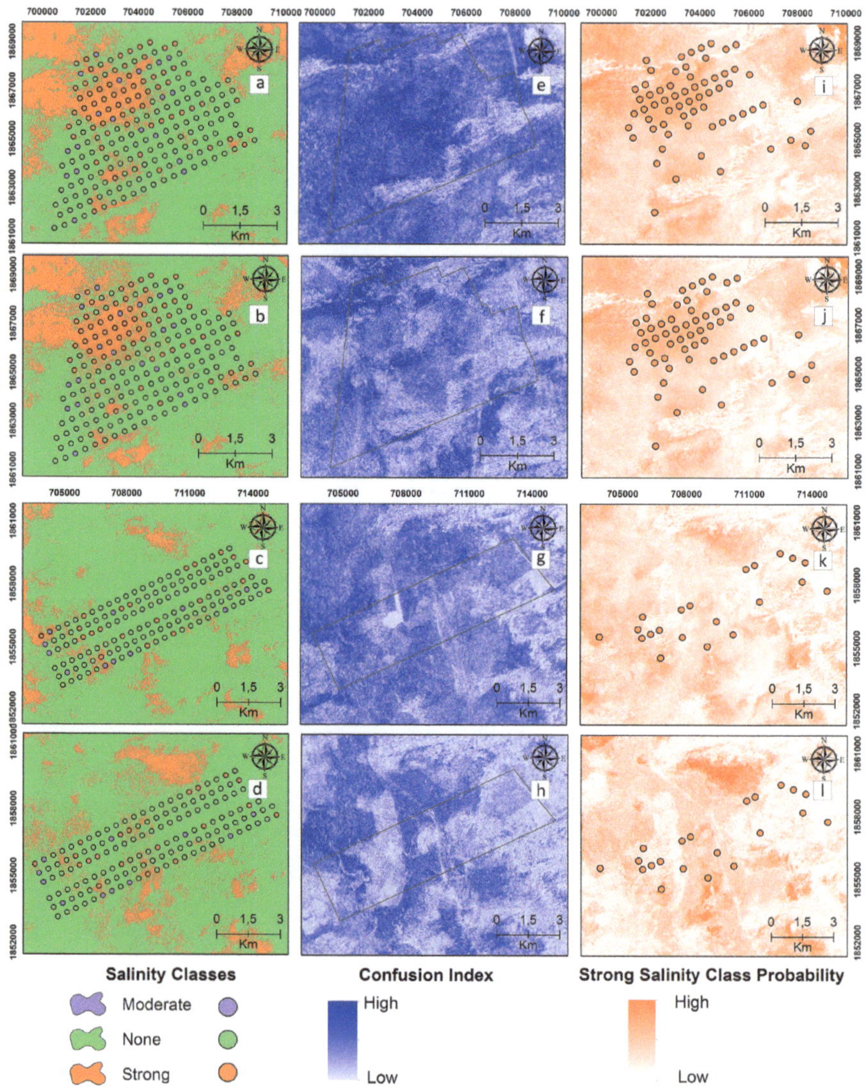


**Figure 7.** Digital maps of salinity classes (60–90 cm). Generated by applying the MNLR (**a**) and GBM (**b**) models for the reference area as well as the MNLR (**c**) and GBM (**d**) models for the target area. For the reference area, confusion index maps for MNLR (**e**) and GBM (**f**) as well as MNLR (**g**) and GBM (**h**) for the target area. Probability map of the "Strong" salinity class in the reference area obtained by applying the MNLR model (**i**) and the GBM model (**j**) as well as the MNLR model (**k**) and the GBM model (**l**) for the target area.

Considering the three different depths of soil salinity classes in the digital maps (Figures 5–7), the presence of a "strong" class increases with depth in the northwest of the reference area, both in the samples taken and in the predicted maps. The current salinity risk in the northwest region of the reference area was found to be high, and high-resolution (10 m) digital maps can play an effective role in defining the management zones for salinity.

### 3.5. Importance of Environmental Variables

Figure 8 shows the relative importance of the environmental variables used in modeling soil salinity classes at the three different depths. In both models, the salinity class of surface soils is determined by the indices produced from optical-based satellite images (Figure 8a,d). In the MNLR model, the relative importance of SAR data increased in the modeling of 30–60 and 60–90 samples (Figure 8b,c). In the GBM model, the increase is not as noticeable as in MNLR (Figure 8e,f). In arid areas, salinity and soil-based indices seem to be relatively more important for the models than vegetation indices.

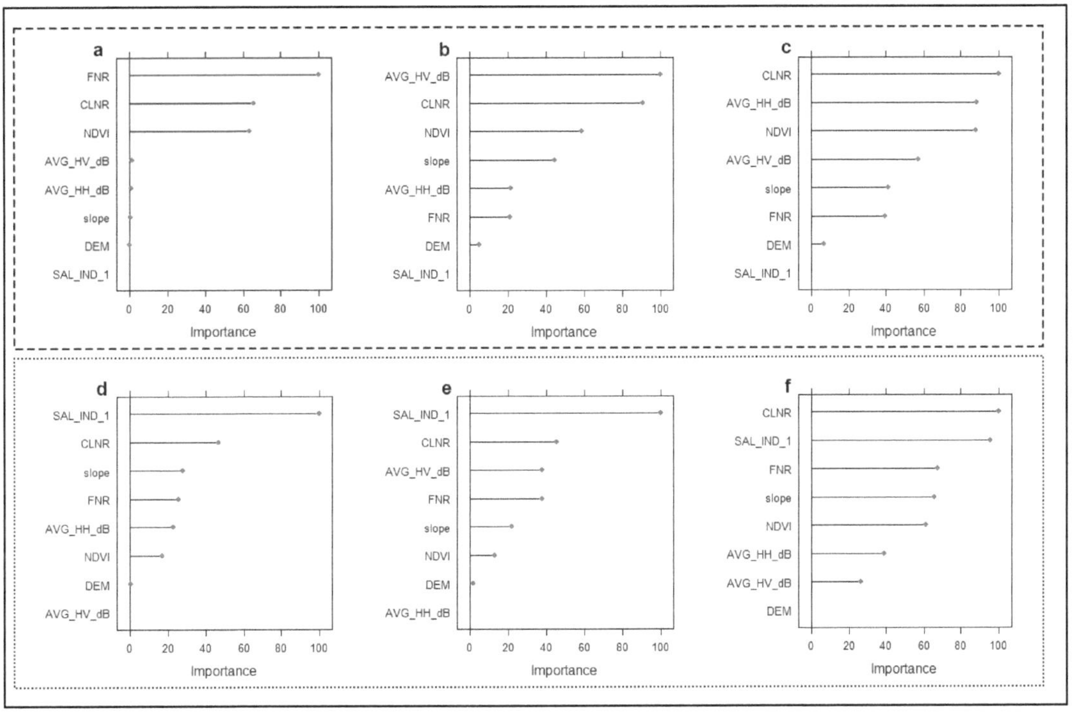


**Figure 8.** Importance of environmental variables in predicting soil salinity classes using different algorithms. 0–30 cm (**a**), 30–60 cm (**b**), and 60–90 cm (**c**) for MNLR (Multinomial Logistic Regression). 0–30 cm (**d**), 30–60 cm (**e**), and 60–90 cm (**f**) for GBM (Gradient Boosting Machine).

## 4. Discussion

The most accurate spatial determination and subsequent monitoring of soil salinity are crucial for sustainable agriculture and food security [3,6]. Up-to-date, reliable, and accurate assessments of soil salinity are important for land use planners and managers. In our study, a three-class estimation process was carried out, and Tau index values were found to be very similar to the Tau value of 0.74 reported by Omuto et al. [57] in Northwestern Sudan. Differences in the relative overall accuracy or Tau index values in the literature comparisons of classification results may be due to the number of salinity classes. For example, Kumar et al. [88] mapped the salt-affected areas with the logistic regression model in their study in the part of the Indo–Gangetic plain affected by soil salinity, with an overall accuracy of 81%.

Since soil salinity is a dynamic environmental problem, it is critical to monitor temporal and spatial changes [57]. Considering the temporal variability of soil salinity, the use of advanced sensor technologies for precision agriculture applications in the future [89], both in the study area and in similar regions, can be used to optimize the growing conditions [90]. Especially in arid regions where irrigated agriculture is practiced, Zhu et al. [91] emphasized the importance of creating soil salinity maps in terms of changes in soil salinity during the irrigated or non-irrigated period to understand the main mechanisms causing soil salinity.

Two ML algorithms that make predictions based on relatively different mathematical calculations presented the results of a comparative study in an arid region of Sudan to assess the transferability of salinity classes using selected covariates. The majority of misclassified and unpredicted cases were found within the moderate salinity class in both the reference and target areas. However, the primary objective of this study is not centered around maximizing the predictive accuracy of the models; rather, it aims to provide initial insights into the transferability of soil salinity models with relevance to agronomic applications. Although the reference and target areas are characterized by very similar climates and topographies [92], there may be concerns about quantifying the degree of similarity between them based on more quantitative results just before model transfer. Therefore, it is recommended for future studies to present comparative results of different mathematical bases such as the Gower similarity index [93] and dissimilarity index [94]. Enhancing the predictive accuracy of transferability related to soil salinity can involve the exploration of specific geographical stratifications [53], such as physiography or topography (slope-aspect categories), as well as the consideration of land use factors.

The low variation in elevation and the homogeneity of the climate in the study area may have caused the elevation digital covariate to be ineffective in the modeling. The effectiveness of optical satellite-based salinity indices [27,32,35,95] and SAR data are consistent with the literature [5,29]. Nevertheless, our effort has been to leverage remote sensing data for the purpose of transferring salinity class models to research areas characterized by quantified similarity analysis. The study outcomes have revealed the substantial transferability of satellite-based radar and optical environmental variables within an arid region, substantiating their potential for generating beneficial outputs. For transferability of soil salinity levels in arid regions using ML algorithms, the PlanetScope satellite [96] can offer important opportunities to capture the spatial variability of salinity [97].

The ultimate aim is to produce useful insights as a result of the models. Among the salinity class maps resulting from the study, special attention should be paid to the spatial distribution of the "strong" class. Our study includes not only defining the problem but also searching for solutions. In this regard, Soil-Improving Cropping Systems, which aim to prevent, mitigate, or ameliorate the adverse effects of soil salinity and improve associated soil functions and ecosystem services related to agricultural production, should be given importance [98]. Sugarcane, which is an important crop in Sudan [44,99], is a very sensitive plant to salinity in terms of cultivation [8,100]. In this study, the cultivation of this deep-rooted plant in areas with increasing "strong" class probability and especially in areas where the danger of salinity increases with depth may experience negative effects. It is important to select plants with relatively high resistance in saline environments [101,102]. As a matter of fact, the study area right next to the Atbara River should be subjected to evaluations such as irrigated land classification [103,104] in a wider perspective for its effective use in irrigated agriculture activities.

Future work should center on assessing the temporal and spatial transferability of remote sensing, including its capability to detect fluctuations within soil salinity classes. While the determination of large-scale soil limitations with DSM methodology is an important objective, ML models are increasingly being used for this purpose. However, it is well known that tree-based algorithms are sensitive to extrapolation, i.e., transferability [105]. In tree-based learners (GBM in our study), any split threshold within the nodes for the "y" (dependent) variable is limited by the minimum and maximum value range of the particular feature in the training dataset [105]. Therefore, when the algorithm encounters a value of "y" that is outside the bounds of the training dataset, it applies the closest corresponding dependent variable value from the training dataset in the mathematical prediction process. Therefore, it will be more difficult to extrapolate regression-based estimates of transferability for tree-based learners. Soil scientists skilled in the mathematics of the models are important at the point of applicability of ML to soil data [106]. It enables more applicable methodologies for transferability by harmonizing the EC values into salinity classes (categorical data) that adhere to international standards for continuous data types.

Furthermore, future studies should focus on measuring the transferability risk associated with MLAs for soil salinity prediction while also focusing on research that will help assess the reliability of their predictions [107]. These studies can reveal valuable information regarding the integration of ML model predictions into the decision/support system [108]. It can be recommended in research for predictive models to provide information at the reconnaissance scale [109].

## 5. Conclusions

In this study, we integrated indices generated from long-term optical Sentinel data and PALSAR-2 radar imagery through GEE for digital mapping of high-resolution regional-scale soil salinity classes in Sudan. We also addressed the transferability of ML-based soil salinity classes in arid areas and used MESS before transferring from the reference area to the target area. This paper presents transfer learning techniques for fast and accurate soil salinity mapping using open-access digital data and machine learning algorithms. In this process, soil scientists should be well-skilled in the mathematical basis of algorithms for integrating soil data to be transferred by modeling into the ML. The spatial information on soil salinity generated in this study can provide remarkable insights into decision-making processes that are compatible with the growing need for soil information for future sustainable development goals.

**Author Contributions:** Conceptualization, F.K. and M.M.S.; methodology, F.K. and M.M.S.; software, F.K.; formal analysis, F.K.; investigation, M.M.S. and M.A.E.; resources, M.M.S. and M.A.E.; data curation, F.K. and M.M.S.; writing—original draft preparation, F.K. and M.M.S.; writing—review and editing, M.M.S., F.K., M.A.E., L.B. and R.F.; visualization, F.K. All authors have read and agreed to the published version of the manuscript.

**Funding:** This research received no external funding.

**Data Availability Statement:** The data presented in this study are available on request from the first author.

**Conflicts of Interest:** The authors declare no conflict of interest.

## Appendix A

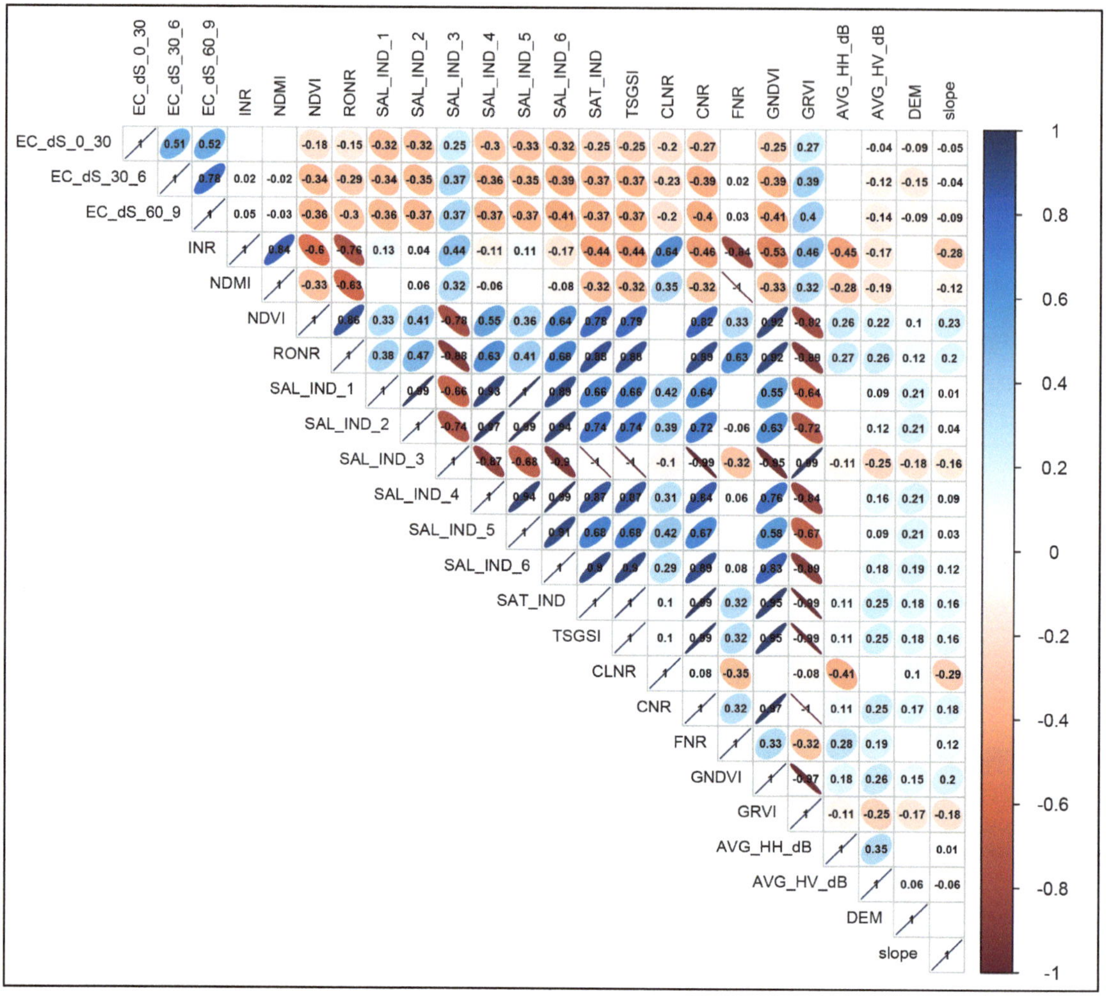

**Figure A1.** Spearman correlation analysis of environmental variables. Blue color indicates a positive correlation, while red color exhibits a negative correlation at $p < 0.05$. Empty boxes indicate no correlation between parameters.

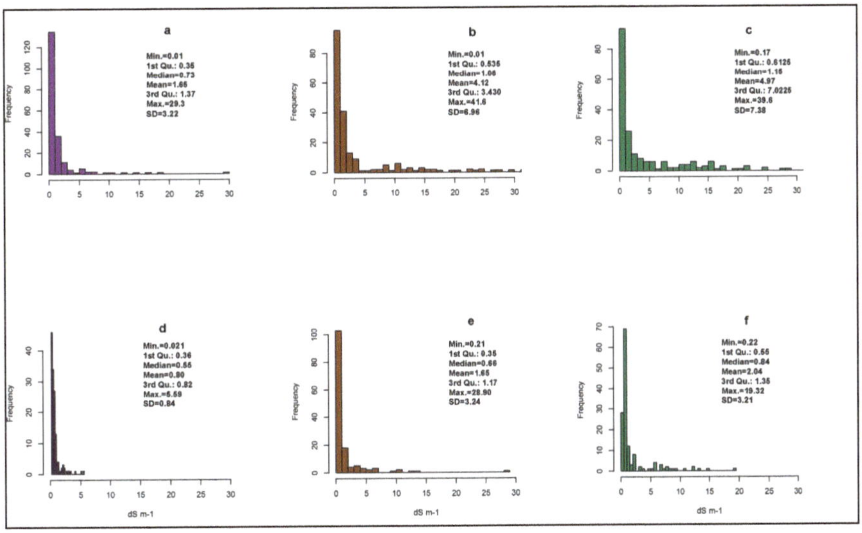


**Figure A2.** Histograms of measured EC at reference and target sites and some descriptive statistics. 0–30 cm (**a**), 30–60 cm (**b**), and 60–90 cm (**c**) for the reference area. 0–30 cm (**d**), 30–60 cm (**e**), and 60–90 cm (**f**) for the target area.

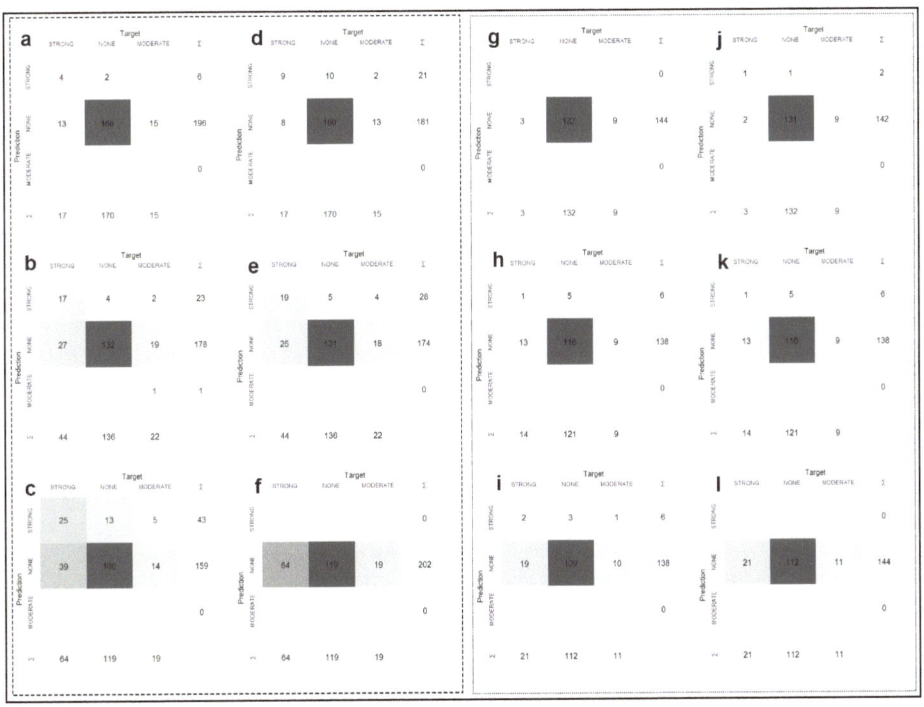

**Figure A3.** Confusion matrix results of the classification performances of the models for reference and target areas. Confusion matrices of 0–30 cm (**a**), 30–60 cm (**b**), 60–90 cm (**c**) maps produced by applying the MNLR model for reference area and 0–30 cm (**d**), 30–60 cm (**e**), 60–90 cm (**f**) maps produced by applying the GBM model. Confusion matrices of 0–30 cm (**g**), 30–60 cm (**h**), 60–90 cm (**i**) maps produced by applying the MNLR model and 0–30 cm (**j**), 30–60 cm (**k**), 60–90 cm (**l**) maps produced by applying the GBM model for the target area.

## References

1. FAO. GSASmap v1.0, Global Map of Salt-Affected Soils. Available online: https://www.fao.org/3/cb7247en/cb7247en.pdf (accessed on 7 September 2022).
2. Kaya, F.; Schillaci, C.; Keshavarzi, A.; Basayigit, L. Predictive Mapping of Electrical Conductivity and Assessment of Soil Salinity in a Western Türkiye Alluvial Plain. *Land* **2022**, *11*, 2148. [CrossRef]
3. Negacz, K.; Vellinga, P.; Barrett-Lennard, E.; Choukr-Allah, R.; Elzenga, T. (Eds.) *Future of Sustainable Agriculture in Saline Environments*; CRC Press: Boca Raton, FL, USA, 2021.
4. Hazbavi, Z.; Zabihi Silabi, M. Innovations of the 21st Century in the Management of Iranian Salt-Affected Lands. In *Future of Sustainable Agriculture in Saline Environments*; Negacz, K., Vellinga, P., Barrett-Lennard, E., Choukr-Allah, R., Elzenga, T., Eds.; CRC Press: Boca Raton, FL, USA, 2021; pp. 147–170.
5. Zdruli, P.; Zucca, C. *Restoring Land and Soil Health to Ensure Sustainable and Resilient Agriculture in the Near East and North Africa Region—State of Land and Water Resources for Food and Agriculture Thematic Paper*; Zdruli, P., Zucca, C., Eds.; FAO: Cairo, Egypt, 2023.
6. Negacz, K.; Malek, Ž.; de Vos, A.; Vellinga, P. Saline Soils Worldwide: Identifying the Most Promising Areas for Saline Agriculture. *J. Arid Environ.* **2022**, *203*, 104775. [CrossRef]
7. Mukhopadhyay, R.; Sarkar, B.; Jat, H.S.; Sharma, P.C.; Bolan, N.S. Soil Salinity under Climate Change: Challenges for Sustainable Agriculture and Food Security. *J. Environ. Manag.* **2021**, *280*, 111736. [CrossRef] [PubMed]
8. Singh, A. Soil Salinity: A Global Threat to Sustainable Development. *Soil Use Manag.* **2022**, *38*, 39–67. [CrossRef]
9. Singh, A. Soil Salinization Management for Sustainable Development: A Review. *J. Environ. Manag.* **2021**, *277*, 111383. [CrossRef]
10. Basak, N.; Rai, A.K.; Barman, A.; Mandal, S.; Sundha, P.; Bedwal, S.; Kumar, S.; Yadav, R.K.; Sharma, P.C. Salt Affected Soils: Global Perspectives. In *Soil Health and Environmental Sustainability*; Shit, P.K., Adhikary, P.P., Bhunia, G.S., Sengupta, D., Eds.; Springer Science and Business Media Deutschland GmbH: Berlin/Heidelberg, Germany, 2022; pp. 107–129.
11. Devkota, K.P.; Devkota, M.; Rezaei, M.; Oosterbaan, R. Managing Salinity for Sustainable Agricultural Production in Salt-Affected Soils of Irrigated Drylands. *Agric. Syst.* **2022**, *198*, 103390. [CrossRef]
12. Cf, O.D.D.S. *United Nations General Assembly Transforming Our World: The 2030 Agenda for Sustainable Development*; United Nations: New York, NY, USA, 2015.
13. Tomaz, A.; Palma, P.; Alvarenga, P.; Gonçalves, M.C. Soil Salinity Risk in a Climate Change Scenario and Its Effect on Crop Yield. In *Climate Change and Soil Interactions*; Elsevier: Amsterdam, The Netherlands, 2020; pp. 351–396. [CrossRef]
14. Okur, B.; Örçen, N. Soil Salinization and Climate Change. In *Climate Change and Soil Interactions*; Elsevier: Amsterdam, The Netherlands, 2020; pp. 331–350. [CrossRef]
15. Keshavarzi, A.; del Árbol, M.Á.S.; Kaya, F.; Gyasi-Agyei, Y.; Rodrigo-Comino, J. Digital Mapping of Soil Texture Classes for Efficient Land Management in the Piedmont Plain of Iran. *Soil Use Manag.* **2022**, *38*, 1705–1735. [CrossRef]
16. Tziolas, N.; Tsakiridis, N.; Ogen, Y.; Kalopesa, E.; Ben-Dor, E.; Theocharis, J.; Zalidis, G. An Integrated Methodology Using Open Soil Spectral Libraries and Earth Observation Data for Soil Organic Carbon Estimations in Support of Soil-Related SDGs. *Remote Sens. Environ.* **2020**, *244*, 111793. [CrossRef]
17. Wang, J.; Zhen, J.; Hu, W.; Chen, S.; Lizaga, I.; Zeraatpisheh, M.; Yang, X. Remote Sensing of Soil Degradation: Progress and Perspective. *Int. Soil Water Conserv. Res.* **2023**, *11*, 429–454. [CrossRef]
18. Bennett, J.M.L.; Roberton, S.D.; Ghahramani, A.; McKenzie, D.C. Operationalising Soil Security by Making Soil Data Useful: Digital Soil Mapping, Assessment and Return-on-Investment. *Soil Secur.* **2021**, *4*, 100010. [CrossRef]
19. Malone, B.; Arrouays, D.; Poggio, L.; Minasny, B.; McBratney, A. Digital Soil Mapping: Evolution, Current State and Future Directions of the Science. *Ref. Modul. Earth Syst. Environ. Sci.* **2022**, *4*, 684–695. [CrossRef]
20. Miller, B.A. Digital Soil Mapping and Pedometrics. In *International Encyclopedia of Geography: People, the Earth, Environment and Technology*; John Wiley & Sons, Ltd.: Oxford, UK, 2017; pp. 1–8.
21. European Commission. *Soil Monitoring and Resilience (Soil Monitoring Law)*; European Commission: Brussels, Belgium, 2023.
22. Rezaei, M.; Mousavi, S.R.; Rahmani, A.; Zeraatpisheh, M.; Rahmati, M.; Pakparvar, M.; Jahandideh Mahjenabadi, V.A.; Seuntjens, P.; Cornelis, W. Incorporating Machine Learning Models and Remote Sensing to Assess the Spatial Distribution of Saturated Hydraulic Conductivity in a Light-Textured Soil. *Comput. Electron. Agric.* **2023**, *209*, 107821. [CrossRef]
23. Mousavi, S.R.; Sarmadian, F.; Omid, M.; Bogaert, P. Three-Dimensional Mapping of Soil Organic Carbon Using Soil and Environmental Covariates in an Arid and Semi-Arid Region of Iran. *Measurement* **2022**, *201*, 111706. [CrossRef]
24. Sahbeni, G.; Ngabire, M.; Musyimi, P.K.; Székely, B. Challenges and Opportunities in Remote Sensing for Soil Salinization Mapping and Monitoring: A Review. *Remote Sens.* **2023**, *15*, 2540. [CrossRef]
25. Mohamed, S.A.; Metwaly, M.M.; Metwalli, M.R.; AbdelRahman, M.A.E.; Badreldin, N. Integrating Active and Passive Remote Sensing Data for Mapping Soil Salinity Using Machine Learning and Feature Selection Approaches in Arid Regions. *Remote Sens.* **2023**, *15*, 1751. [CrossRef]
26. Mousavi, S.R.; Sarmadian, F.; Omid, M.; Bogaert, P. Digital Modeling of Three-Dimensional Soil Salinity Variation Using Machine Learning Algorithms in Arid and Semi-Arid Lands of Qazvin Plain. *Iran. J. Soil Water Res.* **2021**, *52*, 1915–1929. [CrossRef]
27. Kaplan, G.; Gašparović, M.; Alqasemi, A.S.; Aldhaheri, A.; Abuelgasim, A.; Ibrahim, M. Soil Salinity Prediction Using Machine Learning and Sentinel—2 Remote Sensing Data in Hyper—Arid Areas. *Phys. Chem. Earth Parts A/B/C* **2023**, *130*, 103400. [CrossRef]
28. Merembayev, T.; Amirgaliyev, Y.; Saurov, S.; Wójcik, W. Soil Salinity Classification Using Machine Learning Algorithms and Radar Data in the Case from the South of Kazakhstan. *J. Ecol. Eng.* **2022**, *23*, 61–67. [CrossRef]

29. Ma, G.; Ding, J.; Han, L.; Zhang, Z.; Ran, S. Digital Mapping of Soil Salinization Based on Sentinel-1 and Sentinel-2 Data Combined with Machine Learning Algorithms. *Reg. Sustain.* **2021**, *2*, 177–188. [CrossRef]
30. He, Y.; Zhang, Z.; Xiang, R.; Ding, B.; Du, R.; Yin, H.; Chen, Y.; Ba, Y. Monitoring Salinity in Bare Soil Based on Sentinel-1/2 Image Fusion and Machine Learning. *Infrared Phys. Technol.* **2023**, *131*, 104656. [CrossRef]
31. Wang, J.; Wu, F.; Shang, J.; Zhou, Q.; Ahmad, I.; Zhou, G. Saline Soil Moisture Mapping Using Sentinel-1A Synthetic Aperture Radar Data and Machine Learning Algorithms in Humid Region of China's East Coast. *Catena (Amst.)* **2022**, *213*, 106189. [CrossRef]
32. Kaya, F.; Ferhatoglu, C.; Turgut, Y.Ş.; Başayiğit, L. State of Art Approaches, Insights, and Challenges for Digital Mapping of Electrical Conductivity as a Dynamic Soil Property. In *Proceedings of the 7th International Scientific Meeting as Soil Science Symposium on "Soil Science & Plant Nutrition"*; Kizilkaya, R., Gülser, C., Dengiz, O., Eds.; Federation of Eurasian Soil Science Societi: Samsun, Turkey, 2022; pp. 75–82.
33. Hassani, A.; Azapagic, A.; Shokri, N. Predicting Long-Term Dynamics of Soil Salinity and Sodicity on a Global Scale. *Proc. Natl. Acad. Sci.* **2020**, *117*, 33017–33027. [CrossRef] [PubMed]
34. Omuto, C.T.; Vargas, R.R.; El Mobarak, A.M.; Mohamed, N.; Viatkin, K.; Yigini, Y. *Mapping of Salt-Affected Soils: Technical Manual*, 1st ed.; FAO: Rome, Italy, 2020.
35. Avdan, U.; Kaplan, G.; Küçük Matcı, D.; Yiğit Avdan, Z.; Erdem, F.; Tuğba Mızık, E.; Demirtaş, İ. Soil Salinity Prediction Models Constructed by Different Remote Sensors. *Phys. Chem. Earth Parts A/B/C* **2022**, *128*, 103230. [CrossRef]
36. Foronda, D.A.; Colinet, G. Prediction of Soil Salinity/Sodicity and Salt-Affected Soil Classes from Salt Soluble Ions Using Machine Learning Algorithms. *Soil Syst.* **2023**, *7*, 47. [CrossRef]
37. Guo, Y.; Shi, Z.; Li, H.Y.; Triantafilis, J. Application of Digital Soil Mapping Methods for Identifying Salinity Management Classes Based on a Study on Coastal Central China. *Soil Use Manag.* **2013**, *29*, 445–456. [CrossRef]
38. Konyushkova, M.; Krenke, A.; Khasankhanova, G.; Mamutov, N.; Statov, V.; Kontoboytseva, A.; Pankova, Y. An Approach to Monitoring of Salt-Affected Croplands Using Remote Sensing Data. In *Future of Sustainable Agriculture in Saline Environments*; Negacz, K., Vellinga, P., Barrett-Lennard, E., Choukr-Allah, R., Elzenga, T., Eds.; CRC Press: Boca Raton, FL, USA, 2021; pp. 171–180.
39. Golestani, M.; Mosleh Ghahfarokhi, Z.; Esfandiarpour-Boroujeni, I.; Shirani, H. Evaluating the Spatiotemporal Variations of Soil Salinity in Sirjan Playa, Iran Using Sentinel-2A and Landsat-8 OLI Imagery. *Catena (Amst.)* **2023**, *231*, 107375. [CrossRef]
40. Kabiraj, S.; Jayanthi, M.; Vijayakumar, S.; Duraisamy, M. Comparative Assessment of Satellite Images Spectral Characteristics in Identifying the Different Levels of Soil Salinization Using Machine Learning Techniques in Google Earth Engine. *Earth Sci. Inform.* **2022**, *15*, 2275–2288. [CrossRef]
41. Lekka, C.; Petropoulos, G.P.; Triantakonstantis, D.; Detsikas, S.E.; Chalkias, C. Exploring the Spatial Patterns of Soil Salinity and Organic Carbon in Agricultural Areas of Lesvos Island, Greece, Using Geoinformation Technologies. *Environ. Monit. Assess* **2023**, *195*, 391. [CrossRef]
42. Nenkam, A.M.; Wadoux, A.M.J.C.; Minasny, B.; McBratney, A.B.; Traore, P.C.S.; Falconnier, G.N.; Whitbread, A.M. Using Homosoils for Quantitative Extrapolation of Soil Mapping Models. *Eur. J. Soil Sci.* **2022**, *73*, e13285. [CrossRef]
43. Ibrahim, T.S.; Workneh, T.S. Identification of Technical Factors That Influence Sugar Productivity of Factories in Sudan. *Afr. J. Sci. Technol. Innov. Dev.* **2022**, *14*, 234–246. [CrossRef]
44. Mahgoub, F. *Current Status of Agriculture and Future Challenges in Sudan*; Nordiska Afrikainstitutet: Uppsala, Sweden, 2014; ISBN 9789171067487.
45. Liu, L.; Ji, M.; Buchroithner, M. Transfer Learning for Soil Spectroscopy Based on Convolutional Neural Networks and Its Application in Soil Clay Content Mapping Using Hyperspectral Imagery. *Sensors* **2018**, *18*, 3169. [CrossRef] [PubMed]
46. Padarian, J.; Minasny, B.; McBratney, A.B. Transfer Learning to Localise a Continental Soil Vis-NIR Calibration Model. *Geoderma* **2019**, *340*, 279–288. [CrossRef]
47. Francos, N.; Heller-Pearlshtien, D.; Dematte, J.A.M.; Van Wesemael, B.; Milewski, R.; Chabrillat, S.; Tziolas, N.; Sanz Diaz, A.; Yague Ballester, M.J.; Gholizadeh, A.; et al. A Spectral Transfer Function to Harmonize Existing Soil Spectral Libraries Generated by Different Protocols. *Appl. Environ. Soil Sci.* **2023**, *2023*, 4155390. [CrossRef]
48. Lemercier, B.; Lacoste, M.; Loum, M.; Walter, C. Extrapolation at Regional Scale of Local Soil Knowledge Using Boosted Classification Trees: A Two-Step Approach. *Geoderma* **2012**, *171–172*, 75–84. [CrossRef]
49. Du, L.; McCarty, G.W.; Li, X.; Rabenhorst, M.C.; Wang, Q.; Lee, S.; Hinson, A.L.; Zou, Z. Spatial Extrapolation of Topographic Models for Mapping Soil Organic Carbon Using Local Samples. *Geoderma* **2021**, *404*, 115290. [CrossRef]
50. Neyestani, M.; Sarmadian, F.; Jafari, A.; Keshavarzi, A.; Sharififar, A. Digital Mapping of Soil Classes Using Spatial Extrapolation with Imbalanced Data. *Geoderma Regional.* **2021**, *26*, e00422. [CrossRef]
51. Abbaszadeh Afshar, F.; Ayoubi, S.; Jafari, A. The Extrapolation of Soil Great Groups Using Multinomial Logistic Regression at Regional Scale in Arid Regions of Iran. *Geoderma* **2018**, *315*, 36–48. [CrossRef]
52. Broeg, T.; Blaschek, M.; Seitz, S.; Taghizadeh-Mehrjardi, R.; Zepp, S.; Scholten, T. Transferability of Covariates to Predict Soil Organic Carbon in Cropland Soils. *Remote Sens.* **2023**, *15*, 876. [CrossRef]
53. Mirzaeitalarposhti, R.; Shafizadeh-Moghadam, H.; Taghizadeh-Mehrjardi, R.; Demyan, M.S. Digital Soil Texture Mapping and Spatial Transferability of Machine Learning Models Using Sentinel-1, Sentinel-2, and Terrain-Derived Covariates. *Remote Sens.* **2022**, *14*, 5909. [CrossRef]

54. Soil Survey Staff. *Keys to Soil Taxonomy*, 12th ed.; USDA-Natural Resources Conservation Service: Washington, DC, USA, 2014.
55. Ditzler, C.; Scheffe, K.; Monger, H.C. (Eds.) *Soil Science Division Staff Soil Survey Manual*; USDA Handbook 18; Government Printing Office: Washington, DC, USA, 2017.
56. FAO. *Standard Operating Procedure for Saturated Soil Paste Extract*; FAO: Rome, Italy, 2021.
57. Omuto, C.T.; Vargas, R.R.; Elmobarak, A.A.; Mapeshoane, B.E.; Koetlisi, K.A.; Ahmadzai, H.; Abdalla Mohamed, N. Digital Soil Assessment in Support of a Soil Information System for Monitoring Salinization and Sodification in Agricultural Areas. *Land Degrad. Dev.* **2022**, *33*, 1204–1218. [CrossRef]
58. Handbook, S.U.; Tools, E. *Sentinel-2 User Handbook, Version 1.2.*; European Space Agency (ESA): Paris, France, 2015.
59. ALOS PALSAR Dataset: © JAXA/METI ALOS PALSAR L1.0 2007. Available online: https://asf.alaska.edu/ (accessed on 12 February 2023).
60. Gorelick, N.; Hancher, M.; Dixon, M.; Ilyushchenko, S.; Thau, D.; Moore, R. Google Earth Engine: Planetary-Scale Geospatial Analysis for Everyone. *Remote Sens. Environ.* **2017**, *202*, 18–27. [CrossRef]
61. Japan Aerospace Exploration Agency (JAXA); Earth Observation Research Center (EORC). *Global 25 m Resolution PALSAR-2 Mosaic (Ver.2.1.2).* Dataset Description; ALOS. Available online: https://www.eorc.jaxa.jp/ALOS/en/dataset/fnf_e.htm (accessed on 12 February 2023).
62. Shimada, M.; Itoh, T.; Motooka, T.; Watanabe, M.; Shiraishi, T.; Thapa, R.; Lucas, R. Global PALSAR-2/PALSAR Yearly Mosaic, Version 1. Available online: https://developers.google.com/earth-engine/datasets/catalog/JAXA_ALOS_PALSAR_YEARLY_SAR (accessed on 12 February 2023).
63. Franceschini, G.; Ali, M. *Introductory Course to Google Earth Engine*; FAO: Rome, Italy, 2022.
64. ESRI. ArcGIS 2021. Available online: https://www.arcgis.com/index.html (accessed on 12 February 2023).
65. NASA JPL NASADEM Merged DEM Global 1 Arc Second V001 [Dataset]. NASA EOSDIS Land Processes DAAC. Available online: https://cmr.earthdata.nasa.gov/search/concepts/C1546314043-LPDAAC_ECS.html (accessed on 21 March 2023).
66. Brown, K.S.; Libohova, Z.; Boettinger, J. Digital Soil Mapping. In *Soil Survey Manual*; USDA Handbook 18; Ditzler, C., Scheffe, K., Monger, H.C., Eds.; Government Printing Office: Washington, DC, USA, 2017.
67. Kaya, F.; Başayiğit, L. Using Machine Learning Algorithms to Mapping of the Soil Macronutrient Elements Variability with Digital Environmental Data in an Alluvial Plain. In *Artificial Intelligence and Smart Agriculture Applications*; Kose, U., Prasath, V.B.S., Mondal, M.R.H., Podder, P., Subrato, B., Eds.; Auerbach Publications: Boca Raton, FL, USA, 2022; pp. 107–136.
68. Kuhn, M. Building Predictive Models in R Using the Caret Package. *J. Stat. Softw.* **2008**, *28*, 1–26. [CrossRef]
69. Liu, F.; Wu, H.; Zhao, Y.; Li, D.; Yang, J.L.; Song, X.; Shi, Z.; Zhu, A.X.; Zhang, G.L. Mapping High Resolution National Soil Information Grids of China. *Sci. Bull.* **2022**, *67*, 328–340. [CrossRef] [PubMed]
70. Van der Westhuizen, S.; Heuvelink, G.B.M.; Hofmeyr, D.P. Multivariate Random Forest for Digital Soil Mapping. *Geoderma* **2023**, *431*, 116365. [CrossRef]
71. Hosmer, D.W.; Lemeshow, S.; Sturdivant, R.X. Logistic Regression Models for Multinomial and Ordinal Outcomes. In *Applied Logistic Regression*; John Wiley & Sons, Ltd.: Hoboken, NJ, USA, 2013; pp. 269–311.
72. Venables, W.N.; Ripley, B.D. *Modern Applied Statistics with S.*, 4th ed.; Springer: New York, NY, USA, 2002.
73. Friedman, J.H. Stochastic Gradient Boosting. *Comput. Stat. Data Anal.* **2002**, *38*, 367–378. [CrossRef]
74. Greenwell, B.; Boehmke, B.; Cunningham, J.; Developers, G.B.M. Gbm: Generalized Boosted Regression Models. 2022. Available online: https://cran.r-project.org/web/packages/gbm/index.html (accessed on 15 February 2023).
75. R Core Team. R: A Language and Environment for Statistical Computing. R Foundation for Statistical Computing. Available online: https://www.r-project.org/index.html (accessed on 15 February 2023).
76. Kaya, F.; Başayiğit, L. Spatial Prediction and Digital Mapping of Soil Texture Classes in a Floodplain Using Multinomial Logistic Regression. In *Intelligent and Fuzzy Techniques for Emerging Conditions and Digital Transformation*; Kahraman, C., Cebi, S., Onar Cevik, S., Oztaysi, B., Tolga, A.C., Sari, I.U., Eds.; Springer International Publishing: Berlin/Heidelberg, Germany, 2022; pp. 463–473.
77. Friedman, J.H. Greedy Function Approximation: A Gradient Boosting Machine. In *The Annals of Statistic*; JSTOR: New York, NY, USA, 2001; p. 29. [CrossRef]
78. Estevez, V.; Beucher, A.; Mattback, S.; Boman, A.; Auri, J.; Bjork, K.-M.; Osterholm, P. Machine Learning Techniques for Acid Sulfate Soil Mapping in southeastern Finland. *Geoderma* **2022**, *406*, 115446. [CrossRef]
79. RStudio Team. *RStudio: Integrated Development for R*; PBC: Boston, MA, USA, 2023.
80. Hijmans, R.J.; Phillips, S.; Leathwick, J.; Elith, J. Dismo: Species Distribution Modeling; 2022. Available online: https://cran.r-project.org/web/packages/dismo/index.html (accessed on 15 February 2023).
81. Camera, C.; Zomeni, Z.; Noller, J.S.; Zissimos, A.M.; Christoforou, I.C.; Bruggeman, A. A High Resolution Map of Soil Types and Physical Properties for Cyprus: A Digital Soil Mapping Optimization. *Geoderma* **2017**, *285*, 35–49. [CrossRef]
82. Silva, B.P.C.; Silva, M.L.N.; Avalos, F.A.P.; de Menezes, M.D.; Curi, N. Digital Soil Mapping Including Additional Point Sampling in Posses Ecosystem Services Pilot Watershed, Southeastern Brazil. *Sci. Rep.* **2019**, *9*, 13763. [CrossRef]
83. Olsen, L.R.; Zachariae, H.B. Cvms: Cross-Validation for Model Selection 2023. Available online: https://cran.r-project.org/web/packages/cvms/cvms.pdf (accessed on 15 February 2023).
84. Rossiter, D.G.; Zeng, R.; Zhang, G.L. Accounting for Taxonomic Distance in Accuracy Assessment of Soil Class Predictions. *Geoderma* **2017**, *292*, 118–127. [CrossRef]

85. Beaudette, D.; Roudier, P.; Brown, A. Aqp: Algorithms for Quantitative Pedology 2022. Available online: https://cran.r-project.org/web/packages/aqp/aqp.pdf (accessed on 15 February 2023).
86. Burrough, P.A.; Van Gaans, P.F.M.; Hootsmans, R. Continuous Classification in Soil Survey: Spatial Correlation, Confusion and Boundaries. *Geoderma* **1997**, *77*, 115–135. [CrossRef]
87. Flynn, T.; Rozanov, A.; Ellis, F.; de Clercq, W.; Clarke, C. Farm-Scale Digital Soil Mapping of Soil Classes in South Africa. *S. Afr. J. Plant Soil* **2022**, *39*, 175–186. [CrossRef]
88. Kumar, N.; Singh, S.K.; Reddy, G.P.O.; Naitam, R.K. Developing Logistic Regression Models to Identify Salt-Affected Soils Using Optical Remote Sensing. In *Interdisciplinary Approaches to Information Systems and Software Engineering*; IGI Global: Hershey, PA, USA, 2019; pp. 233–256.
89. García, L.; Parra, L.; Jimenez, J.M.; Parra, M.; Lloret, J.; Mauri, P.V.; Lorenz, P. Deployment Strategies of Soil Monitoring WSN for Precision Agriculture Irrigation Scheduling in Rural Areas. *Sensors* **2021**, *21*, 1693. [CrossRef]
90. Yin, H.; Cao, Y.; Marelli, B.; Zeng, X.; Mason, A.J.; Cao, C. Soil Sensors and Plant Wearables for Smart and Precision Agriculture. *Adv. Mater.* **2021**, *33*, 2007764. [CrossRef] [PubMed]
91. Zhu, C.; Ding, J.; Zhang, Z.; Wang, J.; Chen, X.; Han, L.; Shi, H.; Wang, J. Soil Salinity Dynamics in Arid Oases during Irrigated and Non-Irrigated Seasons. In *Land Degradation & Development*; Wiley Online Library: Hoboken, NJ, USA, 2023. [CrossRef]
92. Khosravani, P.; Baghernejad, M.; Moosavi, A.A.; FallahShamsi, S.R. Digital Mapping to Extrapolate the Selected Soil Fertility Attributes in Calcareous Soils of a Semiarid Region in Iran. *J. Soils Sediments* **2023**, 1–23. [CrossRef]
93. Gower, J.C. A General Coefficient of Similarity and Some of Its Properties. *Biometrics* **1971**, *27*, 857. [CrossRef]
94. Meyer, H.; Pebesma, E. Predicting into Unknown Space? Estimating the Area of Applicability of Spatial Prediction Models. *Methods Ecol. Evol.* **2021**, *12*, 1620–1633. [CrossRef]
95. Ma, S.; He, B.; Xie, B.; Ge, X.; Han, L. Investigation of the Spatial and Temporal Variation of Soil Salinity Using Google Earth Engine: A Case Study at Werigan–Kuqa Oasis, West China. *Sci. Rep.* **2023**, *13*, 1–16. [CrossRef]
96. Planet Team Planet Application Program Interface. Available online: https://www.planet.com/explorer/ (accessed on 8 September 2022).
97. Tan, J.; Ding, J.; Han, L.; Ge, X.; Wang, X.; Wang, J.; Wang, R.; Qin, S.; Zhang, Z.; Li, Y. Exploring PlanetScope Satellite Capabilities for Soil Salinity Estimation and Mapping in Arid Regions Oases. *Remote Sens.* **2023**, *15*, 1066. [CrossRef]
98. Cuevas, J.; Daliakopoulos, I.N.; Del Moral, F.; Hueso, J.J.; Tsanis, I.K. A Review of Soil-Improving Cropping Systems for Soil Salinization. *Agronomy* **2019**, *9*, 295. [CrossRef]
99. Ibrahim, T.S.; Workneh, T.S. Development and Current Status of the Sugar Industry in Sudan. *Sugar Ind.* **2019**, *144*, 655–659. [CrossRef]
100. Kumar, R.; Dhansu, P.; Kulshreshtha, N.; Meena, M.R.; Kumaraswamy, M.H.; Appunu, C.; Chhabra, M.L.; Pandey, S.K. Identification of Salinity Tolerant Stable Sugarcane Cultivars Using AMMI, GGE and Some Other Stability Parameters under Multi Environments of Salinity Stress. *Sustainability* **2023**, *15*, 1119. [CrossRef]
101. Tedeschi, A. Irrigated Agriculture on Saline Soils: A Perspective. *Agronomy* **2020**, *10*, 1630. [CrossRef]
102. Tedeschi, A.; Schillaci, M.; Balestrini, R. Mitigating the Impact of Soil Salinity: Recent Developments and Future Strategies. *Ital. J. Agron* **2023**. [CrossRef]
103. FAO. *Land Evaluation for Irrigated Agriculture*; Food and Agriculture Organization of the United Nations: Rome, Italy, 1985; ISBN 92-5-102243-7.
104. Kau, A.S.; Gramlich, R.; Sewilam, H. Modelling Land Suitability to Evaluate the Potential for Irrigated Agriculture in the Nile Region in Sudan. *Sustain. Water Resour. Manag.* **2023**, *9*, 1–17. [CrossRef]
105. Malistov, A.; Trushin, A. Gradient Boosted Trees with Extrapolation. In Proceedings of the 18th IEEE International Conference on Machine Learning and Applications, ICMLA, Boca Raton, FL, USA, 16–19 December 2019; pp. 783–789. [CrossRef]
106. Kaya, F.; Basayigit, L. The Effect of Spatial Resolution of Environmental Variables on the Performance of Machine Learning Models in Digital Mapping of Soil Phosphorus. In Proceedings of the IEEE Mediterranean and Middle-East Geoscience and Remote Sensing Symposium (M2GARSS), Istanbul, Türkiye, 7–9 March 2022; pp. 169–172. [CrossRef]
107. Gutzwiller, K.J.; Serno, K.M. Using the Risk of Spatial Extrapolation by Machine-Learning Models to Assess the Reliability of Model Predictions for Conservation. *Landsc. Ecol.* **2023**, *38*, 1363–1372. [CrossRef]
108. Lark, R.M.; Chagumaira, C.; Milne, A.E. Decisions, Uncertainty and Spatial Information. *Spat. Stat.* **2022**, *50*, 100619. [CrossRef]
109. Keshavarzi, A.; Kaya, F.; Levent, B.; Gyasi-Agyei, Y.; Rodrigo-Comino, J.; Caballero-Calvo, A. Spatial Prediction of Soil Micronutrients Using Machine Learning Algorithms Integrated with Multiple Digital Covariates. *Nutr. Cycl. Agroecosystems* **2023**, 1–17. [CrossRef]

*land*

MDPI

*Article*

# Bibliometric and Visualization Analysis of the Literature on the Remote Sensing Inversion of Soil Salinization from 2000 to 2023

**Chengshen Yin [1], Quanming Liu [1,2,*], Teng Ma [1], Yanru Shi [3] and Fuqiang Wang [1]**

1 College of Water Conservancy and Civil Engineering, Inner Mongolia Agricultural University, Hohhot 010018, China; cheng20192020@emails.imau.edu.cn (C.Y.)
2 Autonomous Regional Collaborative Innovation Center for Integrated Management of Water Resources and Water Environment in the Inner Mongolia Reaches of the Yellow River, Hohhot 010018, China
3 Taipusi Banner Meteorological Bureau, Taipusi Banner 027000, China
* Correspondence: nndlqm@imau.edu.cn

**Abstract:** Tracing the historical development of soil salinization and monitoring its current status are crucial for understanding the driving forces behind it, proposing strategies to improve soil quality, and predicting future trends. To comprehensively understand the evolution of research on the remote sensing inversion of soil salinity, a scientific bibliometric analysis was conducted on research from the past two decades indexed in the core scientific databases. This article analyzes the field from various perspectives, including the number of publications, authors, research institutions and countries, research fields, study areas, and keywords, in order to reveal the current state-of-the-art and cutting-edge research in this domain. Special attention was given to topics such as machine learning, data assimilation methods, unmanned aerial vehicle (UAV) remote sensing technology, soil inversion under vegetation cover, salt ion inversion, and remote sensing model construction methods. The results indicate an overall increase in the volume of publications, with key authors such as Metternicht, Gi and Zhao, Gengxing, and major research institutions including the International Institute for Geoinformatics Science and Earth Observation and the Chinese Academy of Sciences making significant contributions. Notably, China and the USA have made substantial contributions to this field, with research areas extending from Inner Mongolia's Hetao irrigation district to the Mediterranean region. Research in the remote sensing domain focuses on various methods, including hyperspectral imaging for salinized soil inversion, with an increasing emphasis on machine learning. This study enriches researchers' knowledge of the current trends and future directions of remote sensing inversion of soil salinization.

**Keywords:** remote sensing technology; soil salinity; machine learning; data assimilation; summarize; bibliometric analysis

**Citation:** Yin, C.; Liu, Q.; Ma, T.; Shi, Y.; Wang, F. Bibliometric and Visualization Analysis of the Literature on the Remote Sensing Inversion of Soil Salinization from 2000 to 2023. *Land* **2024**, *13*, 659. https://doi.org/10.3390/land13050659

Academic Editors: Tiago Brito Ramos, Maria da Conceição Gonçalves and Mohammad Farzamian

Received: 2 April 2024
Revised: 8 May 2024
Accepted: 10 May 2024
Published: 11 May 2024



## 1. Introduction

The global map of saline soils published by the Food and Agriculture Organization of the United Nations (FAO) in 2021 estimates that saline soils are widespread in more than 100 countries and regions of the world, covering more than 833 million hectares, or 8.7% of the Earth's surface, most of which are found in naturally arid or semi-arid zones in Asia, Africa, and Latin America. Presently, the situation of arable land resources is dire, with low land quality. Soil salinization and secondary salinization have emerged as primary contributors to the desertification and degradation of land [1]. Consequently, due to various factors, portions of arable land have experienced phenomena such as salt and alkali accumulation, leading to severe land degradation and abandonment [2]. Thus, for agriculture to flourish sustainably, the accurate prediction and effective monitoring of soil salinity and alkalinity at a broad regional scale are crucial.

Soil salinization is a global issue, and traditional field soil sampling methods require a lengthy process of data accumulation and handling, often lagging behind the cyclical natural processes and failing to meet the demands of agricultural research. Since remote sensing technology is always evolving, using it to monitor and study saline–alkaline soil ensures the availability of information needed for both the sustainable growth of agriculture and the management of this type of land. One of the main methods used to monitor soil salinization is remote sensing [3]. International efforts to monitor soil salinity using remote sensing began in the 1970s [4–6]. Satellite remote sensing technology has acquired nearly 40 years of observational data, laying the foundation for the mapping, quantitative inversion, and dynamic monitoring of soil salinity. These observational data record solar reflection signals (visible and near-infrared bands), thermal infrared emission signals (thermal infrared bands), and microwave emission and radar scattering signals (microwave bands) from the land surface. These signals contain complex information such as soil moisture, vegetation cover, and surface roughness. Separating the soil salinity signal from this information is the main task of remote sensing inversion. With the support of ground sampling data, various methods for soil salinity inversion have been developed. China started employing visible-light bands from remote sensing in the 1980s to reverse the salinity of soil. Because spectral reflectance and saline alkaline soil characteristics have a complicated and nonlinear relationship [7], artificial neural networks have gradually been used to determine the saline alkaline parameters [8–10]. Based on the salt content value and radar backscatter coefficient, Liu et al. [11] constructed a quantitative inversion model for soil salt content in the Jiefangzha irrigation district using a back propagation artificial neural network (BP ANN), which was consistent with the ground validation results. Sun [12] took the Songnen Plain in western Jilin Province as the study area, and after the Box Cox transform of the measured EC of the sampled soils satisfied the normal distribution, the optimal spectral parameters were constructed based on the correlation between the spectral reflectance of EC and Sentinel-2 MSI in various bands, and then, a support vector machine (SVM) was used to build the linear regression model, while regression tree (RT), Gaussian process regression (GPR), and ensemble tree (ET) methods were used to build the EC inversion models. The test results of the model on the validation set showed that the GPR model performed optimally ($R^2$ = 0.66, RMSE = 0.48 m S/cm, MAE = 0.52 m S/cm), and the $R^2$ was improved by 29.04% compared with the traditional linear regression model. There is currently a dearth of thorough literature that summarizes and analyzes the state of the research in this field as well as the dynamic shifts in remote sensing inversion of soil salinity, both domestically and internationally. Traditional literature summarization methods such as one-by-one reading, tracing back, and qualitative summarization reveal low efficiency, low timeliness, and serious deficiencies in objectivity and intuitiveness when facing massive literature targets [13,14].

Bibliometric analysis is a kind of cross-science that applies mathematical and statistical methods to quantitatively analyze all knowledge carriers; it integrates several disciplines, such as mathematics, statistics, and bibliography, to form a comprehensive knowledge system focusing on quantification. The main measurement objects of bibliometric analysis include the volume of literature (including various publications, especially journal articles, and citations), the number of authors (including individuals, collectives, or groups), and the vocabulary (various literature markers, such as narratives, etc.). The essential feature of this method of analysis is that its output must be "quantitative", i.e., when studying the use of a particular teaching methodology, it is necessary to collect a wide range of relevant literature and to count the number of such documents, their sources, etc. This refers to the analytical method of researching the collected literature on a particular aspect in order to determine the nature and condition of the subject of study and to derive one's own viewpoints from it, so as to provide a systematic and comprehensive account and commentary on the research results and progress of a certain discipline or topic within a certain period of time, after summarizing, organizing, analyzing, and identifying them [15]. Bibliometric analysis is a powerful tool that can help researchers gain a deeper understanding and analyze the development of knowledge domains from a quantitative perspective. Citation management software tools for information visualization are primarily used to measure and analyze data from the scientific literature [16]. They can also be used to identify important literature, hot research topics, and frontier directions in a particular scientific field, as well as to draw knowledge maps of how science and technology have developed, and to intuitively display information panoramas of scientific knowledge fields [17]. The mapping methods used in bibliometric analysis mainly rely on a variety of visualization tools and techniques that can present complex bibliographic data in an intuitive and easy-to-understand graphical form. The following are some common mapping methods for bibliometric analysis: citation network analysis, co-occurrence analysis, cluster analysis, temporal order analysis, and word cloud analysis, to name but a few. In order to carry out these mapping operations, researchers can use a variety of professional bibliometric analysis software platforms or tools; for example, CiteSpace and VOSviewer are professional bibliometric information analysis software platforms that can be combined with SPSS and other software to analyze the main authors of the research in relevant fields, along with the issuing organizations, keywords, and other information, which can be used to understand the focus of the research in related fields through keyword clustering analysis. These tools usually have powerful data import, cleaning, analysis, and visualization functions, which can help researchers easily complete the mapping work needed for bibliometric analysis.

When conducting scientific research, scientists must sift through a vast volume of material. The first challenge to be addressed before beginning any research is how to identify the research hotspots within the literature. CiteSpace and VOSviewer, as a mainstream software for knowledge graph analysis and bibliometric mining, plays a vital role in domestic and foreign review research. They can show the relationships between papers as a scientific knowledge graph, which can be used to both organize the history of previous studies and show the links between documents. Additionally, it can help give us a rough idea of the potential for future study. Many scholars, at home and abroad, have conducted detailed studies and overviews from different perspectives, scales, and regions, but there is a lack of overview studies on the remote sensing inversion of soil salinity based on bibliometric analysis.

In order to fulfill our research objectives, we asked the following research questions: Q1. What are the trends in the scientific literature on the remotely sensed inversion of soil salinity? Q2. What are the future research trends? The specific objectives of this study were as follows: (1) to obtain bibliometric information on scientific studies extracted from the Web of Science (WoS) Core Collection database as a data source; (2) to use bibliometric analysis methods to transform and analyze the quantitative data of the selected articles; (3) to use the total number of citations to identify the main authors, countries, institutions, etc., in this field of study; and (4) to use keywords to analyze the research history and current research hotspots. The remainder of this paper is organized as follows: Section 2 describes the data and methods used in the bibliometric analysis. Section 3 describes the basic features and research hotspots of remote sensing inverse soil salinization studies. Section 4 discusses the research hotspots in the field of the remote sensing inversion of soil salinization and discusses future trends based on the analysis.

## 2. Materials and Methods

### 2.1. Methods

(1) This work presents a visual analysis of the past 20 years of research on soil salinity inversion from remote sensing by examining keywords, author groups, institutions, and publication volume, among other things; it makes use of data mining and knowledge mapping with analytical tools such as CiteSpace and VOSviewer to visually display the emergence of knowledge clusters and disclose the structural dynamics of this sector [18,19]. The cartographic method is detailed in [20].

(2) Analysis of the authors of publications: The core author group refers to a collection of authors who have published a considerable number of influential papers in the field's relevant journals [21]. The core author group serves as a compass and the foundation of

the literature flow. The following formula was used to determine the number of papers published by core authors in accordance with Price's hypothesis [22]:

$$N = 0.747\sqrt{N_{max}} \quad (1)$$

where N represents the minimum number of papers that a core author should have published, while $N_{max}$ represents the maximum number of papers published by an author during the study period.

Only authors who have published at least N papers can be considered to be core authors.

(3) Keyword analysis: The clustering of knowledge graphs in this study was conducted using the VOSviewer [21] software, while the layout was carried out using the Pajek [22] software. Different colors are used in this study to represent the average year of appearance of a particular keyword. The calculation of the average year T is as follows:

$$T = \frac{\sum(year \cdot C_{year})}{\sum C_{year}} \quad (2)$$

where year represents the year of the keyword's appearance, and while $C_{year}$ represents the frequency of the keyword's appearance in that year.

(4) Keyword emergence method: The term "emergence" describes a quick rise that occurs over time. One can comprehend the evolution and shifts in research hotspots, trends, and frontiers over a certain period of time by detecting keyword emergence. In order to identify emergent words with high-frequency change rates and rapid growth rates, emergent word analysis looks at the temporal distribution of keywords. This allows for the investigation of the discipline's frontier areas and development patterns.

### *2.2. Data Sources*

The Web of Science database provided the data, which were retrieved between 1 January 2000 and 31 December 2023. A search was conducted in the WOS core database using "salt", "remote sensing", and "soil" as the main themes, with the language set to English. After refining and filtering the retrieved data, all records were exported to a plain text file, including "full records and cited references". Following deduplication upon importation into the analysis software, a total of 300 relevant and valid documents pertaining to the focus of this study were obtained.

## 3. Analysis of the Fundamental Characteristics of Research on Remote Sensing Inversion of Soil Salinization

A summary of bibliometric statistics was provided by the preliminary results of the bibliometric analysis. Subsequently, we delved into the details of the literature, including authors, institutions, journals, and countries of origin.

### *3.1. Descriptive Bibliometric Analysis*

The overall direction of research and the advancement of academic issues can be reflected in the fluctuations in the number of publications over time [23]. An essential measure of a research direction's development process and potential future trends is its annual publication volume. The pattern of variations in cumulative publication output can be used to determine the phases of development and potential future directions for research.

Figure 1 presents the scientific achievements during the study period. From 2000 to 2007, the research was in its initial stages, with few achievements. Then, there was a period of erratic growth from 2008 to 2014. There were 227 publications in the ten years following 2014, accounting for 75.67% of all publications.

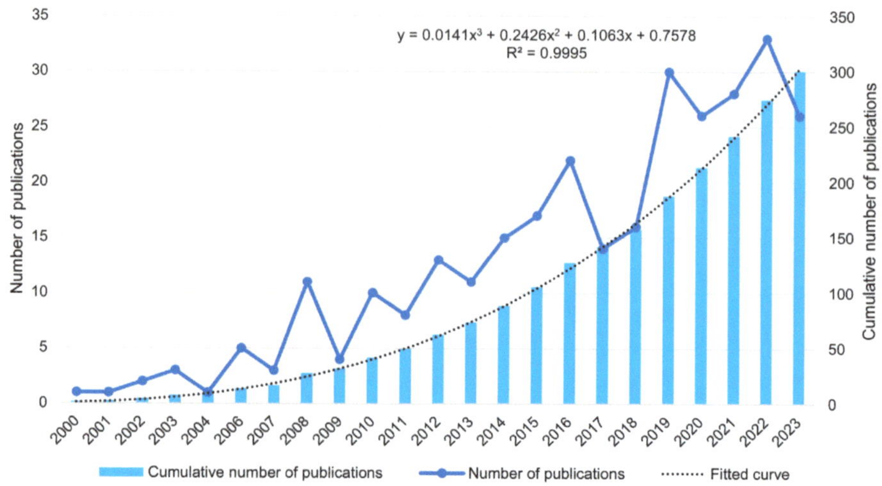


**Figure 1.** Number of scientific publications on remote sensing inversion of soil salinity from 2000 to 2023.

Regression analysis of the data from 2000 to 2023 yielded a well-fitting index curve of $y = 0.0141x^3 + 0.2426x^2 + 0.1063x + 0.7578$ ($R^2 = 0.9995$). The fitting curve indicates that the field of remote sensing inversion of soil salinization is still in its early stages and is predicted to expand quickly in the coming years. A growing number of academics are beginning to concentrate on the topic of remote sensing monitoring of soil salinization, which has been growing actively and is in a comparable developmental stage. The pattern of studies indicates that future publications will continue to be published in greater numbers.

### 3.2. Analysis of Primary Authors

According to the statistics, the maximum number of publications by an author is eight, denoted by $N_{max} = 8$. Hence, N = 2.11 articles, rounded to the nearest integer (3), define authors who have published at least three articles as core authors in this field. Therefore, 36 authors were identified as core contributors to this research.

From the author collaboration network diagram (Figure 2) and the top 10 authors by publication volume, it can be observed that the three authors with the largest nodes are Zhao, Gengxing with eight papers; Zhang, Fei with six papers; and Wu, Jingwei with five papers, all hailing from China. Based on the optimization of the cooperation network [24] using pathfinding algorithms, the main groups of authors consist of three concentrated author clusters: the Zhao, Gengxing team; the Zhang, Fei and Ding, Jianli team; and the Wu, Jingwei team. Among them, the research area of Zhao, Gengxing's team mainly focuses on the Yellow River Delta in China [25], Ding, Jianli's team primarily focuses on Xinjiang, China [26], and Wu, Jingwei's team concentrates on the Hetao Plain in China. There are also many cooperation networks composed of small nodes and scattered independent authors, indicating that the concentration of authors is not high and the research teams are dispersed. The loose connections between different research teams suggest weak citation relationships among them, which could lead to academic barriers over time, adversely affecting sustainable development research.

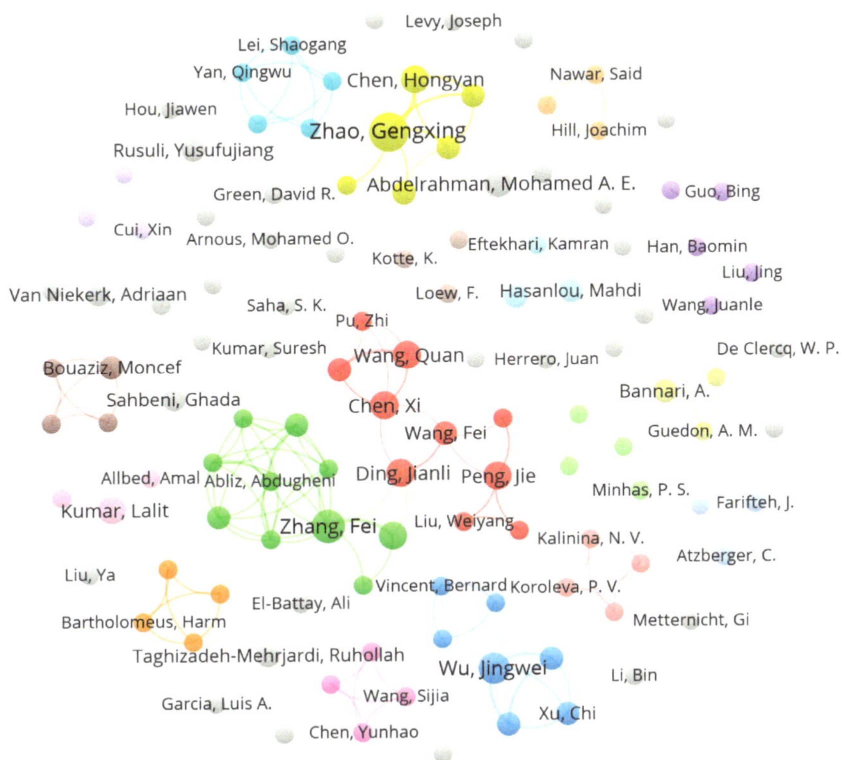


**Figure 2.** Collaborative network of authors.

The top ten authors by citation frequency are listed in Table 1. Ding Jianli has the highest citation frequency and can be considered the most influential author. Among the ten most influential researchers, three are from China, and the rest are from Australia, Germany, Poland, India, Tunisia, and the United States. The two authors with the most publications—Zhao Gengxing (eight papers) and Wu, Jingwei (five papers)—are not listed in Table 1, because of their comparatively low citation counts.

**Table 1.** Top ten authors ranked by citation frequency.

| Author | TA | TC | Country |
|---|---|---|---|
| Metternicht, Gi | 2 | 729 | Australia |
| Atzberger, C. | 2 | 437 | Netherlands |
| Farifteh, J. | 2 | 437 | Netherlands |
| Van Der Meer, F. | 2 | 437 | Netherlands |
| Ding, Jianli | 4 | 381 | China |
| Kumar, Lalit | 4 | 336 | Australia |
| Nicolas, Herve | 2 | 313 | France |
| Walter, Christian | 2 | 313 | France |
| Allbed, Amal | 2 | 302 | Australia |
| Zhang, Fei | 6 | 260 | China |

Abbreviations: TA stands for total article count; TC stands for Web of Science Core Collection times cited count.

### 3.3. *Research Institutions and Countries*

Research institutions with more than two published papers were selected as the research objects to identify the top ten institutions by citation frequency, as shown in Table 2. The three institutions with the highest citation frequencies are the International Institute for Geoinformatics Science and Earth Observation, Curtin University of Technology, and Xinjiang University. Among the top ten, five institutions are based in China, indicating a significant contribution of Chinese research in the field of saline alkaline land, both nationally and globally. Globally, China, the Netherlands, the United States, Australia, and Egypt are the leading nations in the field of remote sensing soil salinity research, setting the stage for future research endeavors. Although institutions such as Shandong Agricultural University (with 11 publications and 176 citations) and China Agricultural University (with 8 publications and 176 citations) have relatively few citations compared to the top ten, their substantial publication outputs have exerted significant influence on this research. The visualization of the institutions (Figure 3) reveals a network where larger core institutions connect with smaller ones, indicating close collaborations among Chinese institutions.

**Table 2.** The institutions ranked in the top ten by publication volume.

| Institution | TA | TC | Country |
|---|---|---|---|
| International Institute for Geoinformatics Science and Earth Observation | 2 | 1006 | Netherlands |
| Curtin University of Technology | 2 | 729 | Australia |
| Xinjiang University | 18 | 701 | China |
| Chinese Academy of Sciences | 41 | 651 | China |
| Wageningen University and Research | 2 | 417 | Netherlands |
| University of Chinese Academy of Sciences | 20 | 390 | China |
| University of New England | 4 | 336 | USA |
| Tel Aviv University | 2 | 315 | Israel |
| Beijing Normal University | 6 | 313 | China |
| Wuhan University | 12 | 246 | China |

Abbreviations: TC stands for total citations; TA stands for total articles.

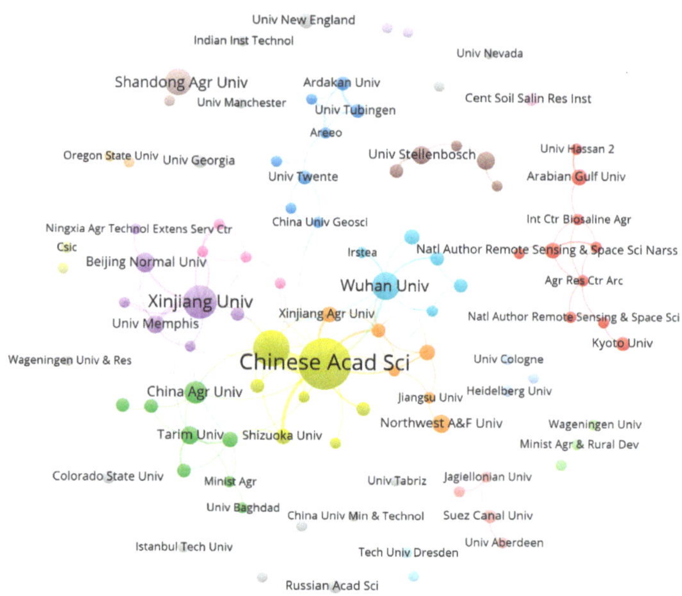


**Figure 3.** The visual representation of institutional networks.

According to the analysis (Figure 4), China ranks first in publication output, contributing 28% of the total publications. Given China's vast saline alkaline lands and considerable development potential, along with the intensifying trend of land salinization in some regions, numerous studies reflect that the highest research output in this field is from Chinese institutions and scholars, showcasing China's growing research prowess. The United States ranks second in publication output, accounting for 11%, followed by India and Australia in the third and fourth positions, both at 5%. The countries with the highest citation frequencies are China (2884 citations), the Netherlands (2040 citations), Australia (1778 citations), the United States (1703 citations), and Germany (1070 citations). An analysis of country relationships (Figure 5) indicates close collaborations between China and the United States with various other countries, particularly with Australia and Italy, among others. Considering both publication output and citation frequency at the national and institutional levels, China and the United States are the leaders in the research field of remote sensing for soil salinity.

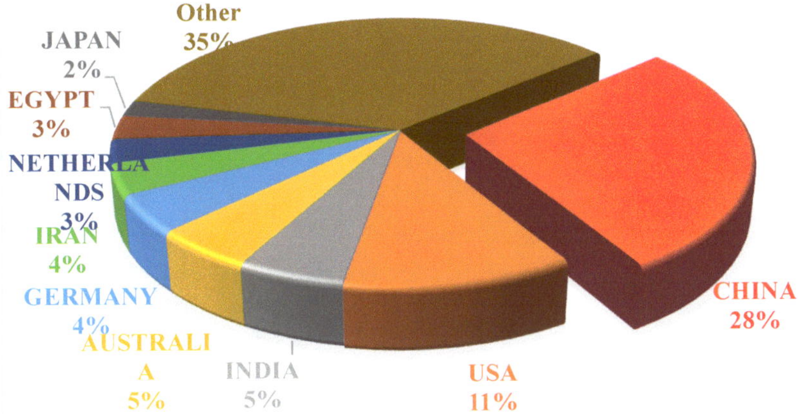


**Figure 4.** Pie chart of document volume by country.

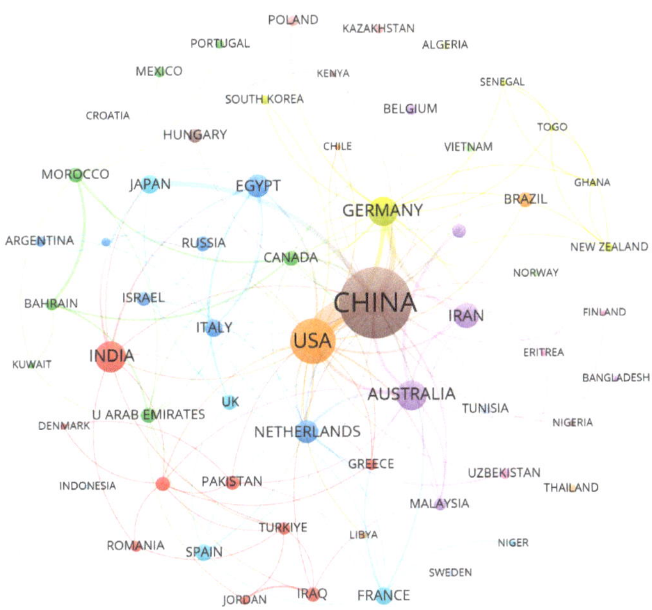


**Figure 5.** Graph depicting international collaboration among nations.

### 3.4. The Most Influential Journals

The top ten journals by citation frequency published 105 papers (35%), as shown in Table 3. The five journals with the highest citation frequencies are "Geoderma" (publishing 14 papers), "Remote Sensing" (publishing 47 papers), and the "Remote Sensing of Environment" (publishing three papers). These journals played significant roles in research on remote sensing for soil salinity retrieval.

**Table 3.** The top ten journals presented by citation frequency.

| Sources | TC | TA |
|---|---|---|
| Geoderma | 1591 | 14 |
| Remote sensing | 1110 | 47 |
| Remote sensing of environment | 1019 | 3 |
| Science of the total environment | 629 | 6 |
| Ecological indicators | 389 | 7 |
| International journal of remote sensing | 356 | 14 |
| Land degradation & development | 349 | 5 |
| Advances in agronomy | 293 | 1 |
| Journal of arid environments | 218 | 2 |
| International journal of applied earth observation and geoinformation | 184 | 6 |

Abbreviations: TC stands for total citations; TA stands for total articles.

### 3.5. Field of Research

When conducting research field analysis in CiteSpace, the software categorizes research areas based on the features the of literature data, such as Web of Science categories and research directions. In the knowledge graph of research fields, one can observe the frequency of occurrence of 39 research areas and the cross-connections between them (Figure 6 and Table 4). Among these, environmental science has the highest frequency of publications and intermediary centrality, which is related to the predominant focus of this field on environmental issues such as soil salinization—a topic relevant to the remote sensing monitoring of soil salinity. Remote sensing ranks second in publication frequency but exhibits relatively lower intermediary centrality, indicating limited connections with other fields. Water resources and instruments and instrumentation show higher intermediary centrality, suggesting closer connections with other research domains. Since changes in soil moisture significantly affect soil salinity, it is reasonable that water resources—ranking second in intermediary centrality, with a value of 0.37—are closely associated with the study of soil salinity. The other top 10 research fields include chemistry, analytics, geosciences, and multidisciplinary, among others.

**Table 4.** The top 10 research fields ranked by citation frequency and their intermediary centrality.

| Research Field | Number of Published Papers | Centrality |
|---|---|---|
| Environmental sciences | 154 | 0.43 |
| Remote sensing | 102 | 0.11 |
| Imaging science and photographic technology | 82 | 0.11 |
| Geosciences, multidisciplinary | 76 | 0.23 |
| Soil science | 41 | 0.15 |
| Water resources | 26 | 0.37 |
| Engineering, electrical and electronic | 19 | 0.43 |
| Agronomy | 15 | 0.16 |
| Chemistry, analytical | 15 | 0.26 |
| Biodiversity conservation | 12 | 0.01 |

Abbreviations: TA and TC stand for total articles and citations, respectively.

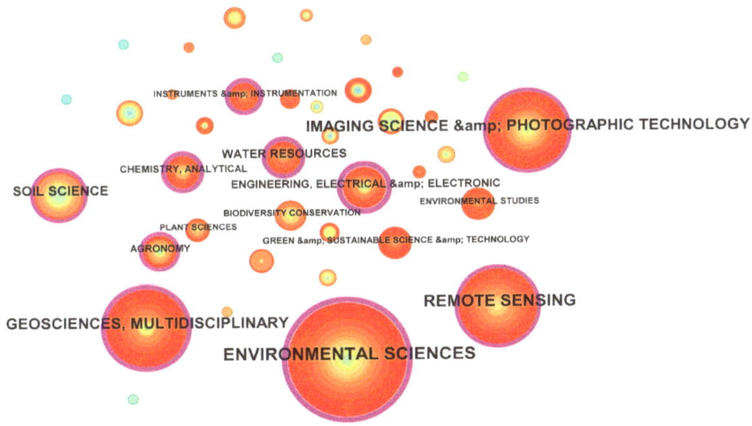


**Figure 6.** The main areas of study for soil salinization and its relationships via remote sensing monitoring.

## 3.6. Study Region

With the increasing trend of global warming, the problem of soil salinization in middle- and low-latitude regions will become more and more obvious, becoming especially pronounced in China, the United States of America, Hungary, and Australia, as well as becoming more serious in the north and east of Africa, South America, the Middle East, Central Asia, and South Asia. Global soil salinization hotspots include numerous Central and West Asian nations, as well as Pakistan, China, the United States, India, Argentina, and Sudan [27]. The three primary saline alkaline zones on Earth are the Songnen Plain in China, Victoria in Australia, and California in the United States.

China has a diverse range of saline soils with abundant resources spread across vast territories. For the study of saline soils in China, there are many different types to consider, including coastal saline soils in the eastern low plains, soda saline soils in the Songnen Plain in Northeast China, oasis saline soils in Xinjiang, silty saline soils in the irrigation area of the Hebei Plain (Figure 6), tidal saline soils in the Huang-Huai-Hai Plain, alkaline saline soils in the Hexi Corridor, acidic sulfate saline soils in the southern coastal areas, and saline soils in the extremely arid regions of Qinghai and Xinjiang [28].

Saline soils in China are mainly distributed in the northern arid and semi-arid areas, with a total area of 36.3 million hectares, accounting for nearly 4.88% of China's available land area—significantly higher than the global average [29]. Among them, saline alkaline land covers 7.6 million hectares of arable land, accounting for 6% of the arable land area, and is widely distributed in coastal and inland areas. A large part of this saline alkaline land, serving as reserve land, remains undeveloped. Through an analysis of research the literature, Inner Mongolia's Hetao irrigation district [30], Shandong's Yellow River Delta [31], the oasis and Aibi Lake in the Weigan River Kuqa River Basin in Xinjiang [32], Yinchuan in Ningxia [33], and the Songnen Plain in Northeast China were identified among the areas of high research interest, with secondary salinization being a hot topic. This is consistent with the findings of the China Geological Survey's 2020 distribution survey of saline alkaline land in China. For further information, see the distribution map of China's saline alkaline land in the He Jin article "The Earth's Misery—Saline-Alkali Land" on the China Geological Survey's website (http://www.chegs.cgs.gov.cn/, accessed on 1 January 2024).

Outside of China, research areas include the Aral Sea in Central Asia [34], the Nile River Basin in East Africa [35], the Mediterranean coast of Europe (affected by factors such as groundwater extraction and rising sea levels), and Western Australia, which are consistent with the most influential countries. For more details, view the 2021 World Soil Salinization Distribution Map from the Food and Agriculture Organization of the United Nations.

### 3.7. Keyword Co-Occurrence Analysis

The keyword co-occurrence analysis method uses terms or noun phrases that appear together in a body of literature to determine the relationships between various topics in the field of reference. An association network composed of topic term pairs can be constructed by calculating their frequency of occurrence in a specific document within a corpus of literature. The number of nodes in the graph represents the total number of keywords, while the number of edges in the graph represents the number of links between keywords. A relationship exists between two keywords as long as they are found in the same document. Figure 7 illustrates the clustering results of keywords in this research over the past 20 years, where keywords belonging to the same cluster are arranged in vertical columns. Moreover, the color gradient, from blue to yellow, represents the keywords' average year of occurrence, from old to new, aiding in identifying the evolution of research hotspots within each cluster.

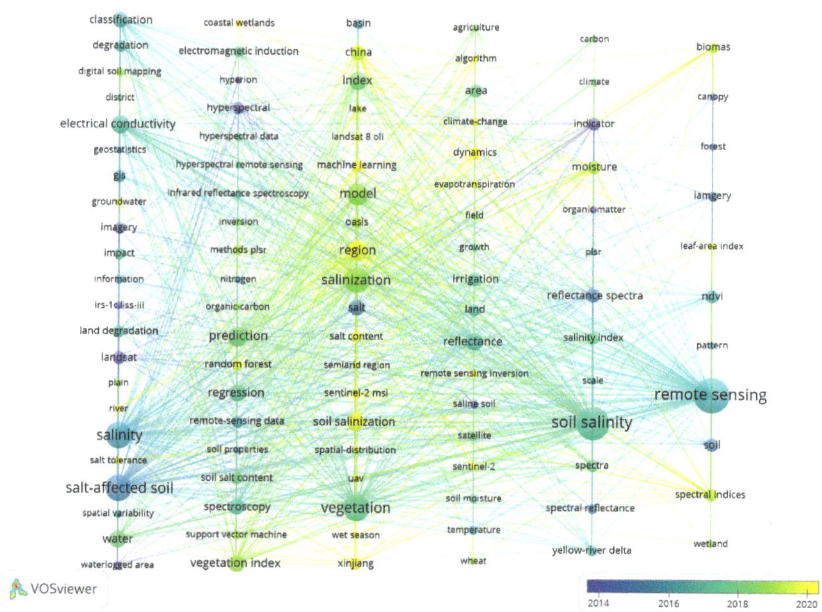


**Figure 7.** Co-occurrence graph of keywords.

Based on the color characteristics of the keyword co-occurrence graph for the WoS database, the key terms that appeared during the initial phase from 2000 to 2007 included remote sensing, saline soil, Landsat, geographic information system (GIS), reflectance spectroscopy, etc. The amount of salt present in soil serves as a significant indicator of its salinization level, and remote sensing technology has increasingly been used to track soil salinization. During this period, the main focus was on identifying saline soils using optical satellites and GIS methods to understand the extent, area, and degree of soil salinization in a particular region. Among them, the main node was "remote sensing", followed by "soil salinity". Remote sensing extends several prominent paths from the node, with close connections to vegetation indices, spectral indices, factor analysis, and spectral reflectance, which are used in remote sensing studies for retrieving soil salinity information. Particularly emphasized is the keyword "water". Significant differences in commonly used spectral indices over various moisture gradients are caused by the influence of soil moisture content on spectral reflectance in the near-infrared and infrared bands. Thus, soil moisture is seen as an important factor influencing the accuracy of soil salinity monitoring [36]. During the period from 2007 to 2017, which was characterized by development, high-frequency keywords included vegetation, model, index, indices, spectroscopy, reflectance, salinity index, land degradation, vegetation index, water, area, Yellow River Delta, biomass, etc. The research started to shift from qualitative to quantitative during this time, and studies began looking at the connections between soil salinity, hyperspectral data, and land satellite data. During this time, researchers began examining the connection between spectral reflection characteristics and soil's physico-chemical qualities (e.g., salinity index, conductivity), with the goal of using remote sensing images to dynamically monitor soil salinity. Alternative remote sensing indices for determining soil salinity in agricultural fields have been those linked to water stress or vegetation health. Plants' health is hampered by salt stress, which in manifests symptoms akin to those of a water shortage. Significant growth patterns, spectral characteristics of salt-tolerant plants, or salt crusts and efflorescence that might be present in bare soil can all be used to infer high salt concentrations. A variety of salinity indices are available for the detection and mapping of soil salinity using multispectral and hyperspectral satellite sensors, much like vegetation indices. On the other hand, the color, roughness, salinity, and soil moisture all have significant impacts on surface reflectivity. Choosing one index may not be appropriate in every situation, because these indices do not always produce outcomes. In summary, the normalized difference vegetation index (NDVI), a reliable and error-free indicator, appears to be able to quickly evaluate the spatial patterns of vegetation health. Statistical techniques can be applied to link soil parameters with various indicators based on this criterion. Moreover, there are drawbacks when describing the productivity of saline alkaline soils in various locations using the productivity indicators obtained from remote sensing. During the period from 2018 to 2023, which experienced rapid development, high-frequency keywords included random forest, Sentinel 2, machine learning, climate change, moisture, etc. This indicates that machine learning and random forest methods combined with Sentinel-2 data are being applied in the remote sensing monitoring of soil salinity.

### 3.8. Analysis of Emerging Trends in Frontier-Stage Research

Keyword emergence analysis, aided by the burst detection function of CiteSpace software, divides time into one-year intervals to identify burst keywords. The red range indicates the period with the greatest frequency change, during which the keywords have the most significant impact [37]. Figure 8 displays the first appearance year (Year), burst strength (Strength), burst start year (Begin), burst end year (End), and their positions on the timeline (with the red portion indicating the burst year). The keywords are arranged in chronological order according to their burst start time, with burst strength arranged from largest to smallest.

From Figure 8, it can be observed that, for a rather long portion of the early and quite long period of research, soil salinity was inferred using remote sensing satellites such as Landsat. In 2014, the Yellow River Delta emerged as the main research area and entered the spotlight. In this section, the focus is on the keywords emerging in recent years, with the aim of identifying the research frontier. In order to increase the accuracy of inference, researchers in this phase started to account for the impact of external factors such as surface vegetation, soil moisture, and climate change on the outcomes of a remote sensing soil salinity retrieval. This is indicated by the appearance of keywords such as "biomass", "moisture", and "climate change" in 2016, 2018, and 2021, respectively. The term "machine learning" began to appear in 2021–2023, which suggests the developing trends and frontier dynamics of applying machine learning techniques to the remote sensing monitoring of soil salinization. It is clear from the keyword emergence analysis that machine learning and remote sensing are gradually becoming more and more important in research on remote sensing soil salinity inference. Among the keywords, those that appeared earlier but emerged in recent years may correspond to the emergence of new technologies or methods.

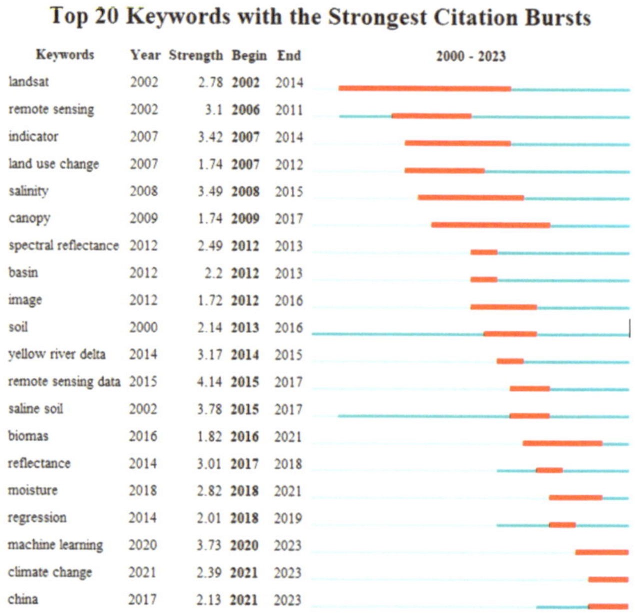


**Figure 8.** Keyword burst graph (The red part is the year in which the keyword appears).

### 3.9. The Most Influential Articles

Based on the quantity of citations, the most important works from 2000 to 2023 are listed in this section. Interestingly, the top ten papers by number of citations in all databases match those that are exclusive to the Web of Science (WoS) database.

According to Table 5, the paper by Metternicht and Zinck, which ranks first in citation count, discusses the potential and limitations of the remote sensing inversion of soil salinity. In this field of study, its early publication date has drawn attention. The article reviews a variety of sensors and techniques for the remote identification and mapping of salt-affected areas, including aerial photography, satellites, airborne multispectral sensors, microwave sensors, video imaging, airborne geophysics, hyperspectral sensors, and electromagnetic induction meters [38]. The spectral behavior of salts, their spatial distribution on the terrain surface, temporal fluctuations in salinity, vegetation interference, and spectral confusion with other terrain surfaces are the difficulties associated with employing remote sensing data to map areas affected by salt. Spectral decomposition, maximum likelihood classification, fuzzy classification, band ratios, principal component analysis, and pertinent equations are some of the techniques that are covered. Finally, the paper presents an integrated method to simulate the spatiotemporal variability of salinity using various data fusion and data integration techniques. Daliakopoulos et al. [27] outlined the drivers and pressures of soil salinity, key indicators, and the latest advances in monitoring, modeling, and mapping methods. Their report discusses how salinization affects soil functions, and it concludes by outlining Europe's salinization situation. In order to support policies and strategies for safeguarding European soils, future research in the field of soil salinization should concentrate on the carbon dynamics of saline soils, further investigate soil properties through remote sensing, and coordinate and enhance soil salinity maps throughout Europe.

Farifteh et al. [5] explored the possibility of predicting the soil's salt concentration using partial least squares regression (PLSR) and artificial neural networks (ANNs) at three different scales, employing different datasets in four distinct study areas. The findings show that there is considerable potential for both approaches in terms of mapping and quantifying soil salinity. Performance measures indicate strong similarities between the two approaches, with PLSR showing a minor advantage. This implies that a linear function can be used to approximate the relationship between soil salinity and soil reflectance. Douaoui et al. [39] analyzed the logistic regression method by combining remote sensing data with European Community ground measurement data in a plain plot in Algeria. Compared to pure regression methods, this approach significantly improved the accuracy of salinity estimation. By integrating remote sensing data with a ground monitoring network, this method allows for the more precise spatiotemporal monitoring of soil salinity in arid regions. Ben-Dor [40] primarily expressed the potential and possibilities of using hyperspectral remote sensing technology for the quantitative analysis of soil properties, while also discussing the challenges faced and their possible resolution strategies. Through specific case studies, the paper validated the application value of this technology for monitoring soil salinity and other key soil properties, while also expressing expectations and directions for future technological developments in this field. Farifteh et al. [4] outlined a conceptual framework for a method that combines optical remote sensing data with simulation models and geophysical survey results to economically and effectively predict different degrees (low, medium, severe) of salinization/alkalization. In this integrated method, combining data not only categorizes existing saline soils but also tracks salinization as a soil-forming process. Since the goal of this approach is to integrate data of various scales and types, data fusion and upgrading are crucial. Fan et al. [41] studied the Yellow River Delta area, assessing the distribution maps of salinized soil over the previous 20 years by combining spatial models, field data obtained at three different times, and remote sensing imagery. Using the Kriging interpolation method, they analyzed the spatiotemporal dynamics of groundwater levels and total dissolved solids (TDSs) over nearly 20 years. The Kriging method, utilizing random subsamples of observations as a basis for validation, significantly improved the accuracy of the soil salinity predictions. An examination of spatial data correlation revealed a close relationship between groundwater dynamics and the distribution and evolution of salinized soil, with a higher TDS and rising groundwater levels being associated with worsening soil salinization.

Allbed et al. [42] focused on an oasis in eastern Saudi Arabia, establishing various vegetation indices and soil salinity indices. The methods that worked best for determining the salinity of soil in heavily vegetated farmed fields were the soil-adjusted vegetation index (SAVI), the normalized difference salinity index (NDSI), and the salinity index (SI-T). The NDSI and SI-T had the strongest association connection with salinity in regions with less plant density and bare ground. The salinity of the soil in the area was effectively determined using vegetation and soil salinity indices that were taken from IKONOS satellite photography. Wang et al. [43] examined the potential for utilizing Sentinel-2 MSI spectral bands and generated spectral indices to forecast the soil salinity of wetlands affected by salt in the Xinjiang region around Aibi Lake. The study found significant correlations between the newly proposed NDI and TBI4 spectral indices and soil salinity. By applying different algorithms, such as the random forest partial least squares regression model, the study successfully constructed high-accuracy soil electrical conductivity (EC) prediction model and produced high-resolution soil salinity maps. Abbas et al. [44] focused on the upper Indus River basin in Pakistan, creating and utilizing salinization indices (S1–S4) based on remote sensing data and categorizing photos using the maximum likelihood approach. The majority of the salt-affected soil types in the region were saline soils. Poor irrigation channel management not only led to water resource waste but also caused soil degradation, with soil salinization causing irreversible losses to agricultural productivity and the regional economy.

The first and second papers primarily review the issues and developments in remote sensing inversion of soil salinity, while the seventh, ninth, and tenth papers focus on vegetation indices or salinity indices. Other papers primarily use different methods and remote sensing data to invert soil salinization information in various locations.

**Table 5.** Top ten cited papers.

| Paper | DOI | Year | Times Cited, WoS Core | Times Cited, All Databases |
|---|---|---|---|---|
| Metternicht, GI; Remote sensing of environment | [38] | 2003 | 698 | 885 |
| Daliakopoulos, IN; Science of the total environment | [27] | 2016 | 402 | 433 |
| Farifteh, J; Remote sensing of environment | [5] | 2007 | 306 | 385 |
| Douaoui, AEK; Geoderma | [39] | 2006 | 302 | 344 |
| Ben-Dor, E; Advances in agronomy | [40] | 2002 | 291 | 334 |
| Farifteh, J;Geoderma | [4] | 2006 | 204 | 262 |
| Allbed, A; Geoderma | [42] | 2014 | 199 | 259 |
| Fan, X; Land degradation and development | [41] | 2011 | 175 | 195 |
| Wang, JZ; Geoderma | [43] | 2019 | 170 | 191 |
| Abbas, A; Physics and chemistry of the earth | [44] | 2013 | 159 | 189 |

## 4. Discussion

Since the 1990s, soil salinity retrieval by remote sensing has undergone tremendous developments. Initially, satellite remote sensing technology was utilized to provide evidence for the on-site measurements of soil salinization [45]. Subsequently, advancements were made in utilizing optical satellite band information, spectral reflectance, and other remote sensing data for retrieval purposes. Additionally, retrieval accuracy has been further improved by integrating several remote sensing sources, such as microwave radar satellites and drones, and by improving the models and algorithms [46,47]. A bibliometric analysis of the literature related to the remote sensing retrieval of soil salinity indicated that universities and research institutions in China are leading in this field, with high research activity. In this section, we explore the research hotspots from six perspectives based on the keywords identified in Section 3.

### *4.1. Utilizing Machine Learning Methods for Soil Salinity Retrieval*

Machine learning has outstanding advantages in screening soil salt sensitive variables and mining hidden information in massive data. Researchers have begun to pay extensive attention to the application of machine learning in the remote sensing monitoring of soil salinization, which is consistent with the research results of Wang et al. [48]. Currently, the machine learning methods used for soil salinity retrieval primarily include BP neural networks and random forests [49]. Machine learning methods have been employed to construct estimation models for spectral characteristic parameters obtained by unmanned aerial vehicles (UAVs). The models with the best estimation accuracy were those that combined the elastic net regression algorithm with the extreme learning machine algorithm at a depth of 10–20 cm [50]. The results indicated that their root-mean-square-error (RMSE) was 0.141% and their highest cross-validation coefficient of determination ($R^2$) was 0.783. Meanwhile, $R^2$ was 0.66 and the relative percent difference (RPD) was 2.59 for an inversion model based on the first-order fractional differentiation of optical remote sensing spectral bands [51]. Additionally, a machine learning method combined with microwave data was utilized to flip the soil salinity model at a 0–10 cm depth, with the cubist model exhibiting the highest accuracy, with a validation set $R^2$ of 0.822 and an RMSE of 3.064 [52]. The inversion of diverse remote sensing data can achieve a maximum inversion depth of 40 cm. In order to predict soil salinity in the Fraser Valley of British Columbia, Canada, Heung et al. [53] assessed and contrasted machine learning algorithms, such as artificial neural networks, random forests, support vector machines, and multivariate logistic regression. Their study found that these machine learning models have advantages in predicting soil salinity, with the support vector machine algorithm achieving the highest accuracy of 72% for the large ROS group. In areas with abundant vegetation and mild-to- moderate salinization, machine learning techniques have proven to be dependable for the digital mapping of soil salinity [54]. Particularly, the random forest regression model produced better estimation accuracy results [55], with a validation set $R^2$ of 0.86, an RMSE of 1.83, and an RPD of 2.7. It is important to remember that machine learning depends on a huge number of training samples. Overfitting is a condition in which the training samples' fitness is noticeably higher that of validation samples, and this must be taken into account when choosing the model and structural parameters. The advantages of machine learning in data fitting allow the quantitative relationship between soil salinity and multi-source remote sensing and GIS data to be fitted using suitable models. This will increase the accuracy of soil salinity assessment. Subsequent investigations may concentrate on the ongoing integration of machine learning with several remote sensing data sources for inversion [56,57].

### 4.2. Soil Salinity Retrieval Based on UAV Remote Sensing

For unmanned aerial vehicle (UAV) data, object-oriented classification methods can be employed to improve classification accuracy. Soil salinity may now be estimated in this way due to advances in quantitative remote sensing. The primary technique for determining soil salinity is the use of spectral indices, which have proven effective in removing background and noise effects, minimizing interference from internal and external sources, and improving the extraction of spectral absorption features. They also play an important role in fully and accurately exploiting spectral information and constructing high-precision, robust models [58]. Yao et al. [59] used machine learning to invert mulch-covered farmland using multispectral UAV data. With a validation set $R^2$ of 0.717 and an RMSE of 0.171, the model built with an extreme learning machine to measure surface soil salinity content at a depth of 0–20 cm fared the best. To create spectral indices for soil salinity at a depth of 0–10 cm, UAV spectral properties were divided into different degrees of mild, moderate, and severe salinization [60]. The optimal model, based on grey relational analysis and a support vector machine, achieved an $R^2$ of 0.692 and an RMSE of 8.562. Zhang er al. [61] utilized UAV satellite remote sensing using the dominant variable weighting method and multiscale transformation through multiple linear regression models to effectively improve the monitoring accuracy of surface soil salinity at a depth of 0–10 cm, with an $R^2$ of 0.420 and an RMSE of 0.219, achieving fusion inversion from UAV to airborne remote sensing. Chen, et al. [62] created an improved TsHARP scale transformation approach to accomplish the upscale UAV satellite remote sensing monitoring of soil salinization. The model was based on GF-1 satellite remote sensing data and UAV multispectral remote sensing data. Most studies focus on surface soil, with a maximum inversion depth of up to 60 cm. Multiple linear regression models were used by Ivushkin, et al. [63] to identify connections between vegetation indices, canopy temperature, and plant height derived from three separate UAV sensors that measured salinity, stomatal conductance, and real plant height, but the overall $R^2$ was low at 0.46. Utilizing UAV electromagnetic interference technology in conjunction with the random forest approach, Hu, et al. [64] quantitatively evaluated the salinity of surface soil at a depth of 0–20 cm. The results showed that the prediction model established employing data from UAVs exceeded the model using GF-2 data, with an RMSE of 1.40 and RPD of 2.98.

In the future, the integration of comprehensive stereo satellites, drones, and ground-based data for surface information monitoring will be explored. Continuous experimental research can be conducted through an unmanned aerial vehicle remote sensing to infer soil salinity, salt ions, machine learning parameters, and model optimization under different vegetation classifications.

### 4.3. Soil Salinity Inversion Based on Data Assimilation

Data assimilation is one of the crucial methods for improving forecast accuracy by incorporating physical mechanisms. By assimilating the multiple sources of remote sensing and ground observation data into soil water/salt models, the simulation process can adjust the model parameters in a timely manner, correct the model trajectories in real-time, and obtain spatiotemporally consistent soil salinity data, thereby reducing the accumulated model errors and better representing the spatiotemporal variations in soil salinization. On the one hand, model simulation compensates for the limitations of remote sensing data, which can monitor but not predict. On the other hand, remote sensing data can provide more accurate input parameters for the model. Leveraging satellite remote sensing technology coupled with data assimilation principles can aid in rapidly determining the crop salinity conditions at the irrigation district scale and further assessing the regional soil salinization levels.

Articles on soil salinity inversion based on data assimilation have been published in recent years. Lin Lin [65], who established equations for simulating solute transport in soil, including convection dispersion equations, proposed the earliest studies on the mechanism of water and salt transport in soil, both domestically and globally. A set of models that can accurately simulate the movement of salt and water in saturated and unsaturated porous media based on different solute transport equations, such as MODFLOW, MODPATH, SWAP, and HYDRUS, has been widely applied in different types of research [66].Using data assimilation methods, with soil water/salt transport models as model operators and large-scale observation data as driving data, observational data are incorporated into the model using assimilation algorithms. For example, Yao et al. [67] utilized ensemble Kalman filter algorithms to assimilate electromagnetic induction data into the HYDRUS-1D model, improving the spatiotemporal dynamic estimation accuracy of soil salinity. Ding Jianli et al. used ensemble Kalman filter methods to assimilate MODIS and Landsat TM data information into HYDRUS-1D. Assimilating remote sensing data for the HYDRUS-1D model and ensemble Kalman filter [68] was superior to the single inversion of the HYDRUS-1D model or ensemble Kalman filter alone, but the assimilation effectiveness decreased with increasing soil sampling depth, with the best assimilation effect at 0–20 cm, and the depth of soil inversion could reach up to 60 cm below ground.

Currently, research on the scale transformation of the soil salinity movement is still immature, mainly because salt migration is based on soil's hydrological processes, which are complex and have temporal and spatial dependencies. Future research should strengthen the integration of observation data at different scales, construct universal scale transformation functions, optimize the construction of assimilation systems [69], and introduce machine learning to improve the models' accuracy, and continuously study the coupling of multiple assimilation algorithms.

### 4.4. Remote Sensing Retrieval of Soil Salinity under Vegetation Cover

Current research on soil salinity inversion is mostly focused on bare soil, which can be inverted using various salinity indices via microwave remote sensing with penetration capability [70]. However, as the distribution and health of vegetation can act as indicators of soil salinity levels, plant cover has an impact on soil salinity monitoring. Studies have shown that vegetation indices can be used to indirectly measure the salinity of soil [71]. By analyzing the spectral information of vegetation under salt stress [72], the optimal model for estimating soil salinity at depths of 0 to 60 cm was found to be the quantile regression model, with an $R^2$ of 0.636 and an RMSE of 0.249. Using particle filtering algorithms based on different vegetation cover percentages [73], the optimal depth for indirect soil salinity inversion was found to be 20–40 cm, with an RMSE of 0.0422. The maximum depth of soil inversion reached 60 cm below ground. Subdividing the vegetation cover significantly improves the accuracy of soil salinity prediction. The best model for soil salinity inversion utilizing vegetation indicators for agricultural land with vegetation cover was determined to be the BPNN (back-propagation neural network) model, with a validation set $R^2$ of 0.836, an RMSE of 0.027, and an RPD of 2.100 [47].

Different regions have different soil environments and varying degrees of vegetation cover, which pose challenges to the universality of models. Studies have shown that soil salinity inversion models suitable for the city of Yichun, Jilin Province, may not be suitable for saline–alkali paddy fields in Hotan County, Xinjiang [74]. Rice fields, wheat fields, vegetable plots, and orchards are mainly distributed in areas with low salinity [75], while drylands, dominated by cotton planting, are distributed in areas with higher salinity. Additionally, research indicates that vegetation cover has a greater impact on reflectance spectra than soil moisture. Furthermore, a number of other variables affect the soil's salinity, including the amount and types of salt minerals present, soil color, surface roughness, soil texture, amount of organic matter, depth of groundwater, groundwater mineralization, geography [76], and climate.

### *4.5. Reversal of Soil Salinity Ions*

Not only can soil salinity be inversely estimated as a whole, but the individual ions responsible for the soil's salinity can also be inferred. Zhang et al. [77] conducted field-based, in situ spectral measurements of the soil and combined them with laboratory-determined pH values, EC, and soluble salt ion measurements. Based on the selection of spectral reflectance sensitive to various salt indicators and the optimal transformation methods, they used partial least squares regression modeling and stepwise regression analysis. The results showed similar patterns in the spectral characteristics of the soils with different types and amounts of salinization in the study area. They achieved the analysis of different salt ions, with the predictive model for $SO_4^{2-}$ based on sensitive bands in the 0–5 cm layer reaching an $R^2$ of 0.9676 using partial least squares regression. However, this study was limited to the bare surface soil. For various soil salt ions, the same inversion model's accuracy varies. Research indicates that soil salinity prediction models yield satisfactory results for total salt content, EC values, and $Na^+$, $K^+$, and $Cl^-$ ions, but the accuracy of predicting pH values and $Ca^{2+}$ ions is not high, and the prediction accuracy for $Mg^{2+}$ ions groundwater depth, groundwater mineralization, topography, and climate is insignificant.

### *4.6. Monitoring Soil Salinization Using Remote Sensing Model Building Techniques*

At the moment, the high-precision monitoring of soil salinization within regions is possible using remote sensing models for soil salinization inversion, which has also produced a wealth of research findings [78]. The contents of such research mainly include the establishment of new spectral indices [79], the coordinated use of different types of remote sensing data and scale conversion [80], the improvement of mathematical modeling methods, and the optimization of model parameters. In order to characterize soil salinization within regions and reflect the interrelationships between the soil salinization status and its influencing factors, the development of remote sensing models for soil salinization monitoring will remain a research hotspot in the field. The establishment of the model typically includes steps such as obtaining measured soil salinity data, acquiring remote sensing images, extracting and selecting modeling factors, establishing the model, and verifying its accuracy, although there may be differences between different studies. The pixel values obtained from remote sensing product data are an important source of modeling factor data. Remote sensing data, climate factors, soil physicochemical properties, terrain factors, spatial locations, and vegetation factors used for modeling can be obtained through band calculation and corresponding remote sensing data acquisition. Gorji et al. [81] used 25 remote sensing images to monitor agricultural land near Lake Tuz in Turkey over multiple time periods, producing a distribution map of soil salinity. They assessed the EC values of soil samples in the field, created five remote sensing indices of soil salinity, and then utilized regression analysis to link the measured data with the salinity indices produced from the remote sensing photos. The findings show how crucial remote sensing technology is for tracking and forecasting land salinization, which supports and ensures agricultural output.

In the modeling process, different factors can be divided into different sets according to the modeling requirements. Their methods for model construction mainly include linear regression models and machine learning models. It is essential to consistently research and develop new model algorithms in order to increase the accuracy of model construction, which needs to be implemented through the corresponding modules of different mathematical analysis software platforms. Numerous research findings show that the modeling accuracy of machine learning models is often higher than that of linear models and that the accuracy of various models varies significantly between experimental trials. Future studies will concentrate on expanding these models' applicability, making them more capable of reflecting the conditions of vegetation cover, and gathering data on the salinity of deep soil. This is consistent with the findings of Wang et al. [82].

The following challenges will affect the development of the remote sensing inversion of soil salinity in the future, as determined by the examination and discussion of the six research hotspots in this section—(1) Machine learning-based soil salinity inversion: Large numbers of training samples are necessary for machine learning, and overfitting—a situation in which the training samples' fitness greatly surpasses that of the validation samples—must be taken into account when choosing the model and its structural parameters; (2) Efficiency and scale of UAV-based remote sensing: UAV-based remote sensing for soil salinity inversion is efficient, more accurate, and cost-effective. Nevertheless, it is limited by problems specific to UAVs, which make it challenging to satisfy the demands of extensive area surveillance; (3) Soil salinity inversion based on data assimilation methods—Research on the scale conversion of soil salinity movement based on data assimilation methods is immature. This is primarily due to the complexity and spatiotemporal dependency of the driving factors of salt migration, which are based on soil hydrological processes; (4) Remote sensing inversion of soil salinity under vegetation cover: Variations in vegetation cover across different regions pose obstacles to the universality of models; (5) Remote sensing inversion of soil salinity ions: Current methods mainly focus on surface soil, and the accuracy of inversion varies for different salt ions. Although some salt ions have ideal results, the accuracy of inversion for certain ions remains unsatisfactory; (6) Construction of soil salinization monitoring models based on remote sensing: The applicability of models is affected by factors such as the research area or season, and a large number of modeling parameters can lead to significant errors. In general, future research priorities will include combining multi-element and multi-scale "space-ground" observation data of soil salinization, removing the impact of variables such as vegetation cover, and applying machine learning algorithms for multi-data fusion assimilation to create high-precision remote sensing models. This will help deduce the spatiotemporal evolution process of salinization and reveal the driving mechanisms behind it.

## 5. Conclusions and Outlook

### 5.1. Conclusions

This study utilized bibliometric visualization software to construct a knowledge map, providing a visual analysis of research on remote sensing inversion of soil salinity from 2000 to 2023. To comprehend the advancements in the remote sensing monitoring of soil salinization, a number of factors were visually analyzed, including the number of publications, authors, institutions, research fields, and keywords. In-depth discussions were conducted on the research hotspots identified through the keyword analysis, leading to the following conclusions:

(1) The overall trend of publication quantity in remote sensing inversion of soil salinity is increasing over time. Zhao, Gengxing was identified as the author with the highest publication quantity, while Metternicht, Gi was the most cited author.

(2) Regarding publication institutions, the International Institute for Geoinformatics Science and Earth Observation and the Chinese Academy of Sciences emerged as the primary publishing institutions. Notably, the United States and China have

made major contributions. Influential journals in this field include "Remote Sensing" and "Geoderma".

(3) Many fields have conducted research on the remote sensing monitoring of soil salinization, with the environmental field publishing the most frequently. This aligns with the environmental focus of the remote sensing monitoring of soil salinization, primarily addressing the environmental issue of soil salinization. Major research areas include the Hetao irrigation district in Inner Mongolia, Xinjiang, the Yellow River Delta, the Nile River Basin in Egypt, and the Mediterranean coast.

(4) Looking at development trends, keywords such as remote sensing and saline soil appeared with the highest frequency, primarily focusing on the identification, mapping, inversion, and prediction of soil salinity. The main objective is to propose a response mechanism for soil salinization issues. Furthermore, the emergence of keywords indicates a shift in the research frontier towards areas such as machine learning.

### 5.2. Outlook

Global soil degradation is primarily caused by soil salinization and secondary salinization, which are both influenced by climate change and human activity. Using bibliometric analysis, we found that since the year 2000, remote sensing technology has been crucial in tracking past changes, providing early warning, and keeping an eye on soil salinization—all of which have produced positive outcomes. Nevertheless, it must be acknowledged that there are still a number of significant gaps in our understanding of how salt affects soil inversion. Specifically, there are still a number of pressing problems that need to be resolved in order to obtain high-precision real-time data and the lengthy historical time series of soil salinity. The following are the future development directions for the remote sensing inversion of soil salinity, driven by advancements in computer modeling approaches and remote sensing technology:

(1) Unmanned aerial vehicle (UAV) technology: The advantages of UAV-based remote sensing include great mobility and high spatial resolution. Equipped with various sensors, UAVs can conduct real-time observations in key areas to obtain high-resolution and high-precision soil salinization monitoring data. Meanwhile, by integrating UAV hyperspectral data, near-ground hyperspectral data, and satellite remote sensing, salinization information can be comprehensively extracted to achieve the dynamic, large-scale, and accurate monitoring of soil salinization over extensive regions.

(2) Multivariate collaborative inversion of soil salinity: A more accurate representation of the spatiotemporal distribution of soil salinization can be achieved through the effective integration of visible near-infrared remote sensing, thermal infrared remote sensing, microwave remote sensing, topography, and meteorological data.

(3) Remote sensing has limited penetration depth. Currently, the deep soil's salinity is mainly estimated through modeling using conductivity meters and surface remote sensing observation data. Surface soil salinity estimated by remote sensing can serve as the upper boundary condition to accurately predict the soil's salinity profile based on soil hydrodynamics models. Additionally, variations in land types, soil wetness, groundwater depth, and soil types should be taken into consideration in studies on the possible use of remote sensing data for predicting soil salinity.

(4) The mechanisms behind variations in soil salinity can be uncovered by the use of long-term time-series soil salinity data. In order to investigate the origins of soil salinity and analyze its seasonal and interannual fluctuations, long-term time series data are essential.

(5) Using platforms such as Google Earth Engine (GEE, USA) and Pixel Information Expert Engine (PIE-Engine, China), it is now possible to monitor soil salinization on a broad spatial and temporal scale due to the advent of remote sensing cloud platforms.

(6) Future research paths for the dynamic monitoring and prediction of soil salinization will be made possible by combining remote sensing monitoring models of salinization with models of soil water and salt transport.

(7) To satisfy the demands of large-scale monitoring, unified quantitative inversion models of soil salinity at national or even worldwide scales must be established.

However, this study still has certain shortcomings. Firstly, the reliance on electronic databases as data sources may have resulted in the omission or the erroneous selection of literature, even when combined with WoS. To address this, it may be beneficial to incorporate more databases and utilize improved screening software. Secondly, the clustering and summarization of topics via the bibliometric software still require subjective judgment. Therefore, there is a need to improve intelligent algorithms to reduce the subjectivity of summarization. Additionally, increasing the flexibility of parameter settings in the visualization process of knowledge graphs is essential for enhancing the quality of the analysis. At the same time, the keyword network used in the bibliometric analysis of the research can be strengthened through the application of deep learning and natural language processing technology, the continuous development of which may help make the construction and analysis of the keyword networks more automated and precise in the future. Deep learning models can help identify keywords in the text and use them to construct the network structure, which in turn reveals the correlation between the documents.

**Author Contributions:** Writing—original draft, data curation, methodology, C.Y.; Writing—original draft; methodology; funding acquisition; validation; Supervision, Q.L.; Writing—original draft; methodology; funding acquisition; validation, T.M.; writing—review and editing, Y.S. and F.W. All authors have read and agreed to the published version of the manuscript.

**Funding:** National Natural Science Foundation of China (52069020). Research Program of science and technology at Universities of Inner Mongolia Autonomous Region (NJZY21498).

**Data Availability Statement:** No new data were created or analyzed in this study. Data sharing is not applicable to this article.

**Acknowledgments:** Thanks to National Natural Science Foundations for supporting this work. Thanks to China Geological Survey website. Thanks to the anonymous reviewers, academic editors and editors for their comments and suggestions.

**Conflicts of Interest:** The authors declare no conflicts of interest.

## References


1. Li, X. Synergistic Inversion of Soil Salinity Multi-Source Remote Sensing Data in Hetao Irrigation District. Master's Thesis, Inner Mongolia Agricultural University, Hohhot, China, 2017.
2. Wang, X. Study on Soil Dielectric Properties and Multisource Remote Sensing Moisture Inversion in Salinization Irrigated Area. Master's Thesis, Inner Mongolia Agricultural University, Hohhot, China, 2017.
3. Zhang, F. The Study on the Saline Soil Spectrum, Spatial Characteristic and Composition in the Arid Regions. Master's Thesis, Xinjiang University, Ürümqi, China, 2007.
4. Farifteh, J.; Farshad, A.; George, R.J. Assessing salt-affected soils using remote sensing, solute modelling, and geophysics. *Geoderma* **2006**, *130*, 191–206. [CrossRef]
5. Farifteh, J.; Van der Meer, F.; Atzberger, C.; Carranza, E.J.M. Quantitative analysis of salt-affected soil reflectance spectra: A comparison of two adaptive methods (PLSR and ANN). *Remote Sens. Environ.* **2007**, *110*, 59–78. [CrossRef]
6. Jimenez, L.O.; Rivera-Medina, J.L.; Rodriguez-Diaz, E.; Arzuaga-Cruz, E.; Ramirez-Velez, M. Integration of spatial and spectral information by means of unsupervised extraction and classification for homogenous objects applied to multispectral and hyperspectral data. *IEEE Trans. Geosci. Remote Sens.* **2005**, *43*, 844–851. [CrossRef]
7. Wang, J.; Liu, X.; Huang, F.; Tang, J.; Zhao, L. Salinity forecasting of saline soil based on ANN and hyperspectral remote sensing. *Trans. Chin. Soc. Agric. Eng.* **2009**, *25*, 161–166. [CrossRef]
8. Valeriano, M.M.; Epiphanio, J.C.N.; Formaggio, A.R.; Oliveira, J.B. Bi-directional reflectance factor of 14 soil classes from Brazil. *Int. J. Remote Sens.* **2007**, *16*, 113–128. [CrossRef]
9. Schaap, M.G.; Leij, F.J. Using neural networks to predict soil water retention and soil hydraulic conductivity. *Soil Tillage Res.* **1998**, *47*, 37–42. [CrossRef]
10. Walthall, C. A comparison of empirical and neural network approaches for estimating corn and soybean leaf area index from Landsat ETM+ imagery*1. *Remote Sens. Environ.* **2004**, *92*, 465–474. [CrossRef]
11. Liu, Q.; Cheng, Q.; Wang, X.; Li, X. Soil salinity inversion in Hetao Irrigation district using microwave radar. *Transactions of the Chin. Soc. Agric. Eng.* **2016**, *32*, 109–114. [CrossRef]

12. Sun, Y. Remote Sensing Retrieval of Electrical Conductivity of Saline Soil in West Jilin Province and Research on Temporal Variation in Thirty Years. Master's Thesis, University of Chinese Academy of Sciences (Northeast Institute of Geography and Agroecology), Jilin, China, 2020. [CrossRef]
13. Li, T.; Cui, L.; Xu, Z.; Hu, R.; Joshi, P.K.; Song, X.; Tang, L.; Xia, A.; Wang, Y.; Guo, D.; et al. Quantitative Analysis of the Research Trends and Areas in Grassland Remote Sensing: A Scientometrics Analysis of Web of Science from 1980 to 2020. *Remote Sens.* **2021**, *13*, 1279. [CrossRef]
14. Zhang, N.; Zhang, S.; Yang, H.; Zhang, J. Visualized Quantitative Research of Soil Pollution in the Guangdong-Hong KongMacao Greater Bay Area. *Environ. Sci.* **2019**, *40*, 5581–5592. [CrossRef]
15. Wu, X.; Wulantuya; Zhang, S.; Wurilige; Anggelima; Sun, D. Visualization Analysis on Grassland Soil Salinization in Arid Regions of China. *J. Inn. Mong. Norm. Univ.* **2024**, *53*, 120–128. [CrossRef]
16. Hou, J.; Hu, Z. Review on the Application of CiteSpace at Home and Abroad. *J. Mod. Inf.* **2013**, *33*, 99–103. [CrossRef]
17. Chen, C. CiteSpace II: Detecting and visualizing emerging trends and transient patterns in scientific literature. *J. Am. Soc. Inf. Sci. Technol.* **2005**, *57*, 359–377. [CrossRef]
18. Yan, S.; Gao, Y.; Wang, H.; LI, J.; Wang, X. Research status of agricultural brackish water irrigation based on CiteSpace. *Sci. Soil Water Conserv.* **2021**, *19*, 132–141. [CrossRef]
19. Ke, L.; Yin, S.; Liu, W. A bibliometric analysis of China's marine ecological economy based on CiteSpace. *Acta Ecol. Sin.* **2018**, *38*, 5602–5610. [CrossRef]
20. Chen, Y.; Chen, C.; Liu, Z.; Hu, Z.; Wang, X. The methodology function of Cite Space mapping knowledge domains. *Stud. Sci. Sci.* **2015**, *33*, 242–253. [CrossRef]
21. Quan, H.; Li, M. Analysis of the Authors'Group for Scientia Geographica Sinica. *Sci. Geogr. Sin.* **2001**, *21*, 570–574. [CrossRef]
22. Dong, H.; Liu, S.; Damdinsuren, B.; Hou, X. CiteSpace-based domestic ecological compensation study. *Acta Ecol. Sin.* **2022**, *42*, 8521–8529.
23. Chen, C.; Ibekwe-SanJuan, F.; Hou, J. The Structure and Dynamics of Co-Citation Clusters: A Multiple-Perspective Co-Citation Analysis. *J. Am. Soc. Inf. Sci. Technol.* **2010**, *61*, 1386–1409. [CrossRef]
24. Luo, J.; Zhao, W.; Li, W.; Zhao, S.; Lv, S.; Huang, L.; Sun, X. Using CiteSpace to Compare the Research Effort of China on Drinking Water in Comparison with the Rest of the World. *J. Irrig. Drain.* **2022**, *41*, 109–119. [CrossRef]
25. An, D.; Zhao, G.; Chang, C.; Wang, Z.; Li, P.; Zhang, T.; Jia, J. Hyperspectral field estimation and remote-sensing inversion of salt content in coastal saline soils of the Yellow River Delta. *Int. J. Remote Sens.* **2016**, *37*, 455–470. [CrossRef]
26. Wang, X.; Zhang, F.; Ding, J.; Kung, H.T.; Latif, A.; Johnson, V.C. Estimation of soil salt content (SSC) in the Ebinur Lake Wetland National Nature Reserve (ELWNNR), Northwest China, based on a Bootstrap-BP neural network model and optimal spectral indices. *Sci. Total Environ.* **2018**, *615*, 918–930. [CrossRef] [PubMed]
27. Daliakopoulos, I.N.; Tsanis, I.K.; Koutroulis, A.; Kourgialas, N.N.; Varouchakis, A.E.; Karatzas, G.P.; Ritsema, C.J. The threat of soil salinity: A European scale review. *Sci. Total Environ.* **2016**, *573*, 727–739. [CrossRef] [PubMed]
28. Yang, J.; Yao, R.; Wang, X.; Xie, W.; Zhang, X.; Zhu, W.; Zhang, L.; Sun, R. Research on Salt-affected Soils in China: History, Status Quo and Prospect. *Acta Pedol. Sin.* **2022**, *59*, 10–27. [CrossRef]
29. Yuan, G.; Chen, D.; Xu, Y.; Meng, D.; Zhang, Y.; Wang, X. Summary of methods for extracting soil salinization information. *J. North China Univ. Water Resour. Electr. Power (Nat. Sci. Ed.)* **2022**, *43*, 95–101. [CrossRef]
30. Feng, X.; Liu, Q. Regional Soil Salinity Monitoring Based on Multi-source Collaborative Remote Sensing Data. *Trans. Chin. Soc. Agric. Mach.* **2018**, *49*, 127–133. [CrossRef]
31. Huang, J.; Zhao, G.; Xi, X.; Cui, K.; Gao, P. xtraction of soil salinization information by combining spectral and texture data in the Yellow River Delta:A case study in Kenli District, Shandong Province. *J. Agric. Resour. Environ.* **2022**, *39*, 594–601. [CrossRef]
32. Zhao, Q.; Ding, J.; Han, L.; Jin, X.; Hao, J. Exploring the application of MODIS and Landsat spatiotemporal fusion images in soil salinization: A case of Ugan River-Kuqa River Delta Oasis. *Arid. Land Geogr.* **2022**, *45*, 1155–1164. [CrossRef]
33. Jia, P.; Shang, T.; Zhang, J.; Sun, Y. Inversion of soil salinity in dry and wet seasons based on multi-source spectral data in Yinbei area of Ningxia, China. *Trans. Chin. Soc. Agric. Eng.* **2020**, *36*, 125–134. [CrossRef]
34. Duan, Z.; Wang, X.; Sun, L. Monitoring and Mapping of Soil Salinity on the Exposed Seabed of the Aral Sea, Central Asia. *Water* **2022**, *14*, 1438. [CrossRef]
35. Abd El-Kawy, O.R.; Rød, J.K.; Ismail, H.A.; Suliman, A.S. Land use and land cover change detection in the western Nile delta of Egypt using remote sensing data. *Appl. Geogr.* **2011**, *31*, 483–494. [CrossRef]
36. Liu, Q. On Radar Inversion and Simulation of Salty Soil Salinization. *Bull. Surv. Mapp.* **2014**, 43–46. [CrossRef]
37. Zhang, W.; Jiang, L.; Ge, X.; Wang, Y.; Liang, J.; Li, Y. Hot spot analysis and future prospect of rural homestead research in China: Quantitative analysis based on CNKI and CiteSpace. *J. Arid. Land Resour. Environ.* **2022**, *36*, 16–25. [CrossRef]
38. Metternicht, G.I.; Zinck, J.A. Remote sensing of soil salinity: Potentials and constraints. *Remote Sens. Environ.* **2003**, *85*, 1–20. [CrossRef]
39. Douaoui, A.E.K.; Nicolas, H.; Walter, C. Detecting salinity hazards within a semiarid context by means of combining soil and remote-sensing data. *Geoderma* **2006**, *134*, 217–230. [CrossRef]
40. Ben-Dor, E. Quantitative Remote Sensing of Soil Properties. *Adv. Agron.* **2000**, *75*, 173–243. [CrossRef]
41. Fan, X.; Pedroli, B.; Liu, G.; Liu, Q.; Liu, H.; Shu, L. Soil salinity development in the yellow river delta in relation to groundwater dynamics. *Land Degrad. Dev.* **2011**, *23*, 175–189. [CrossRef]

42. Allbed, A.; Kumar, L.; Aldakheel, Y.Y. Assessing soil salinity using soil salinity and vegetation indices derived from IKONOS high-spatial resolution imageries: Applications in a date palm dominated region. *Geoderma* **2014**, *230–231*, 1–8. [CrossRef]
43. Wang, J.; Ding, J.; Yu, D.; Ma, X.; Zhang, Z.; Ge, X.; Teng, D.; Li, X.; Liang, J.; Lizaga, I.; et al. Capability of Sentinel-2 MSI data for monitoring and mapping of soil salinity in dry and wet seasons in the Ebinur Lake region, Xinjiang, China. *Geoderma* **2019**, *353*, 172–187. [CrossRef]
44. Abbas, A.; Khan, S.; Hussain, N.; Hanjra, M.A.; Akbar, S. Characterizing soil salinity in irrigated agriculture using a remote sensing approach. *Phys. Chem. Earth Parts A/B/C* **2013**, *55–57*, 43–52. [CrossRef]
45. Li, K.; Ding, J.; Han, L.; Ge, X.; Gu, Y.; Zhou, Q.; Lyu, Y. Digital mapping of soil salinization in a typical oasis based on PlanetScope images. *Arid. Land Geogr.* **2023**, *46*, 1291–1302.
46. Zhang, Z.; Tai, X.; Yang, N.; Zhang, J.; Hhuang, X.; CHEN, Q. UAV Multispectral Remote Sensing Soil Salinity Inversion Based on Different Fractional Vegetation Coverages. *Trans. Chin. Soc. Agric. Mach.* **2022**, *53*, 220.
47. Zhao, W.; Zhou, C.; Zhou, C.; Ma, H.; Wang, Z. Soil Salinity Inversion Model of Oasis in Arid Area Based on UAV Multispectral Remote Sensing. *Remote Sens.* **2022**, *14*, 1804. [CrossRef]
48. Wang, J.; Ding, J.; Yu, D.; Teng, D.; He, B.; Chen, X.; Ge, X.; Zhang, Z.; Wang, Y.; Yang, X.; et al. Machine learning-based detection of soil salinity in an arid desert region, Northwest China: A comparison between Landsat-8 OLI and Sentinel-2 MSI. *Sci. Total Environ.* **2020**, *707*, 136092. [CrossRef] [PubMed]
49. Tao, J.; Xu, G.; Weng, Y.; Fan, X. Progresses in Remote Sensing of Soil Salinization. *Hans J. Soil Sci.* **2020**, *8*, 190–195. [CrossRef]
50. Yang, N. Research on UAV Multispectral Remote Sensing Model for Estimating Soil Salt Content. Master's Thesis, Northwest Agricultural and Forestry University, Xianyang, China, 2021. [CrossRef]
51. Wang, Z. Soil Salinization Inversion and Risk Assessment Based on Fractional-Order Diff-Erentiation and Machine Learning. Master's Thesis, Xinjiang University, Ürümqi, China, 2021. [CrossRef]
52. Tang, P. Research on Application of Coupling Machine Learning and Microwave Data to Monitoring Soil Water and Salt in Oasis. Master's Thesis, Xinjiang University, Ürümqi, China, 2021. [CrossRef]
53. Heung, B.; Ho, H.C.; Zhang, J.; Knudby, A.; Bulmer, C.E.; Schmidt, M.G. An overview and comparison of machine-learning techniques for classification purposes in digital soil mapping. *Geoderma* **2016**, *265*, 62–77. [CrossRef]
54. Qi, G.; Zhao, G.; Xi, X. Soil Salinity Inversion of Winter Wheat Areas Based on Satellite-Unmanned Aerial Vehicle-Ground Collaborative System in Coastal of the Yellow River Delta. *Sensors* **2020**, *20*, 6521. [CrossRef] [PubMed]
55. Erkin, N.; Zhu, L.; Gu, H.; Tusiyiti, A. Method for predicting soil salinity concentrations in croplands based on machine learning and remote sensing techniques. *J. Appl. Remote Sens.* **2019**, *13*, 034520. [CrossRef]
56. Yan, K.; Chen, H.; Fu, D.; Zeng, Y.; Dong, J.; Li, S.; Wu, Q.; Li, H.; Du, S. Bibliometric visualization analysis related to remote sensing cloud computing platforms. *Natl. Remote Sens. Bull.* **2022**, *26*, 310–323. [CrossRef]
57. Lifu, Z.; Peng, M.; Xuejian, S.; Cen, Y.; Tong, Q. Progress and bibliometric analysis of remote sensing data fusion methods (1992–2018). *J. Remote Sens.* **2019**, *23*, 603–619. [CrossRef]
58. Mashimbye, Z.E.; Cho, M.A.; Nell, J.P.; De Clercq, W.P.; Van Niekerk, A.; Turner, D.P. Model-Based Integrated Methods for Quantitative Estimation of Soil Salinity from Hyperspectral Remote Sensing Data: A Case Study of Selected South African Soils. *Pedosphere* **2012**, *22*, 640–649. [CrossRef]
59. Yao, Z.; Chen, J.; Zhang, Z.; Tan, C.; Wei, G.; Wang, X. Effect of plastic film mulching on soil salinity inversion by using UAV multispectral remote sensing. *Trans. Chin. Soc. Agric. Eng.* **2019**, *35*, 89–97.
60. Wang, D. Quantitative Inversion of Soil Moisture and Salinity According to Different Salinization Grades in Coastal Saline Soil of Yellow River Delta. Master's Thesis, Shandong Agricultural University, Taian, China, 2020. [CrossRef]
61. Zhang, Z.; Chen, Q.; Huang, X.; Song, Z.; Zhang, J.; Tai, X. UAV-Satellite Remote Sensing Scale-up Monitoring Model of Soil Salinity Based on Dominant Class Variability-weighted Method. *Trans. Chin. Soc. Agric. Mach.* **2022**, *53*, 226–238+251. [CrossRef]
62. Chen, J.; Wang, X.; Zhang, Z.; Han, J.; Yao, Z.; Wei, G. Soil Salinization Monitoring Method Based on UAV-Satellite Remote Sensing Scale-up. *Trans. Chin. Soc. Agric. Mach.* **2019**, *50*, 161–169. [CrossRef]
63. Ivushkin, K.; Bartholomeus, H.; Bregt, A.K.; Pulatov, A.; Franceschini, M.H.D.; Kramer, H.; van Loo, E.N.; Jaramillo Roman, V.; Finkers, R. UAV based soil salinity assessment of cropland. *Geoderma* **2019**, *338*, 502–512. [CrossRef]
64. Hu, J.; Peng, J.; Zhou, Y.; Xu, D.; Zhao, R.; Jiang, Q.; Fu, T.; Wang, F.; Shi, Z. Quantitative Estimation of Soil Salinity Using UAV-Borne Hyperspectral and Satellite Multispectral Images. *Remote Sens.* **2019**, *11*, 736. [CrossRef]
65. Lin, L.; Yang, J.; Shi, L.; Zhou, F. Simplified model of solute transport in regional saturated-unsaturated porous media. *J. Hydraul. Eng.* **2007**, *38*, 342–348. [CrossRef]
66. Hassanli, M.; Ebrahimian, H.; Mohammadi, E.; Rahimi, A.; Shokouhi, A. Simulating maize yields when irrigating with saline water, using the AquaCrop, SALTMED, and SWAP models. *Agric. Water Manag.* **2016**, *176*, 91–99. [CrossRef]
67. Yao, R.; Yang, J.; Zheng, F.; Wang, X.; Xie, W.; Zhang, X.; Shang, H. Estimation of soil salinity by assimilating apparent electrical conductivity data into HYDRUS model. *Trans. Chin. Soc. Agric. Eng. (Trans. CSAE)* **2019**, *35*, 90–101. [CrossRef]
68. Zhang, Z.; Huang, X.; Chen, Q.; Zhang, J.; Tai, X.; Han, J. Estimation Method of Soil Salinity Based on Remote Sensing Data Assimilation. *Trans. Chin. Soc. Agric. Mach.* **2022**, *53*, 197–207.
69. Han, J. Research on Monitoring Method of Soil Salinization Based on Satellite Remote Sensing Data Assimilation. Master's Thesis, Northwest Agricultural and Forestry University, Xianyang, China, 2020. [CrossRef]

70. Guan, H.; Huang, J.; Li, L.; Li, X.; Miao, S.; Su, W.; Ma, Y.; Niu, Q.; Huang, H. Improved Gaussian mixture model to map the flooded crops of VV and VH polarization data. *Remote Sens. Environ.* **2023**, *295*, 113714. [CrossRef]
71. Wang, X.-m.; Zhou, X.-h. Estimation and inversion modeling of salinity of cotton field soil using remote sensing in the Delta Oasis of Weigan and Kuqa Rivers. *Agric. Res. Arid. Areas* **2018**, *36*, 250–254+262. [CrossRef]
72. Qiu, Y.; Chen, C.; Han, J.; Wang, X.; Wei, S.; Zhang, Z. Satellite Remote Sensing Estimation Model of Soil Salinity in Jiefangzha Irrigation under V egetation Coverage. *Water Sav. Irrig.* **2019**, 108–112.
73. Zhang, J. GF-1 Satellite Remote Sensing Monitoring Method for Soil Salinization Based on Data Assimilation. Master's Thesis, Northwest Agricultural and Forestry University, Xianyang, China, 2021. [CrossRef]
74. Peng, J.; Liu, H.; Shi, Z.; Xiang, H.; Chi, C. Regional heterogeneity of hyperspectral characteristics of salt-affected soil and salinity inversion. *Trans. Chin. Soc. Agric. Eng.* **2014**, *30*, 167–174.
75. Jia, J.; Zhao, G.; Gao, M.; Wang, Z.; Chang, C.; Jiang, S.; Li, J. Study on the relationship between winter wheat sowing area changes and soil salinity in the typical area of the Yellow River Delta. *J. Plant Nutr. Fertil.* **2015**, *21*, 1200–1208. [CrossRef]
76. Hu, Q.; Zhao, Y.; Hu, X.; Qi, J.; Suo, L.; Pan, Y.; Song, B.; Chen, X. Effect of saline land reclamation by constructing the "Raised Field-Shallow Trench" pattern on agroecosystems in Yellow River Delta. *Agric. Water Manag.* **2022**, *261*, 107345. [CrossRef]
77. Zhang, J.; Jia, P.; Sun, Y.; Jia, L. Prediction of salinity ion content in different soil layers based on hyperspectral data. *Trans. Chin. Soc. Agric. Eng.* **2019**, *35*, 106–115. [CrossRef]
78. Dong, Y.; Pan, H.; Wang, L.; Tang, Z. Bibliometric Visualization Analysis of Soil Salinization Remote Sensing. *Chin. J. Agric. Resour. Reg. Plan.* **2024**, 1–16. Available online: http://kns.cnki.net/kcms/detail/11.3513.S.20240129.1751.002.html (accessed on 1 May 2024).
79. Ding, J.; Wu, M.; Liu, h.; Li, Z. Study on the Soil Salinization Monitoring Based on Synthetical Hyperspectral Index. *Spectrosc. Spectr. Anal.* **2012**, *32*, 1918–1922. [CrossRef]
80. Xiao, S.; Ilyas, N.; Nuerbiye, M.; Zhao, J.; Adilai, A. Spatial and temporal analysis of soil salinity in Yutian Oasis by combined optical and radar multi-source remote sensing. *Arid. Zone Res.* **2023**, *40*, 59–68. [CrossRef]
81. Gorji, T.; Sertel, E.; Tanik, A. Monitoring soil salinity via remote sensing technology under data scarce conditions: A case study from Turkey. *Ecol. Indic.* **2017**, *74*, 384–391. [CrossRef]
82. Li, X.; Zhang, F.; Wang, Z. Present situation and development trend in building remote sensing monitoring models of soil salinization. *Remote Sens. Nat. Resour.* **2022**, *34*, 11–21. [CrossRef]

land

*Article*

# Optimization of Irrigation of Wine Grapes with Brackish Water for Managing Soil Salinization

**Vinod Phogat [1,2,3], Tim Pitt [1,2,3], Paul Petrie [1,2,3,4], Jirka Šimůnek [5,*] and Michael Cutting [6]**

1 Crop Sciences, South Australian Research and Development Institute, GPO Box 397, Adelaide, SA 5001, Australia; vinod.phogat@sa.gov.au (V.P.); tim.pitt@sa.gov.au (T.P.); paul.petrie@sa.gov.au (P.P.)
2 School of Agriculture, Food and Wine, The University of Adelaide, PMB No.1, Adelaide, SA 5064, Australia
3 College of Science and Engineering, Flinders University, Adelaide, SA 5042, Australia
4 School of Mechanical and Manufacturing Engineering, The University of New South Wales, Sydney, NSW 2052, Australia
5 Department of Environmental Sciences, University of California, Riverside, CA 92521, USA
6 Murraylands and Riverland Landscape Board, Murray Bridge, SA 5253, Australia; michael.cutting@sa.gov.au
* Correspondence: jiri.simunek@ucr.edu

**Abstract:** Water scarcity and quality are critical impediments to sustainable crop production. In this study, HYDRUS-2D was calibrated using field measurements of water contents and salinities in the soil under wine grapes irrigated with river water (*Rw*, 0.32 dS/m). The calibrated model was then used to evaluate the impact of (a) four different water qualities ranging from 0.32 (*Rw*) to 3.2 dS/m (brackish water, *Gw*) including blended (*Mix*) and monthly alternating (*Alt*) irrigation modes; (b) two rainfall conditions (normal and 20% below normal); and (c) two leaching options (with and without 30 mm spring leaching irrigation) during the 2017–2022 growing seasons. Irrigation water quality greatly impacted root water uptake (*RWU*) by wine grapes and other water balance components. Irrigation with brackish water reduced average *RWU* by 18.7% compared to river water. Irrigation with blended water or from alternating water sources reduced *RWU* by 8.8 and 7%, respectively. Relatively small (2.8–8.2%) average annual drainage (*Dr*) in different scenarios produced a very low (0.05–0.16) leaching fraction. Modeling scenarios showed a tremendous impact of water quality on the salts build-up in the soil. The average electrical conductivity of the saturated soil extract ($EC_e$) increased three times with *Gw* irrigation compared to *Rw* (current practices). Blended and alternate irrigation scenarios showed a 21 and 28% reduction in $EC_e$, respectively, compared to *Gw*. Irrigation water quality substantially impacted site-specific actual basal ($K_{cb\,act}$) and single ($K_{c\,act}$) crop coefficients of grapevine. Threshold leaching efficiency estimated in terms of the salt mass leached vs. added ($LE_s$; kg/kg) for salinity control ($LE_s > 1$) was achieved with *LFs* of 0.07, 0.12, 0.12, and 0.15 for the *Rw*, *Mix*, *Alt*, and *Gw* irrigations, respectively. Applying annual leaching irrigation (30 mm) before bud burst (spring) in the Mix and Alt with *Rw* and *Gw* scenarios was found to be the best strategy for managing irrigation-induced salinity in the root zone, lowering the $EC_e$ to levels comparable to irrigation with *Rw*. Modeling scenarios suggested that judicious use of water resources and continuous root zone monitoring could be key for salinity management under adverse climate and low water allocation conditions.

**Keywords:** grapevine; water quality; water balance; soil salinity; leaching; crop coefficients; drought seasons rainfall; leaf area index; HYDRUS-2D

**Citation:** Phogat, V.; Pitt, T.; Petrie, P.; Šimůnek, J.; Cutting, M. Optimization of Irrigation of Wine Grapes with Brackish Water for Managing Soil Salinization. *Land* **2023**, *12*, 1947. https://doi.org/10.3390/land12101947

Academic Editors: Tiago Brito Ramos, Maria da Conceição Gonçalves and Mohammad Farzamian

Received: 14 September 2023
Revised: 10 October 2023
Accepted: 18 October 2023
Published: 20 October 2023



## 1. Introduction

Water scarcity is a common occurrence and severely impacts irrigated agriculture in arid and semi-arid regions where rainfall totals are far less than the evapotranspiration demand. This skewed relationship is further widened by rapid population growth, industrialization, and increased living standards that compete directly with irrigated agriculture for supplementary water sources [1–3]. Water scarcity is increasing and is expected to become more severe with an unabated rise in greenhouse gas emissions, rising temperatures, frequent droughts, and changing precipitation patterns [4]. A rise in temperatures, for instance, could lead to soil moisture deficits and a growing risk of vegetation desiccation due to increased evapotranspiration [5,6], requiring more irrigation water for crop production [7,8] and reaching yield potentials [9]. Agroecological and economic consequences of climate change are expected to vary widely [10–12], and uncertainty in water availability will remain a crucial factor influencing irrigated agriculture, which already consumes 69% of freshwater resources on the planet [13].

Water allocations for irrigation in the river basins worldwide depend on rainfall, runoff generation, and resulting flows in the river system. Climate variability and uncertainty in irrigation water allocations can severely impact irrigated crops' sustainability, including viticulture. For example, in Australia, water allocations for irrigation during the Millennium drought (2002–2009) were severely reduced to as low as 18% of normal allocations, severely affecting the sustainable production and resilience of vineyards and other irrigated crops. Similar conditions of unreliable and limited water availability for irrigation also occur in other arid and semi-arid regions worldwide. Judicious integration of other water sources, such as brackish water, into irrigation schedules would enhance water security and increase the resilience of cropping systems to climate variability, resulting in long-term sustainability and economic viability of irrigated production systems. This additional natural resource is often poorly understood, undervalued, mismanaged, or abused [14].

Irrigation with marginal quality and brackish water has been practiced on various crops with varied impacts on yield and associated environments depending on site-specific climate, soil, water quality, crop sensitivity, and management practices [15–22]. Numerous wine-grape growing regions use groundwater or recycled water of varying quality (0.8–3.5 dS/m) for irrigation [23–26]. Saline/brackish water irrigation with unsuitable management practices can increase soil salinity [21,25–28]. Pitt and Stevens [29] reported that rootzone electrical conductivity of the soil saturation extract ($EC_e$) in the soil under the vine was four times higher than mid-row (1.5 dS/m) in a vineyard that has received long-term groundwater irrigation. Other studies also reported the development of saline and sodic conditions in the soil under grapevines [27,30,31] and other crops [15,18,32–34], which negatively impacts the soil physicochemical conditions and the growth and yield of crops. However, judicious site-specific use of brackish water following appropriate ameliorative management techniques can help reduce the harmful impacts on soils and provide long-term sustainable production of irrigated crops [21].

Grapevine growth, yield, and berry composition are affected to a varied extent by the degree and duration of salinity stress imposed by saline and brackish water irrigation [35]. Adverse impacts of high salinity on grapevines are two-pronged [36]: (a) reduced water uptake due to high osmotic pressure [35,37] and (b) vine mortality induced by high $Na^+$ and $Cl^-$ concentrations in the leaves. Numerous studies have reported high $Na^+$ and $Cl^-$ concentrations in leaves, juice, and wine [23,25,26,38], depending on the cultivar, rootstock, salt concentrations, and exposure time to saline conditions. High salt concentrations in the wine can exceed maximum residue limits, preventing sales in specific markets and producing adverse sensory outcomes [39]. However, Martínez-Moreno et al. [40,41] reported that deficit irrigation of grapevine with chloride- and sulfate-dominated saline water of 5 dS/m over five years showed an insignificant impact on vine growth and yield. Additionally, the wine received the best sensory scores in well-drained soil, highlighting the importance of drainage for salinity control. They concluded that site-specific soil type, climate, and drainage conditions created a favorable environment for rapid leaching of salts and reduced the detrimental impact on vine performance.

The use of saline water for irrigation can be managed through appropriate leaching of salts from the root zone depending on the level of crop salt tolerance [42]. The traditional irrigation management strategy is to provide extra water so that the ratio of the actual drainage depth to the irrigation depth, i.e., the leaching fraction ($LF$), satisfies the leaching requirement (*LR*) [42], maintaining the root zone salinity below the crops' tolerance threshold. Due to the complexities of the interactions between water, soil, and plant uptake, estimating and matching *LF* with *LR* is not always straightforward [43]. The unavailability of good-quality water and the complexity of implementing proper *LR* strategies for salinity control under drip irrigation systems may lead to enormous amounts of salt deposition in the soils [27,44].

Numerous studies describe multiple approaches to maintain the root zone salinity below the crop threshold, such as (a) the intra/inter-seasonal or physiological stage alternated use of fresh and saline waters for irrigation [26,31,45,46], (b) blending waters from multiple water sources [47–49], and (c) applying additional leaching irrigation if rainfall is insufficient to manage excessive salts [27,43,44,50]. Typically, these options have varied responses for managing the accumulation of salts in the root zone and require site-specific approaches, which are a function of soil type, climate, crop, and the composition of the irrigation water. Complex soil heterogeneity, widespread stratification (as it exists in Australian grapevine growing areas), and poor drainage conditions may further confound the efforts to ward off the salts buildup in the crop root zone [27,51,52].

Long-term field experiments involving numerous variables under varied soil, water, and climate conditions are expansive and labor-intensive. Mathematical models are excellent, cost-effective tools for studying the impact of climate, soil, water, and crop variables on water and solute transport in the soil [21]. Among available agro-hydrological models, HYDRUS (2D/3D) [53] has been widely used to simulate the water movement and salinity dynamics under drip irrigation [27,54–58]. The model became popular because of its flexibility to accommodate different types of complex boundary conditions, consider root water uptake, and account for the coupled impact of water and salinity stresses and its ease of use due to its user-friendly graphical interface.

The objectives of this investigation are two-pronged, viz., (1) to estimate the water balance components and actual single crop ($K_{c\,act}$), basal crop ($K_{cb\,act}$), and evaporation ($K_e$) coefficients for improving site-specific irrigation practices; and (2) to optimize the integration of brackish water in the irrigation schedule of wine grapes while managing the root zone salinity for sustainable production. To achieve these objectives, HYDRUS-2D was first calibrated for water balance and salinity dynamics in the soil under wine grapes during one season (2021–22). The calibrated model was then used to evaluate the impact of sixteen scenarios involving two qualities of irrigation water (river water—*Rw*, $EC_{iw}$ = 0.4 dS/m and brackish water *Gw*, $EC_{iw}$ = 3.2 dS/m); two modes of application [blended *Rw* and *Gw* in a 1:1 ratio, $EC_{iw}$ = 1.87 dS/m and monthly alternate use of *Rw* and *Gw*)]; two types of rainfall occurance [normal (*n*) and 20% reduced (*d*) annual rainfall]; and two leaching irrigation options [no leaching irrigation and annual leaching irrigation (*l*, 30 mm)] over multiple seasons (2017–2022) in heterogeneous soil. Subsequently, the model-predicted water and salt balance components were used to estimate the leaching fractions (*LF*) and leaching efficiency (*LE*) under different scenarios. Thus, the assessment of the entire growing system and potential water quality options aims to evaluate the incorporation of brackish water in the wine grapes' irrigation schedule while controlling soil salinization to maintain sustainable production with effective leaching strategies.

## 2. Materials and Methods

### 2.1. Description of the Experimental Site

An experimental site was established at Kimbolton Wines (Lat −35.32681, Long 139.06178) in the Langhorne Creek (LHC) Geographic indication in South Australia, between the Mount Lofty Ranges and Lake Alexandrina in the flood plains of the Angas and Bremer rivers. The Cabernet Sauvignon wine grapes were planted in 2001 on the Teleki 5C rootstock at a spacing of 2 m between vines and 3 m between vine lines. The vineyards were trained on a single wire double cordon spur pruned system. Vineyard rows are oriented in a North-South direction, facilitating better absorption of photosynthetically active radiation. A drip irrigation system was laid out to irrigate the wine grapes, placing a dripline along the vine line with pressure-compensated emitters spaced at 75 cm with a discharge rate of 2.4 L/h. Irrigation was scheduled based on the estimated soil water regime and local climate conditions in the Langhorne Creek region. Measurements of soil, water, and crop parameters were carried out over the entire growing season (July 2021 to June 2022) as described in the following sections.

Climate data were obtained from the nearest Landscape SA weather station (Langhorne Creek Central). Rainfall during the simulation period (2017–18 to 2021–22) varied from 283 to 504 mm, with an average of 372 mm, which is close to the long-term average rainfall (390 mm) at Langhorne Creek [59]. Most rain events during this period were smaller than 5 mm, which tended to wet the soil surface only and evaporate back into the atmosphere. Above-normal rainfall years potentially generate precipitation surplus during winter months when most rainfall occurs because the LHC region has a Mediterranean climate with hot and dry summers and cool and moist winters. Large rain events (>20 mm) play a key role in regulating the salts in the crop root zone. There were 1, 2, 5, 3, and 0 rainfall events of >20 mm during the 2017–18, 2018–19, 2019–20, 2020–21, and 2021–22 growing seasons, respectively. Seasonal reference crop evapotranspiration ($ET_0$) values estimated by a modified Penman-Monteith approach [60] varied in a narrow range (1015–1096 mm), showing small year-to-year variation.

### 2.2. Soil Water and Canopy Growth Measurements

Soil water distributions were monitored using in situ calibrated capacitance probes installed close to an emitter in the vine row by the vineyard manager with sensors every 10 cm to a depth of 120 cm. The soil moisture probe was installed to support irrigation scheduling for the vineyard. Soil solution extractors were installed at 30, 60, and 90 cm depths at two locations in the vine row near the lateral and adjoining the moisture probe. Soil solutions were collected at fortnightly intervals to estimate soil solution electrical conductivity ($EC_{sw}$). Soil solution samples could not be collected during the summer when deficit irrigation was imposed. Alternatively, salinity ($EC$) in the 1:5 soil-to-water solution was measured on soil samples collected from 0–15, 15–30, 30–60, and 60–90 cm soil depths near the extractors during the summer season. These measurements were used to supplement the calibration of the model.

A high-definition (1080 p) time-lapse camera was installed between two vines to monitor the canopy growth (Figure 1a). The camera was located at ground level and orientated upward to ensure that the sky was equally visible on each side of the image. Images were collected during the early morning to avoid overexposure of the image sections by the sun. Collected images were stored on an SD card and emailed using the 4G network so that the camera operations could be confirmed.

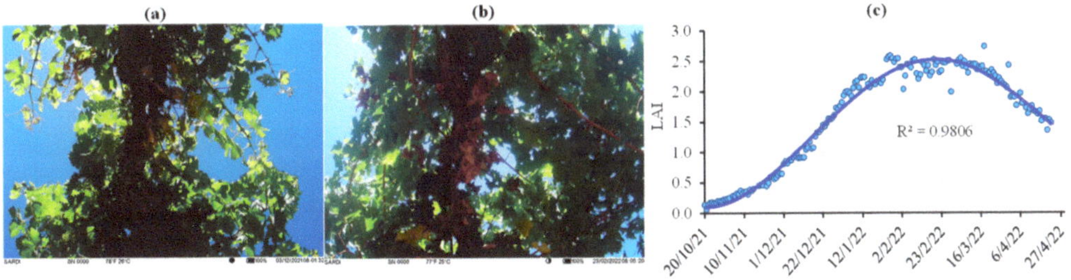


**Figure 1.** Grapevine canopy images (**a**,**b**) taken by an HD camera and (**c**) leaf area index (*LAI*) estimated by image analysis.

Images were analyzed using the ImageJ software (https://imagej.nih.gov/ij/index.html version 1.53k/Java1.8.0_172; accessed 27 August 2021). The canopy size for the entire growing season was indirectly assessed as the leaf area index (*LAI*), improving the methodology described in previous studies [61–64]. The algorithm used in these studies defines three fractions, which are the fractions of foliage of the projective cover ($f_f$), the crown cover ($f_c$), and the crown porosity ($\Phi$). The field view of the full canopy captured by the image is shown in Figure 1a. However, as the canopy keeps growing, the field view of the canopy eventually outgrows the vertical field view of the camera (Figure 1b). Under such conditions, fractional covers ($f_f$ and $f_c$) are not estimated correctly using the partial image of the canopy and are usually ignored in the LAI analysis [61]. For such situations, an approximate correction factor for the crown cover can be applied, assuming the $f_f/f_c$ ratio of the extended canopy is similar to the measurements of the field view by the camera. Therefore, Equation (A3) in [63] can be rewritten as

$$LAI_M = -f_c \frac{ln\Phi}{k} \left( \frac{W_{cci}}{W_c} \right) \quad (1)$$

where $W_c$ is the width of the field view of the canopy captured by the camera, $W_{cc}$ is the maximum width of the canopy measured at the cordon, $W_{cci}$ is the daily interpolation of width protruding the field view of the camera, and $LAI_M$ is the measured $LAI$, which corresponds to the ground area covered by the vertical projection of the foliage and branches [62,63]. $LAI_M$ was further corrected to obtain the effective $LAI$ ($LAI_e$), as reported in previous studies [61–63]. However, the estimated $LAI_e$ deviates from the usual definition of grapevine $LAI$, defined as the ratio of total green surfaces (leaves, shoots, and fruits when present) to the unit of the land area allocated to each vine [65,66]. Therefore, the second correction factor was applied to the $LAI_e$ to obtain corrected $LAI$ for the vine spacing ($LAI_{es}$) as given below:

$$LAI_{es} = LAI_e \left( \frac{W_{cc}}{W_r} \right) \quad (2)$$

where $W_r$ is the width between vine rows. $LAI_{es}$ estimated using Equation (2) indicated a continuous canopy growth from bud burst to harvest (Figure 1c). These methodology modifications reduced the average $LAI_{es}$ of mid to late grapevine season by half of $LAI_e$, measured using the previous method. The LAI values were comparable to those estimated by the Plant Canopy Analyser (LAI-2200C, LI-COR Biosciences, Lincoln, NE, USA). The $LAI$ values for other seasons (2017–18 to 2020–21) were estimated from the NDVI drawn from the Datafarming portal (https://www.datafarming.com.au/about-us/(accessed 27 August 2021)). A linear relationship ($Y = 4.07X$; $R^2 = 0.83$) was fitted between NDVI and measured $LAI$ during the 2021–22 season. This conversion was applied to the NDVI values from the other seasons (2017–18 to 2020–21) for estimating $LAI$. Similar relationships have been reported for the vineyards across various regions [63] and vineyards in other parts of the globe [67].

The $LAI$ data and other soil, climate, and plant parameters were used to estimate the daily values of potential evaporation and transpiration following the FAO-56 dual crop coefficient approach [60,68]. These values served as inputs for the numerical model HYDRUS-2D [53], which was used to simulate the water and salinity dynamics in the soil.

### 2.3. Irrigation Application and Water Quality

The property manager controlled irrigation application decisions at the study site. Irrigations were applied based on the profile's total available water (TAW) and the capacitance probe data. Profile water availability in the spring is key in initiating irrigation at the start of a new season. In-season irrigations were applied based on the phenological stage of the crop, climate conditions, and water deficit in the soil profile. Irrigation was applied when the stored water in the profile declined to a level lower than the readily available water (Table A1).

In-season irrigation water was collected in a catch can installed at the terminal point of the lateral irrigating vines at the study site. Water samples were collected fortnightly to measure electrical conductivity ($EC_{iw}$). The measured $EC_{iw}$ values during 2021–22 varied from 0.17 to 0.38 dS/m and served as the current irrigation water quality for calibrating the model. For the rest of the simulated period (2017–21), the daily $EC_{iw}$ values of the River Murray water ($Rw$) measured at the Langhorne Creek pipeline pumping station were used. Estimated $EC_{iw}$ values ranged from 0.16 to 0.61 dS/m, lower than the tolerance threshold (0.95 dS/m) of salinity-sensitive crops in Australia and New Zealand [69].

The groundwater quality at the Langhorne Creek Prescribed Groundwater Area is predominantly brackish, with salinity ranging between 300 and 30,000 mg/L [59]. The estimated groundwater salinity ($Gw$) in the nearest prescribed well used for irrigation remained around 3.2 dS/m, and this value was adopted in the simulations. The rainfall chemistry analyzed by Cresswell et al. [59] in the study region provides reliable information about rainfall salinity ($EC_{rw}$; 0.16 dS/m) used in simulations.

### 2.4. Estimation of Soil Hydraulic Properties

Soil hydraulic parameters were estimated from water retention curves determined on undisturbed soil samples. These samples were collected in 75 mm diameter and 50 mm tall rings from 0–15, 15–30, 30–50, 50–100, and 100–120 cm soil depths at two locations, representing the site-specific textural heterogeneity. Intact soil rings were first saturated overnight, and then saturated hydraulic conductivity ($K_s$) values were measured using the constant head method [70]. Subsequently, the cores were used to measure the saturated water content ($\theta_s$) and volumetric water contents ($\theta$) at −3 and −6, and −10, −33, −100, −300, and −1500 kPa pressure heads using hanging columns and the pressure plate apparatus, respectively. Measured $K_s$ and water content ($\theta$)—pressure head ($h$) data of each soil layer were used to estimate the soil hydraulic parameters (SHP) according to the van Genuchten-Mualem model [71]. Estimated SHPs were further finetuned by trial and error during model calibration. The final values used in HYDRUS-2D to simulate water movement in the soil are given in Table 1.

**Table 1.** Calibrated soil hydraulic properties ($\theta_s$ = saturated water content, $\theta_r$ = air dry water content, $\alpha$ = inverse of air entry value, $K_s$ = saturated hydraulic conductivity, $\eta$ = pore size distribution parameter, and $l$ = pore connectivity parameter) of different textural layers at the Langhorne Creek study site.

| Soil Depth | Texture | $\theta_r$ | $\theta_s$ | $A$ | $\eta$ | $K_s$ | $l$ |
|---|---|---|---|---|---|---|---|
| (cm) | | $cm^3cm^{-3}$ | | $cm^{-1}$ | | cm $day^{-1}$ | |
| 0–15 | Sandy Loam | 0.041 | 0.4259 | 0.0353 | 1.743 | 141.238 | 0.5 |
| 15–30 | Sandy Loam | 0.053 | 0.347 | 0.058 | 1.774 | 98.61 | 0.5 |
| 30–50 | Sandy Clay Loam | 0.1527 | 0.3942 | 0.0339 | 1.495 | 10.93 | 0.5 |
| 50–100 | Sandy Loam | 0.1044 | 0.3413 | 0.0226 | 1.795 | 7.75 | 0.5 |
| 100–120 | Sandy Clay Loam | 0.1908 | 0.3731 | 0.0403 | 1.363 | 6.93 | 0.5 |

### 2.5. Estimation of Potential Evaporation and Transpiration

HYDRUS-2D requires daily inputs of potential evaporation ($E_s$) and transpiration ($T_p$), representing the evaporative flux from the partially wetted soil surface and the transpiration flux via the canopy, respectively. In this study, these parameters were estimated using the FAO Penman-Monteith dual crop coefficient approach [60,68]:

$$ET_C = (K_{cb} + K_e)\, ET_0 \tag{3}$$

where $ET_C$ is evapotranspiration ($LT^{-1}$), $ET_0$ is reference evapotranspiration ($LT^{-1}$), $K_{cb}$ is the basal crop coefficient, which represents the plant transpiration component, and $K_e$ is the soil evaporation coefficient. Generic values of the grapevine $K_{cb}$ [60] were adjusted for the local climate and crop conditions, considering crop height, wind speed, and minimum relative humidity averages. Other crop and soil-related parameters for estimation of $T_p$ and $E_s$ using the FAO 56 approach, e.g., readily available water ($RAW$), total available water ($TAW$), readily evaporable water ($REW$), total evaporable water ($TEW$), and fractions of wetted and shaded areas ($f_c$, $f_w$), were estimated from the measured data as reported in previous studies [8,72]. Important climate and other parameters used in FAO-56 for different years are provided in the Appendix A (Figure A1 and Table A1).

*2.6. Model Construction, and Initial and Boundary Conditions*

Brief theoretical details on water movement and solute transport processes in the soil adapted in the model are given in the Appendix A. More information on the HYDRUS-2D software (version 5.02.0530) and related references can be found at https://www.pc-progress.com/en/Default.aspx?HYDRUS-3D (accessed 27 August 2021). A two-dimensional model domain was constructed to represent a vertical cross-section perpendicular to the vine line and equally spaced between mid-rows (Figure 2). The simulated domain was 120 cm deep and 300 cm wide (the row spacing at the field site), perpendicular to a vine row. The transport domain was divided into 13,048 finite elements with a very fine grid below the dripper (0.5 cm) and gradually increasing element sizes laterally (up to 3 cm) and vertically (up to 10 cm) from the dripper. Surface drip irrigation was simulated assuming an infinite line source, which was shown previously to be a good representation of the drip irrigation system [44,73].

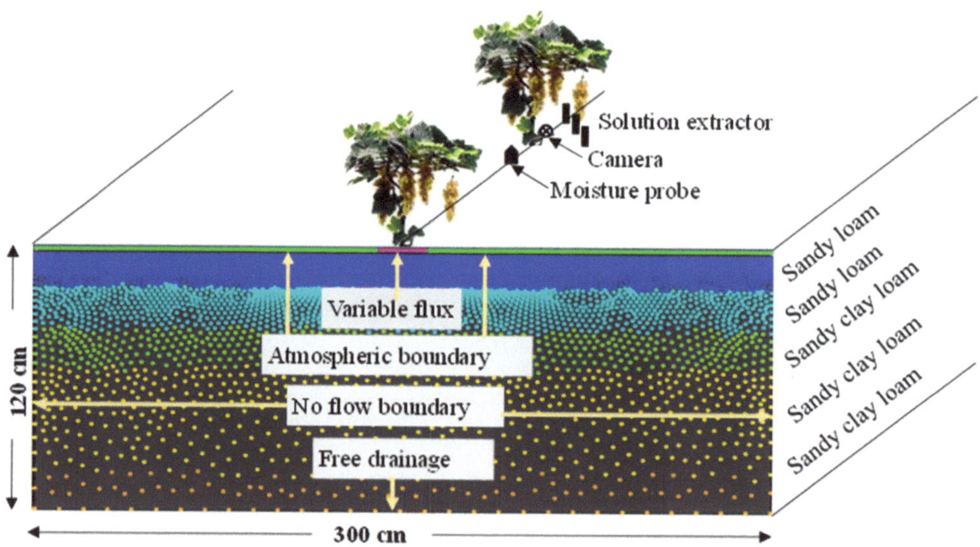


**Figure 2.** Model domain showing field observation setup (camera, moisture probe, and solution extractor), soil textures, and imposed boundary conditions.

Different textural layers (see different colors in Figure 2) were assigned to the domain depending on the measured depths of textural layers at the study site. A drip line runs along the vine line, and irrigation water is applied through the drippers placed on the soil surface. Measured values of the initial water content and soil solution salinity in different textural layers were used as initial conditions for water flow and solute transport in the soil. The soil surface was subjected to the atmospheric boundary condition (BC) and a variable flux BC imposed by dripper discharge (2.3 L/h), resulting in a two-dimensional flow. A free drainage boundary condition was assigned at the bottom, while a no-flow boundary condition was imposed on the sides of the domain. Concentration flux conditions were set as top and bottom boundary conditions for solute transport. Initial conditions reflected the effects of rainfall, irrigation, evaporation, and transpiration on soil salinity before the start of the experiment.

Root water extraction from the soil was computed using the macroscopic model approach [74]. This model adjusts plant root water uptake according to the local soil water pressure head, $h$, at any point in the root zone. It defines how potential transpiration ($T_p$) is reduced when the soil can no longer supply the plant's required water under the prevailing climatic conditions. Values of critical pressure heads for grapevine were taken from previous investigations in South Australia [8,72]. The multiplicative model for osmotic pressure head reductions is considered in this study. The model states that water is extracted at the maximum rate below the crop threshold osmotic head ($EC_e$ = 2.1 dS/m). The slope of the curve determines the fractional reduction of water uptake per unit increase in the osmotic head (12.8%/$EC_e$ unit) beyond the threshold. These parameters for grapevine were obtained from previous regional salinity tolerance studies [75].

Soil solution salinity ($EC_{sw}$) was simulated as a non-reactive solute, similarly as in many previous studies [57,72,76]. These studies showed good correspondence between model predictions and observed soil salinity dynamics in the soil under conditions involving intensive irrigation and fertigation environments. Other solute transport parameters, such as longitudinal and transverse dispersivities, initially assumed to be one-tenth of the modeling domain size and one-tenth of the longitudinal dispersivity, respectively [77], were optimized during the calibration.

Model calibration was performed to optimize the soil hydraulic and solute transport parameters in the soil to mimic the field conditions over the entire growing season. Subsequently, a calibrated model was used to simulate the soil's water balance and salinity dynamics under different climate and water quality scenarios.

### 2.7. Water Quality and Water Management Scenarios

Different scenarios were designed to assess the impact of using two water sources on seasonal evapotranspiration components and salinity dynamics. Water qualities ($EC_{iw}$) of the two sources were involved in various scenarios. For the drought season scenario, a rainfall reduction of 20% was assumed, which matches the future climate projections for South Australia [78]. An annual leaching irrigation ($l$) of 30 mm of $Rw$ was incorporated in various scenarios based on the historical salinity issues in the Langhorne Creek area and a previous leaching study [79]. Sixteen scenarios were designed to evaluate the impact of various quantities and qualities of irrigation water and their use in different application modes, such as mixing $Rw$ and $Gw$ in a 1:1 ratio (*Mix*) and alternate monthly use (*Alt*) of $Rw$ and $Gw$ (Table 2).

**Table 2.** Scenarios involving combinations of different water qualities ($Rw$ = river water, $Gw$ = brackish water, *Mix* = mixed $Rw$ and $Gw$ in a 1:1 ratio, *Alt* = a monthly alternate use of $Rw$ and $Gw$), the amount of annual rainfall (normal, $n$; and 20% less, $d$), and annual pre-season leaching irrigation ($LI$) of 30 mm simulated over the five consecutive grapevine growing seasons (2017–22).

| Water Quality | Normal Rainfall (n) | | Drought Season (d) | |
|---|---|---|---|---|
| | No Leaching | Leaching Irrigation (*l*) | No Leaching | Leaching Irrigation (*l*) |
| River water ($Rw$) | *Rw_n* | *Rw_nl* | *Rw_d* | *Rw_dl* |
| Brackish water ($Gw$) | *Gw_n* | *Gw_nl* | *Gw_d* | *Gw_dl* |
| Mixing $Rw$ and $Gw$ | *Mix_n* | *Mix_nl* | *Mix_d* | *Mix_dl* |
| Alternate use of $Rw$ and $Gw$ | *Alt_n* | *Alt_nl* | *Alt_d* | *Alt_dl* |

The calibrated model was used to understand the impact of different scenarios (Table 2) on the soil's water balance and salinity dynamics over the five seasons (2017–18 to 2021–22), using the growers' actual irrigation schedule.

### 2.8. Statistical Analysis

Model-simulated ($S$) spatiotemporal water content distributions and $EC_{sw}$ values were compared with corresponding measured ($M$) values using three error statistics, i.e., mean error ($ME$), mean absolute errors ($MAE$), and root mean square error ($RMSE$). Relevant equations for the error statistics can be found in [80]. The two-sided Dunnett's analysis test (XLSTAT) was applied to compare the differences in the average values of water balance components, namely, actual evapotranspiration ($ET_{C\,act}$), root water uptake (RWU) or actual transpiration ($T_{p\,act}$), soil evaporation ($E_s$), and deep drainage ($D_r$); leaching fractions (*LF*); and rootzone soil salinities ($EC_e$) obtained for the current irrigation practice and different water quality, drought, season, and leaching irrigation scenarios at a 95% confidence interval.

## 3. Results and Discussion

### 3.1. Model Calibration

HYDRUS-2D was calibrated for spatiotemporal water content distribution and salinity dynamics in the soil over the entire grapevine growing season (2021–22). Statistical errors (*RMSE, ME, MAE*) estimated for water content and salinity dynamics in the soil are shown in Table 3. *RMSE, ME,* and *MAE* values at different depths varied from 0.02 to 0.05, 0.00 to 0.05, and 0.02 to 0.05 $cm^3/cm^3$, respectively, showing slightly higher deviations at the soil surface than deeper in the soil profile. The average values for the entire profile varied in a much narrower range (0.02–0.04 $cm^3/cm^3$), showing a good agreement between measured and model-predicted water contents in the soil (Table 3). Similar error estimates between measured and modeled values have been reported in several studies in different soils, crops, and climate conditions [54,57,72,81,82]. It is well understood that soil water contents measured with sensors such as capacitance probes are not error-free. The magnitude of error estimates observed in the current study is equivalent to those often seen in the field measurements with capacitance probes [83].

**Table 3.** Root mean square error (*RMSE*), mean error (*ME*), and mean absolute error (*MAE*) values for water content dynamics at different soil depths and the average profile salinity.

| Statistics | Water Content ($cm^3/cm^3$) | | | | | | $EC_{sw}$ (dS/m) |
|---|---|---|---|---|---|---|---|
| | 20 cm | 40 cm | 60 cm | 80 cm | 100 cm | Average | |
| *RMSE* | 0.05 | 0.05 | 0.03 | 0.02 | 0.04 | 0.04 | 0.61 |
| *ME/MBE* | 0.05 | 0.04 | 0.00 | 0.00 | 0.02 | 0.02 | −0.13 |
| *MAE* | 0.04 | 0.05 | 0.02 | 0.02 | 0.03 | 0.03 | 0.42 |

The profile-averaged values of *RMSE, ME,* and *MAE* for soil solution salinity ($EC_{sw}$) were 0.61, −0.13, and 0.42 dS/m, respectively (Table 3). Chen et al. [54] reported *RMSE* and *MAE* values ranging between 0.23 and 0.55 dS/m, respectively, while Phogat et al. [84] reported error estimates between 0.35 and 0.86 dS/m for the measured and HYDRUS-2D predicted soil salinities. Similarly, Ramos et al. [57] estimated *ME, MAE,* and *RMSE* in the range of −0.21 to −0.67, 0.6 to 1.25, and 0.85 to 1.76 dS/m, respectively, for similar calibration of HYDRUS-2D for the salinity distribution in the soil. There are varied reasons for the disparity between measured and model-predicted $EC_{sw}$ values, such as measurement errors in the model input parameters, soil solution collection, and measurement errors. Phogat et al. [85] pointed out that the estimation of $EC_{sw}$ of the soil solution extracted from the soil under suction (a region of an uncertain size) may vary from $EC_{sw}$ simulated by the model at a specific point in the soil. The statistical comparison supports the ability of the model to simulate the distribution of water content and salinity dynamics in the soil.

### 3.2. Soil Water Balance

Average annual rainfall (*P*) and seasonal irrigation (*I*) applied during drought season (*d*) and leaching irrigation (*l*) scenarios differed significantly from the current irrigation practices (the *n* scenarios) (Table 4). This means that assumed drought season conditions and ameliorative salt management strategies have been appropriately applied. The average values of wine grape $ET_{C\,act}$ estimated for the sceanarios where leaching irrigation is applied (*Gw_nl, Alt_nl,* and *Mix_nl*) showed insignificant differences compared to the current practice (river water, *Rw_n*). It implies that integration of leaching irrigation (*l*) in the current irrigation schedule (*n*) in the brackish water (*Gw*), 1:1 blending (*Mix*), and monthly alternating (*Alt*) of river and brackish water irrigations can reduce the salt pressure in the root zone and provide a soil environment similar to the current practice. The blending and alternate mode scenarios without leaching irrigation also showed $ET_{C\ act}$ values comparable to current practices.

**Table 4.** Components of the annual water balance (*P* = precipitation/rainfall, *I* = irrigation, *RWU* = root water uptake, *Es* = evaporation, and *Dr* = drainage), leaching fraction (*LF*), and average rootzone salinity ($EC_e$, dS/m) estimated for different irrigation scenarios including the current irrigation practice (*Rw_n*). Statistical significance was estimated using Duncan's two-sided analysis of the difference between *Rw_n* (control) and other scenarios at a 95% confidence interval.

| Scenarios | P | I | $ET_{C\ act}$ | RWU | Es | Dr | LF | $EC_e$ |
|---|---|---|---|---|---|---|---|---|
| | | | (mm) | | | | (%) | (dS/m) |
| *Rw_n* | 373 | 209 | 424 | 218 | 206 | 39 | 0.07 | 2.2 |
| *Rw_d* | 299 * | 209 | 382 * | 201 | 181 * | 15 * | 0.03 * | 2.7 |
| *Rw_nl* | 373 | 237 * | 438 * | 223 | 216 * | 53 | 0.09 | 1.6 |
| *Rw_dl* | 299 * | 237 * | 401 * | 210 | 192 * | 26 | 0.05 | 2.2 |
| *Gw_n* | 374 | 209 | 409 * | 177 * | 232 * | 53 | 0.09 | 4.5 * |
| *Gw_d* | 299 * | 209 | 366 * | 154 * | 213 | 31 | 0.06 | 5.5 * |
| *Gw_nl* | 374 | 237 * | 426 | 188 * | 238 * | 65 * | 0.10 * | 3.9 * |
| *Gw_dl* | 299 * | 237 * | 387 * | 168 * | 219 * | 39 | 0.07 | 4.9 * |
| *Mix_n* | 374 | 209 * | 417 | 198 * | 219 * | 45 | 0.07 | 3.6 * |
| *Mix_d* | 299 * | 209 | 375 * | 177 * | 197 * | 22 | 0.04 | 4.3 * |
| *Mix_nl* | 374 | 237 * | 434 | 208 | 226 * | 58 | 0.09 | 2.9 |
| *Mix_dl* | 299 * | 237 * | 395 * | 190 * | 205 | 31 | 0.06 | 3.8 * |
| *Alt_n* | 374 | 209 | 418 | 202 * | 216 * | 44 | 0.08 | 3.2 |
| *Alt_d* | 299 * | 209 | 376 * | 183 * | 193 * | 21 | 0.05 | 3.9 * |
| *Alt_nl* | 374 | 237 * | 434 | 210 | 224 * | 57 | 0.09 | 2.6 |
| *Alt_dl* | 299 * | 237 * | 396 * | 194 * | 202 | 30 | 0.06 | 3.4 |
| CD (0.05) | 16.5 | 0.005 | 13.5 | 16.3 | 8.9 | 19 | 0.03 | 1.3 |

* Statistically significant at a *p*-value of 0.05. Irrigations with river water (*Rw*, 0.32 dS/m) under normal (*n*) rain (*Rw_n*), brackish water (*Gw*, 3.2 dS/m) under normal rain (*Gw_n*), a blend of *Rw* and *Gw* in a 1:1 ratio under normal rain (*Mix_n*), monthly alternate use of *Rw* and *Gw* under normal rain (*Alt_n*), *Rw* + 20% less rain (*Rw_d*), *Gw* + 20% less rain (*Gw_d*), *Mix* + 20% less rain (*Mix_d*), *Alt* + 20% less rain (*Alt_d*), *Rw_n* + annual leaching irrigation of 30 mm (*l*) before bud burst (*Rw_nl*), *Gw_nl*, *Mix_nl*, *Alt_nl*, *Rw_dl*, *Gw_dl*, *Mix_dl*, and *Alt_dl*.

Seasonal vine water uptake (*RWU*) accounts for 40–51% (154–223 mm) of the total water application, including irrigation and rainfall during the growing season (Table 4). The maximum water uptake was predicted for the *Rw_nl* scenario, representing the most favorable conditions for grapevine growth, i.e., river water irrigation coupled with annual leaching irrigation (*l*) of 30 mm before the bud burst. The simulated daily water balance responded to the diurnal changes in the weather conditions, irrigation, canopy characteristics, soil water regime, and osmotic conditions in the soil (Figure 3). Actual daily root water uptake (RWU) of grapevines under current practices (*Rw_n*) varied between 0.02 and 3.4 mm during the simulation. Note that higher *RWU* values were observed over the December to March period, which coincides with the maximum canopy size and *LAI* (Figure 1). Numerous studies have reported that transpiration in crops without full ground cover, such as grapevine (an energy surplus system), responds to the canopy size, ground cover, leaf area index [86–88], and canopy density [68,89,90] under non-saline conditions.

Rootzone salinity ($EC_e$) is an important factor affecting $ET_{C\ act}$ and *RWU* ($T_{p\ act}$) of grapevines. Both parameters decreased linearly in response to rootzone salinity. However, the impact of $EC_e$ was more pronounced for RWU, which decreased by 6.8% ($R^2$ = 0.96) for a unit increase in the rootzone $EC_e$ (dS/m). The impact of salinity on $ET_{C\ act}$ decreased to less than half (3.1%, $R^2$ = 0.42) as the surface evaporation component increased under brackish water irrigation. Ben-Asher et al. [91] reported a 20 and 65% reduction in potential vine transpiration for 1.8 and 4.8 dS/m saline water irrigation. Such a drastic reduction in vine transpiration occurred due to 4 times higher irrigation applications in different climatic conditions. A linear reduction in $ET_C$ in response to saline water irrigation and rootzone salinity was reported by many [50,92–95].

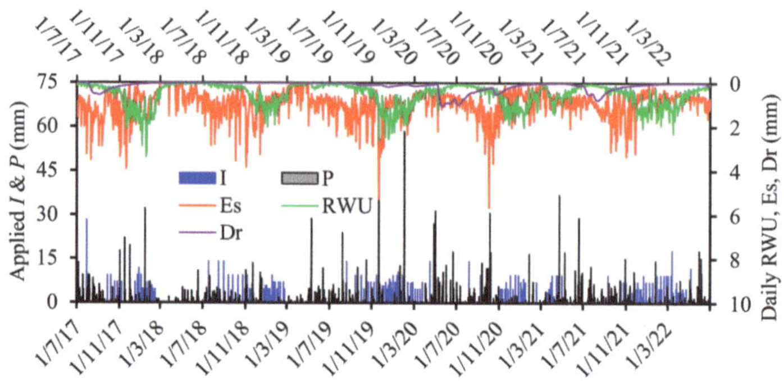


**Figure 3.** Model-predicted daily values of precipitation (*P*), irrigation (*I*), root water uptake (RWU), evaporation (*Es*), and drainage (*Dr*) below the root zone (120 cm) of a grapevine irrigated with river water under normal rainfall conditions (*Rw_n*; current practice) during 2017 to 2022.

Water quality and water stress impacted the seasonal water balance components. The minimum *RWU* was observed in the *Gw_d* scenario, where brackish water of 3.2 dS/m was used for irrigation coupled with 20% less rain (drought season, *d*), representing an adverse soil environment for grapevine growth. Average seasonal *RWU* under the brackish, blending, and alternate mode scenarios showed significant differences compared to the current practices (*Rw_n*), except in the alternate mode with leching irrigation (*Alt_nl*) and 1:1 blending (*Mix_nl*) scenarios (Table 4). This suggests that brackish water irrigation in the blending (*Mix*) and alternate (*Alt*) modes could be a viable irrigation strategy if appropriate leaching of salts from the root zone is maintained with annual leaching irrigation of 30 mm before the bud burst. The *Alt* scenarios, where the application of *Rw* and *Gw* is alternated, performed better in terms of greater RWU than the mixing scenarios (*Mix*), where the two water sources were blended before irrigation. The average seasonal *RWU* for the *Gw*, *Mix*, and *Alt* irrigation scenarios was reduced by 18.7, 8.8, and 7.0%, respectively, compared to *Rw* irrigation. Although the mixing and alternate modes of irrigation applications improved vine water uptake over *Gw* alone, the impact of salts was still visible compared to *Rw* irrigation. The reduction in RWU in the *Gw* and other scenarios can be ascribed to the osmotic stress in the root zone due to salt buildup in the soil, which reduces the ability of vine roots to extract water from the soil [96]. Ben-Asher et al. [65] reported a significant impact of saline water (4.8 dS/m) irrigation on the seasonal transpiration of grapevines. Irrigation with saline water not only induced an osmotic effect on the roots but also severely impacted the green area index, gaseous exchange processes, and photosynthetic activity in the leaves.

Seasonal evaporation losses (*Es*) accounted for 46–56% of the water applied, including rainfall during the growing season (Table 4). Significant differences in *Es* were observed in all scenarios relative to *Rw_n*, except for *Gw_d*, *Mix_dl*, and *Alt_dl*. Leaching irrigation (*l*) scenarios enhanced *Es* losses as 30 mm irrigation was applied in spring when a large portion of the soil surface was exposed. In the *Gw*, *Mix*, and *Alt* scenarios, a reduction in *RWU* due to osmotic stress allowed for an increased *Es* flux from the soil surface. Under reduced rainfall conditions (20% less), *Es* decreased significantly compared to *Rw_n*, although drought season scenarios with leaching irrigation (*l*) promoted surface water evaporation similar to *Rw_n*. Daily evaporation (*Es*) losses across different scenarios ranged between 0.01 and 5.6 mm (Figure 3), and high *Es* losses were observed during spring (August–September) to mid-summer (December–January). During the early growing period of crops with partial ground cover, such as grapevines, post-winter moist soil water regimes provide favorable soil surface conditions for evaporation losses. On the other hand, during mid-summer (December–January), the hot and dry season encourages high *Es* losses.

Seasonal *Es* losses of a similar extent in the current study have been reported in other studies in surface and subsurface drip-irrigated grapevines with similar spacing and climate [72,97]. In contrast, Fandiño et al. [98] reported an *Es* component comprising 8–15% of *ET* under different climate conditions due to using a low wetted fraction (0.01) in their estimation. In sparse vegetation such as grapevine, evaporation often constitutes a large fraction of *ET* due to the considerable area between the mid-rows exposed to the atmosphere [99].

Modeled average seasonal drainage (*Dr*) was remarkably low and varied from 2.8 to 8.2% of the total water application in different scenarios (Table 4). The lowest *Dr* value was observed in the *Rw_d* scenario, while the maximum *Dr* was recorded in the *Gw_nl* scenario. These scenarios significantly differed from the current practice (*Rw_n*) and showed the impact of drought season and annual leaching irrigation, respectively (Table 4). Low *Dr* values cannot leach the rootzone salts and favor the deposition of salts in the soil during the growing season. Daily values of *Dr* below the root zone (120 cm) varied in response to rainfall, irrigation, and leaching conditions during 2017–22 (Figure 4). Daily *Dr* varied from 0 to 1.2 mm and represented a negligible volume of leaching below the root zone over the low rainfall period (November 2017 to November 2019). Winter (June–August) rainfall during 2017 was also much lower (112 mm) than the long-term average (160 mm) in the study area [59]. The extent of winter rainfall plays a crucial role in generating the bottom flux below the root zone, which triggers the leaching of salts and maintains a favorable environment in the soil for water uptake. During the following two vintages (2017–18 and 2018–19), there was an extended period of below-average winter rainfall during late spring, when no leaching occurred (Figure 4). Therefore, most rain and irrigation events during this period were unable to generate drainage surplus below the root zone. This resulted in the rapid accumulation of salts in the root zone, which could impact the normal growth and yield of the vineyard.

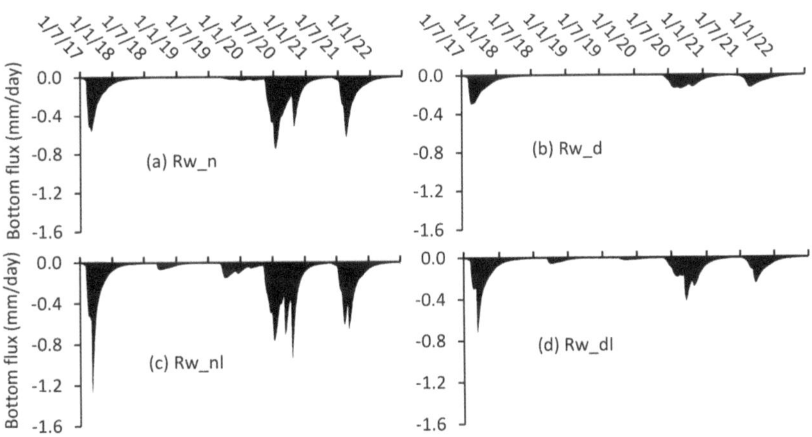


**Figure 4.** Model-predicted daily bottom fluxes (cm/day) below the root zone (120 cm) of a grapevine under (**a**) river water irrigation (*Rw_n*), (**b**) *Rw* + 20% less rain (*Rw_d*), (**c**) *Rw_n* + annual leaching (*Rw_nl*), and (**d**) *Rw_d* + annual leaching (*Rw_dl*) during 2017–22.

Three major leaching events occurred below the root zone over the simulation period from 2017–18 to 2021–22 (Figure 4). These were related to above-average annual rainfall and large rain events during late 2019–20 and throughout 2020–21; this triggered the rapid flushing of salts from the root zone and restored soil conditions. Phogat et al. [85] found that profile establishment is an essential management strategy early in the season to facilitate a favorable environment for root development, canopy growth, and profitable horticultural production in Mediterranean climates. The vineyards in the study region rely heavily on winter rainfall and flood events in the rain-induced local river catchments to flush the salts accumulated in the soil profile [100]. Therefore, to enhance the drainage volume, leaching irrigations play an important role in managing the accumulation of salts in the root zone, reducing the impact of osmotic stress, and providing a congenial medium for plant growth.

### 3.3. Crop Coefficients

The model predicted daily values of $T_{p\,act}$, $E_s$, and $ET_{C\,act}$ for different scenarios were used to compute the monthly actual basal crop coefficients ($K_{cb\,act}$), evaporation coefficients ($K_e$), and actual single crop coefficients ($K_{c\,act}$) of wine grapes under current irrigation practices (*Rw_n*) over the five seasons (Figure 5). The $K_{c\,act}$ values during the initial period ($K_{c\,act\,ini}$) in the current study varied from 0.33 to 0.84 and showed a large variation between growing seasons (Figure 5b). A comparison of average values of $K_{c\,act,ini}$ (0.56) and $K_{cb\,act,ini}$ (0.13) (average of September and October) showed a much higher contribution of evaporation during this period, which corresponds to the bud burst and initial canopy development. Normally, high $K_{c\,act\,ini}$ values and seasonal fluctuations largely reflect the variability in winter rainfall. For example, monthly rainfall of 61 and 65 mm during September and October 2020 (compared to 12 and 9 mm during 2018–19) resulted in a doubling of $K_{c\,act\,ini}$. A high value of $K_{c\,act\,ini}$ (0.33) was also reported by Silva et al. [101]. The maximum $K_{c\,act}$ (0.58) and $K_{cb\,act}$ (0.33) were recorded during November-December, which coincides with the maximum canopy size [102].

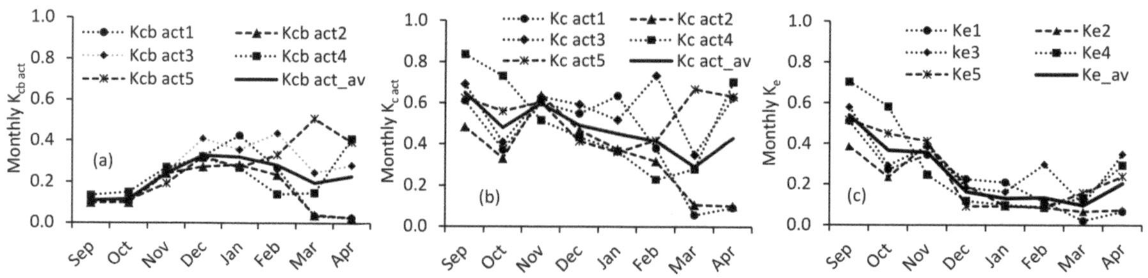


**Figure 5.** Estimated monthly values of (**a**) actual basal crop coefficient ($K_{cb\,act}$), (**b**) single crop coefficient ($K_{c\,act}$), and (**c**) evaporation coefficient ($K_e$) for drip-irrigated grapevines following current irrigation practices (*Rw_n*). Subscripts 1, 2, 3, 4, and 5 represent the values estimated for the 2017–18, 2018–19, 2019–20, 2020–21, and 2021–22 vintages, and '*av*' represents the average coefficient values for five vintages.

During the mid-season (November-February), large variations were recorded in the $K_{c\,act\,mid}$ values (0.23–0.73). The $K_{cb\,act}$ values for the mid-season ($K_{cb\,act\,mid}$) varied from 0.23 to 0.44, with a mean value of 0.31, much lower than the average $K_{cb\,mid}$ (0.6) adjusted for the study site under the prevailing climatic conditions. This indicates that grapevines were irrigated at an almost 50% deficit level. Yunusa et al. [97] and Cancela et al. [103] reported similar $K_{c\,act}$ values during the mid-season to the current study. The mean value of $K_{c\,act}$ during the post-harvest period ($K_{c\,act\,end}$) was, on average, 0.36, although higher in April (0.43). The end season $K_{cb\,act}$ ($K_{cb\,act\,end}$) ranged from 0.02 to 0.51 (average 0.21) and showed immense variability depending on post-harvest irrigation and soil water regime. The average $K_{cb\,act\,end}$ is less than half of the corresponding adjusted value (0.48), indicating deficit moisture regimes adopted by the growers. Very low values of $K_{cb\,act\,end}$ (0.02–0.04) for the 2017–18 and 2018–19 seasons represent extreme moisture-deficit conditions during that period (Figure 5). Significant seasonal fluctuations in the crop coefficients suggest that irrigation decisions made based on the same/standard $K_c$ values across seasons may lead to under- or over-irrigation, affecting the grapevine growth, yield, and efficiency of the irrigation system.

Irrigation water quality had a drastic impact on the crop coefficients. Mean $K_{cb\ ini}$ values for the *Gw* irrigation were reduced by 7–21% compared to the corresponding values for the *Rw* irrigation (Figure 6). The maximum reduction was observed for the drought (*d*) scenario, whereas the *Gw* irrigation with additional annual leaching irrigation (30 mm) showed the lowest impact. The $K_{cb\ mid}$ for the *Gw* irrigation was reduced by 19, 23, 16, and 20% for the normal rain (*n*), drought (*d*), *l* with normal rain (*nl*), and *l* with drought (*dl*) scenarios, respectively, compared to the respective $K_{cb\ mid}$ for the *Rw* scenario (Figure 6b). A similar reduction pattern (19–27%) in the $K_{cb\ end}$ values was observed for the *Gw* irrigation and $K_{cb\ act}$ values with other irrigation water quality scenarios (*Mix* and *Alt*). The $K_{cb\ ini}$, $K_{cb\ mid}$, and $K_{cb\ end}$ for the *Mix* and *Alt* water qualities decreased by 4–12, 7–11, 8–14% and 2–7, 5–8, 8–15%, respectively, compared to the corresponding values for the *Rw* irrigation (Figure 6c,d). The reduction in the $K_{cb\ act}$ values corresponds to the extent of the impact of rootzone salinity on *RWU* by the grapevine, which deviates from the standard conditions [60]. Rosa et al. [104] reported similar responses on the $K_{cb\ act}$ for maize under saline conditions.

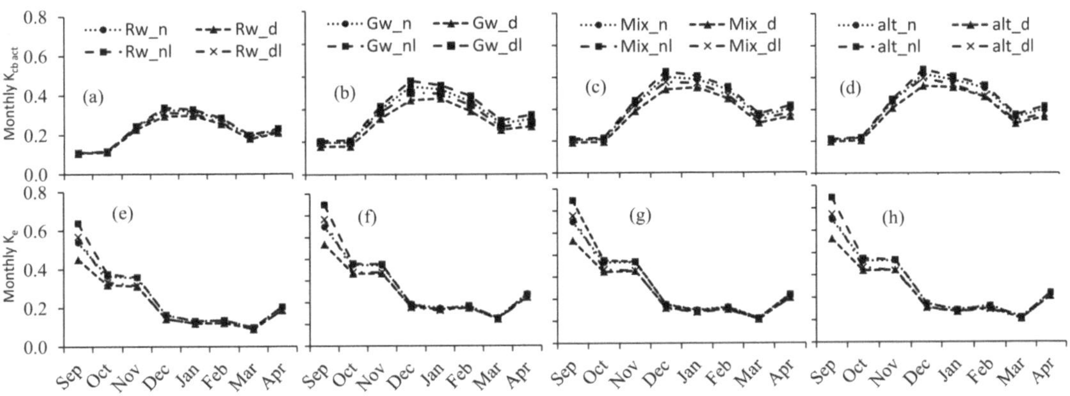


**Figure 6.** Mean values of monthly actual basal crop coefficients ($K_{cb\ act}$) (**a**–**d**) and evaporation coefficients ($K_e$) (**e**–**h**) for a wine grape irrigated with river water (*Rw*) (**a**,**e**), groundwater (*Gw*) (**b**,**f**), 1:1 blending of *Rw* and *Gw* (Mix) (**c**,**g**), and monthly alternate use of *Rw* and *Gw* (*Alt*) (**d**,**h**) under normal irrigation (*n*), drought season (*d*), leaching irrigation with normal rain (*nl*), and leaching irrigation under drought (*dl*).

A wide range of $K_{c\ act}$ and $K_{cb\ act}$ values of grapevines have been reported in many studies reviewed by Rallo et al. [90]. These studies have shown that site-specific growing conditions, including climate, cultivar, training system, ground cover, crop management, and soil conditions, impact evapotranspiration and crop coefficients of grapevines. Timing, variability, and extent of these factors lower the generic crop coefficient values [60,68] estimated by different methods [101,105,106]. Darouich et al. [107] evaluated the crop coefficients of grapevines at three locations in Italy and Portugal using the SIMDualKc model based on FAO-56 [60]. In their study, low values of $K_{cb\ ini}$ (0.09–0.14), $K_{cb\ mid}$ (0.27–0.4), and $K_{cb\ end}$ (0.17–0.32) emphasized the impact of site-specific climate, soil, cultivars, and management conditions. In contrast, some studies [108–110] reported high $K_{c\ mid}$ values (0.8–1.31). Normally, such deviations in crop coefficients from the generic values occur due to variable crop conditions due to insufficient or non-uniform irrigation, variable plant density and canopy cover, soil salinity, and/or agronomic management [111].

A wide variation was observed in the evaporation coefficient ($K_e$) estimated for the current practice (*Rw*) over the five seasons (Figure 6c). During the initial growth period under the current practice (*Rw_n*), large $K_e$ values were recorded (0.23–0.70) because a large portion of the soil surface was wetted by irrigation and rainfall was exposed to the atmospheric evaporative flux. However, the $K_e$ values during the mid-season declined (0.09–0.41; average 0.2) in response to a reduced fraction of the wetted and exposed surface Phogat [112]. $K_e$ towards the end of the season was only 0.15, except after the post-harvest irrigation, when it slightly increased (0.2) during April.

### 3.4. Salinity Dynamics in Soils

The spatial distributions of $EC_e$ in the soil under the *Rw_n* irrigation scenarios were compared at four phenostages of grapevines, i.e., bud burst, flowering, veraison, and harvest (Figure 7). At bud burst, $EC_e$ in the upper 40–50 cm of the soil was <1.5 dS/m, except during the 2018–19 season, which received below-normal rainfall. Without appropriate leaching, pockets of $EC_e$ in the 1.5–3 dS/m range developed in the surface soil alongside the vine row. At deeper depths, $EC_e$ usually varied in the 1.5–3 dS/m range, but some pockets of $EC_e$ in the 3–7 dS/m range were observed. Such variation in $EC_e$ in the soil profile is normal as salt concentrations dynamically vary in space and time in response to irrigation, rainfall, evaporation, and transpiration. Similar salt distributions were observed until flowering/full bloom, which provided favorable conditions for growth and good vine health during the initial stages of development sensitive to salinity.

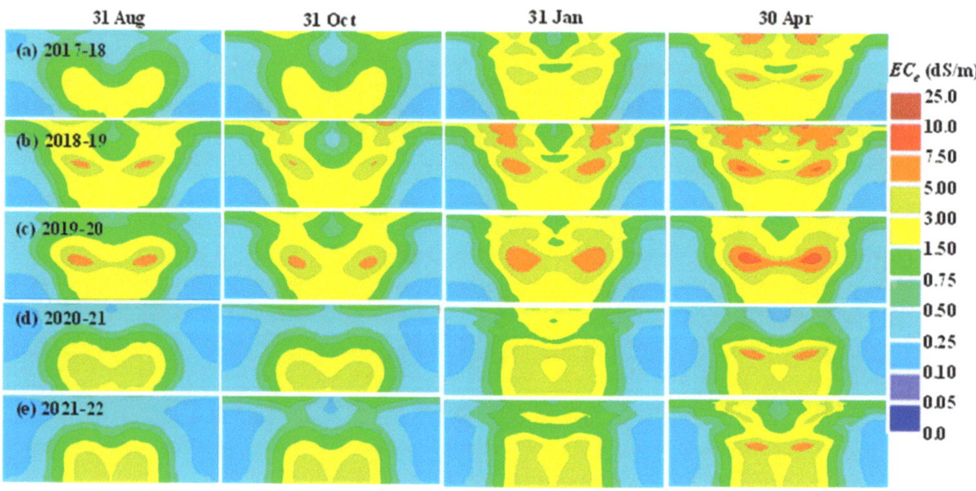


**Figure 7.** Spatiotemporal prediction of salinity dynamics ($EC_e$) in the soil under river water (*Rw*) irrigation at indicated dates (31 August—bud burst, 31 October—flowering, 31 January—veraison, and 30 April—harvest) during (**a**) 2017–18, (**b**) 2018–19, (**c**) 2019–20, (**d**) 2020–21, and (**e**) 2021–22 under a grapevine.

At veraison, the soil region with an $EC_e$ of 1.5–3 dS/m expanded, and pockets of much higher $EC_e$ (3–5 and 5–7.5 dS/m) also developed, especially during 2018–19 and 2019–20. These high salinity zones expanded further during the veraison to harvest period (February-April). During this stage, regulated deficit irrigation was applied to enhance the wine quality traits [113,114]. The imposition of deficit conditions results in the upward movement of water and salts in the soil in response to evaporative fluxes at the soil surface and transpiration fluxes from the root zone. This phenomenon (capillary rise) increases $EC_e$ in the region below the tree line compared to the mid-row region. Numerous studies reported similar observations under varied climate, cultivar, and irrigation conditions [28,31]. The salts deposited in the root zone need to be leached deeper into the soil before the start of the next growing season. Therefore, winter rainfall during the dormant stage plays a critical role in managing irrigation-induced salt accumulation in the soil.

The brackish water (*Gw*) irrigation scenario resulted in a rapid increase in $EC_e$ in the soil (Figure 8). At harvest in the 2017–18 season, $EC_e$ in the soil varied between 3 and 25 dS/m. These pockets of high salinity were normally observed in the upper soil layer (0–40 cm) away from the drip line as the lateral wetting front pushed the salts outward from the vine row [44]. Although subsequent winter rain in the following season (2018–19) reduced the high salinity regions, $EC_e$ remained very high at the time of bud burst and increased continuously during the growing season. In the subsequent seasons (2019–22), high $EC_e$ regions further expanded as there was limited rainfall to aid leaching, and more salts were added through irrigation. Eventually, a large region of high salinity (7.5–25 dS/m) developed during the five years of irrigation with *Gw*, which could impact the sustainability of irrigated vineyards. High soil salinity due to brackish water irrigation of grapevines has been previously reported in the study region [27,79] and other parts of the world under similar climates [36]. These results suggest that grapevine irrigation with Gw is unsustainable and would result in a rapid salt buildup in the soils. Spatiotemporal $EC_e$ distributions in other scenarios are expected to vary between the two extreme examples of irrigation water quality (Figures 6 and 7).

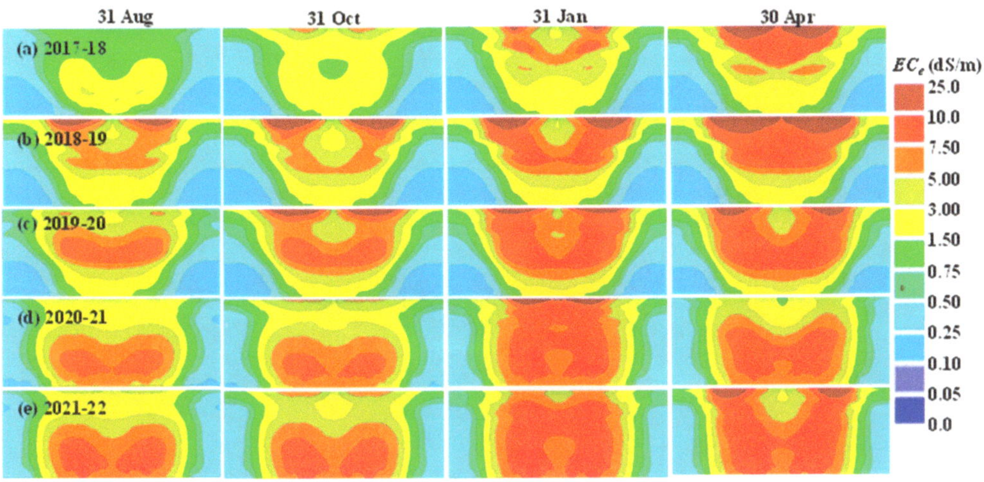


**Figure 8.** Spatiotemporal prediction of salinity dynamics ($EC_e$) in the soil under brackish water (*Gw*) irrigation at indicated dates during (**a**) 2017–18, (**b**) 2018–19, (**c**) 2019–20, (**d**) 2020–21, and (**e**) 2021–22 under a grapevine.

The model-predicted average daily rootzone salinity ($EC_e$) distributions for the 16 scenarios from 2017 to 2022 are compared in Figure 9. The data showed that the $EC_e$ values remained close to or lower than the tolerance threshold for the river water (*Rw*) irrigated scenarios except for reduced rainfall (*Rw_d*) where the salinity was slightly higher due to a rapid reduction in the drainage component. Alternating *Rw* and *Gw* irrigations (*Alt_nl*) and blending *Rw* and *Gw* in a 1:1 ratio (*Mix_nl*) with leaching irrigation scenarios showed rootzone $EC_e$ comparable to the *Rw* irrigation. These results suggest that these scenarios are promising approaches for the adaptation to drought seasons or low water allocations from the river. In all other scenarios, rootzone $EC_e$ increased consistently with time until April 2020. Thereafter, $EC_e$ decreased in response to good winter rainfall and a couple of large rain events (30 and 61 mm) occurring in April 2020, which generated intense leaching, pushing the salts deeper into the soil. Salinity ($EC_e$) increased again during the post-2021 vintage and then decreased gradually during the post-spring season (October 2021). These troughs and peaks in the rootzone salinity are normal occurrences and reflect the extent of leaching triggered by winter rainfall and irrigation. Normally, troughs contribute significantly to salt leaching below the root zone and help reduce the salinity to a great extent. Thus, spatiotemporal salinity dynamics in different scenarios represent a dynamic system that needs constant monitoring and appraisal of drainage, leaching, and soil degradation.

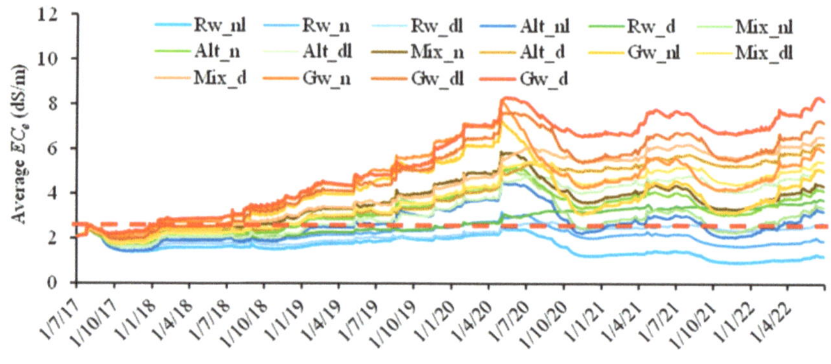


**Figure 9.** Model predicted average daily rootzone salinity ($EC_e$) in the soil under a grapevine irrigated with different water qualities and reduced rainfall scenarios (*Rw* = river water, *Gw* = brackish water, *Alt* = an alternate application of river and brackish water, *Mix* = mixing river and brackish water in a 1:1 proportion, *n* = normal rainfall, *d* = 20% less rainfall, and *l* = 30 mm annual leaching irrigation) during 2017–22. The red dotted line shows the threshold $EC_e$ (2.1 dS/m) for the grapevine salinity tolerance.

The daily average $EC_e$ values predicted by the model for different scenarios are box-plotted in Figure 10. All scenarios with *Rw* irrigation (*Rw_n*, *Rw_nl*, *Rw_d*, and *Rw_dl*) and alternate mode irrigation (*Alt_nl*) showed median rootzone $EC_e$ values close to the grapevine tolerance threshold. Rootzone salinity ($EC_e$) under all *Gw*, *Mix*, and *Alt* scenarios under drought season (*d*) showed significantly higher values than *Rw_n* (current practice). The average $EC_e$ under the brackish water (*Gw*) irrigation with reduced rain (*Gw_d*) increased three times relative to *Rw_n*. The *Mix* and *Alt* scenarios with reduced rainfall (drought season) also resulted in salt concentrations higher than the grapevine tolerance threshold. This indicates increased exposure of grapevine roots to high osmotic pressure, which may restrict plant uptake. Ben-Asher et al. [65] observed a four-fold increase in the rootzone salinity (from 2 to 8 dS/m) in sandy loam soil with an $EC_{iw}$ of 4.8 dS/m irrigation of grapevine over a single growing season.

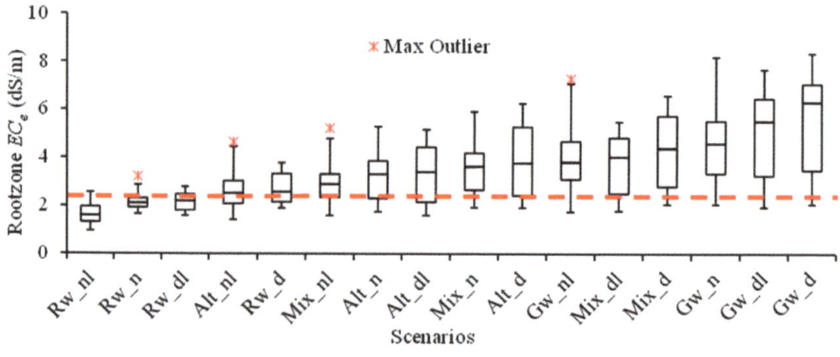


**Figure 10.** Distributions of predicted daily average rootzone salinity ($EC_e$, dS/m) in different scenarios (*Rw* = river water, *Gw* = brackish water, *Mix* = mixing *Rw* and *Gw* in a 1:1 proportion, *Alt* = monthly alternate use of *Rw* and *Gw*, *n* = normal rainfall, *d* = 20% less rainfall, and *l* = 30 mm annual leaching irrigation) during 2017–22. The red dotted line shows the threshold $EC_e$ (2.1 dS/m) for the grapevine salinity tolerance.

Blended water (*Mix_n*) showed a 21% reduction in salinity compared to brackish water irrigation (*Gw_n*). Monthly alternating *Rw* and *Gw* (*Alt_n*) irrigation reduced the salinity by 28% compared to *Gw_n*. Despite a reduction in salinity in these scenarios, the average $EC_e$ values (4.5 and 3.8 dS/m) remained higher than the corresponding $EC_e$ observed under the current practice (*Rw_n*). Average rootzone $EC_e$ values in the *Alt* (*Alt_n*) and *Mix* (*Mix_n*) scenarios under normal rainfall (*n*) were further reduced to 3.3 and 3.6 dS/m, respectively, but were still higher than the crop tolerance level. The *Alt_n* scenario showed non-significant differences in average $EC_e$ compared to the current irrigation practice (*Rw_n*).

The annual leaching irrigation of 30 mm during spring (1st September) showed a drastic reduction in the rootzone salinity values in all water quality scenarios (*Rw*, *Gw*, *Mix*, and *Alt*). Under normal rain (*n*), leaching irrigation (*l*) decreased the average rootzone salinity by 17–24% in different water quality scenarios, and the mean $EC_e$ values decreased to 2.5 and 2.9 dS/m, respectively, in the *Alt_nl* and *Mix_nl* scenarios, which were statistically at par with *Rw_n* and fell in the range of suitable for grapevines [75]. The reduction in $EC_e$ varied from 8 to 17% only under *l* for reduced rain scenarios (*Gw_dl*, *Mix_dl*, and *Alt_dl*) compared to their respective scenarios without leaching (*Gw_d*, *Mix_l*, and *Alt_l*) and $EC_e$ (3.2–4.9 dS/m) remained above the tolerance threshold for grapevines. These results suggest that annual leaching irrigation of more than 30 mm is required for salinity control in the heterogeneous soils at the study site.

### 3.5. Leaching Fraction

The fraction of applied water leaving the root zone (leaching fraction, *LF*) plays a key role in flushing the soluble salts and chemicals out of the root zone and helps facilitate the congenial environment for root water and nutrient uptake essential for plant growth. The estimated annual *LF* in the current study varied from 0.05 to 0.16 in different scenarios over the simulation period (2017–22) (Figure 11). Typically, *LF* smaller than 0.1 may favor the salt buildup in soils, posing a threat to the long-term sustainability of crops. Under the current irrigation practice (*Rw_n*), the *LF* was larger than 0.1 only during 2020–21 due to several large rainfall events (>30 mm) early in the season. It was able to produce an appreciable volume of drainage that transported salts below the root zone. It signifies the importance of large rain events to flush the irrigation-induced salinity below the root zone. However, high-intensity rain events may generate more runoff than infiltration into the soil, depending on the soil type and surface conditions.

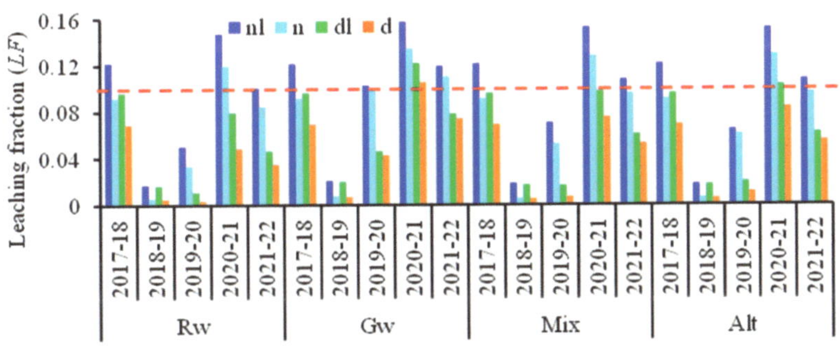


**Figure 11.** Annual leaching fractions (*LF*) below the root zone (120 cm) of a grapevine under different water quality (*Rw* = river water, *Gw* brackish water, *Mix* = mixed *Rw* and *Gw* in a 1:1 ratio, *Alt* = monthly alternate use of *Rw* and *Gw*) and management scenarios (*n* = normal rainfall, *d* = 20% less rainfall, and *l* = 30 mm annual leaching irrigation) during 2017–22 estimated from model predicted water balance components.

Extremely low *LF* values (0.005–0.021) were observed during the 2018–19 season across all scenarios, as annual rainfall (283 mm) was lower than the long-term average of 392 mm [59] in the region, and a few rain events >20 mm (which could generate *LF*) were almost negligible. Application of drought season conditions (20% reduction in the annual rain; the '*d*' scenarios) had a tremendous impact on the drainage volume because the annual

*LF* was reduced by 7–74% during different years. Average *LF* decreased by 33 to 52% in the reduced rainfall scenarios (the '*d*' scenarios) as compared to the normal rainfall scenarios (the '*n*' scenarios) under different water qualities (*Gw*, *Mix*, and *Alt*); the reduction was highest in *Rw* and the lowest in the *Gw* scenarios. The development of high salinity regions in the root zone of the *Gw* irrigation scenarios reduced *RWU* and allowed more water to drain below the root zone. Negative correlations were observed between the reduction in *RWU* and *LF* under *Gw* (−0.98), *Mix* (−0.98), and *Alt* (−0.98) irrigation scenarios. Other studies have also reported a similar negative correlation between transpiration and *LF* [115]. However, numerous other factors, such as soluble chemical constituents, soil physical state, and flow conditions, may also profoundly impact *LF*.

The application of annual leaching irrigation (*l*) of 30 mm in the spring over multiple growing seasons had a tremendous impact on the extent of drainage and leaching of soluble salts in all scenarios (Figure 11). *LF* increased from 3 to 225% in the normal irrigation scenarios (*nl*) and from 9 to 220% in the reduced rainfall scenarios (*dl*). The mean increase in *LF* for river (*Rw*), brackish (*Gw*), 1:1 blending (*Mix*), and monthly alternate (*Alt*) irrigation under normal and reduced rainfall was 31, 18, 26, and 21% and 55, 22, 38, and 33%, respectively, compared to the corresponding no-leaching irrigation scenarios. Among the seasons, 2018–19 recorded a four-fold increase in *LF*, although it was still less than 0.1. Even this small leaching fraction can greatly impact the salt balance in the soil. The *LF* increased above 0.1 during the 2017–18, 2020–21, and 2021–22 seasons, rapidly flushing the salts out of the root zone. The application of annual leaching irrigation (*l*) under *Rw* irrigation (*Rw_nl*) showed an overall depletion of 1.2 t salts/ha over the simulation period (Figure 12a). However, in all other scenarios, salt deposition varied from 2.2 to 16.7 t/ha in response to the salts added through different modes of irrigation application. Therefore, rootzone salt depletion occurred only in the *Rw_nl* scenario over the simulation period. Nonetheless, annual leaching irrigation with the *Gw*, *Mix*, and *Alt* irrigation modes produced a 1.8–2.2 t/ha reduction in the rootzone salts, indicating this as a viable management option to manage the irrigation-induced salts in the soil.

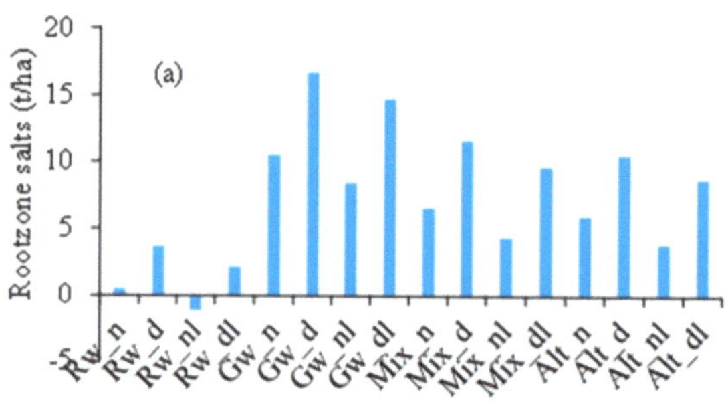


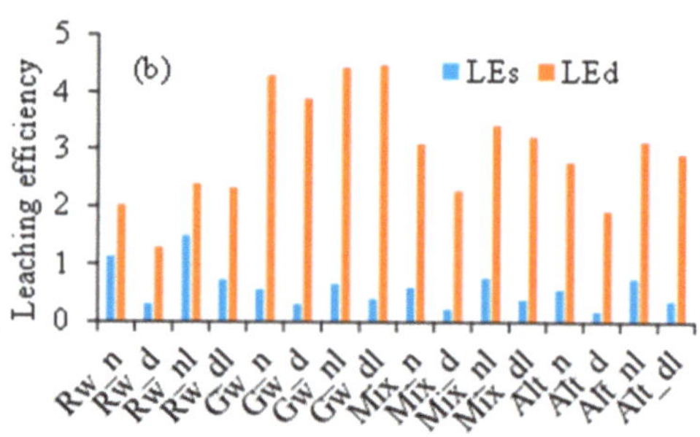


**Figure 12.** Leaching efficiency (**a**) in terms of salts leached/salts applied ($LE_s$, kg/kg), salts leached/drainage volume ($LE_d$, kg/m$^3$), and salts deposited (+)/depleted (−) in the soil root zone (120 cm) at the end simulations (**b**) for different scenarios (*Rw* = river water, *Gw* = brackish water, *Mix* = mixed *Rw* and *Gw* in a 1:1 ratio, *Alt* = monthly alternate use of *Rw* and *Gw*) and managements (under normal *n*, drought season *d*, and leaching irrigation *l* scenarios).

### 3.6. Leaching Efficiency

The leaching efficiency of salts ($LE_s$) is defined as the ratio of the salt mass leached below the crop root zone and the salt mass added through irrigation and rainfall during a given time (e.g., an annual or crop cycle), which offers a more quantitative measure of the salt balance [27,116] under drip irrigation. The maximum value of $LE_s$ (1.5 kg salts leached/kg added) was observed in the river water with the leaching irrigation (*Rw_nl*) scenario, followed by *Rw_n* (1.14 kg salts leached/kg added), while the lowest $LE_s$ values (0.2–0.3 kg salts leached/kg added) were found in the reduced rainfall scenarios (*Gw_d*,

*Rw_d*, *Alt_d*, and *Mix_d*), indicating a profound influence of drought period conditions, which lead to a rapid reduction in the leaching fraction and drainage volume (Figure 12b). Annual leaching irrigation (*l*) (30 mm) increased $LE_s$ by 31, 18, 29, and 34% in the *Rw*, *Gw*, *Mix*, and *Alt* scenarios, respectively, with normal rainfall compared to their respective scenarios without leaching irrigation. The annual leaching irrigation also showed a much higher increase in $LE_s$ (32–139%) in the reduced rain (drought season) scenarios.

Leaching efficiency in terms of the amount of salts leached with a drainage volume ($LE_d$) below the root zone varied from 1.3 to 4.5 kg/m$^3$ between scenarios (Figure 12b). The maximum value was observed in *Gw_dl* whereas the lowest was observed under *Rw_d* because of a higher amount of salts in the former scenario. It suggests that a small leaching fraction in the *Gw_dl* scenario triggered large quantities of salts transported out of the root zone that had accumulated from the Gw irrigation. The average $LE_d$ was larger than 1 in all scenarios, which may not imply that rootzone salinity has been reduced below the crop threshold. For example, even in the *Rw_n* scenario, the salt balance showed an increase of the initial salt mass by 0.53 t of salts/ha out of 4.05 tons/ha of salts added over the simulation period (2017–22) (Figure 12a). It suggests that $LE_d$ represents only the intensity factor for salt leaching and may not provide a valid criterion to estimate the leaching requirement (*LR*) for salinity management.

The *LR* is *LF*, which, when passed through the root zone, reduces the rootzone salts below the crop threshold [117]. Estimation of *LR* following steady state uniform soil conditions typically overestimates *LR* required for heterogeneous soils that have variations in their wetting patterns under drip and micro irrigation systems [118,119]. Hanson et al. [44] observed that, in drip-irrigated systems, it is hard to determine *LR* due to spatially variable soil wetting patterns that lead to localized leaching below the drip line, depositing large amounts of salts laterally along the wetting fronts, as observed in the current study (Figures 7 and 8). Calibrated numerical models that accurately simulate water and salt movement under drip irrigation can help develop better salinity management guidelines [19]. Phogat et al. [27] estimated $LE_s$ in various soils, climatic conditions, water qualities, and irrigation schedules applied to vineyards. They reported large differences in $LE_s$ values obtained for light-textured deep uniform soil (>1 kg salts leached/kg added) compared to heavy-textured heterogeneous soils (0.12–0.13 kg salts leached/kg added), showing a tremendous influence of inherent soil transmission properties. They concluded that in drip irrigation systems, *LF* that corresponds to $LE_s \geq 1$ could potentially leach annually added salts from the root zone. Such *LF* can serve as a good estimate for the leaching of irrigation-induced salts from the root zone. In the current study, $LE_s > 1$ was observed in the *Rw_nl* and *Rw_n* scenarios, which had an average *LF* of 0.09 and 0.07 (Figure 13), respectively, and the average rootzone salinity less than the tolerance threshold of grapevines (Figure 13), maintaining a favorable environment for normal vine growth. This suggests that $LE_s > 1$ proposed in the previous study [27] can serve as a good indicator for sustainable crop production for different soils, climates, and irrigation managements. However, restricted drainage conditions [51], high *SAR* of irrigation water, and high soil *ESP*, especially in the subsoils, have an immense impact on the mobility of salts in the soil [116,120,121] and require different management options for salinity control [21].

Water quality and annual rainfall strongly influence the salt leaching in the soil. A strong positive linear relationship ($R^2$ = 0.99) was found between irrigation water sources (*Rw*, *Gw*, *Mix*, and *Alt*) and the corresponding leaching fractions (*LF*) (Figure 13). A threshold $LE_s$ for salinity control ($LE_s > 1$) was achieved with an *LF* of 0.07, 0.12, 0.12, and 0.15 for *Rw*, *Mix*, *Alt*, and *Gw* water quality irrigation, respectively. This suggests that *LR* increased proportionately to the salts added through irrigation. Irrigating the grapevines below the water requirement added a smaller mass of salts in the soil but could not provide adequate leaching for salinity control in the study region [79]. The $LE_s$ also showed a very weak but positive linear relationship with annual rainfall, suggesting that an increase in annual rainfall influenced the mobilization of salts out of the root zone, depending on the timing, amount, and intensity of the rain events. It is worth noting that under the current irrigation practice (*Rw*), annual rainfall equal to the long-term average (390 mm) can generate $LE_s > 1$, suggesting a typical threshold for maintaining appropriate salt balance in the soil for irrigated viticulture at the study site (Figure 13b).

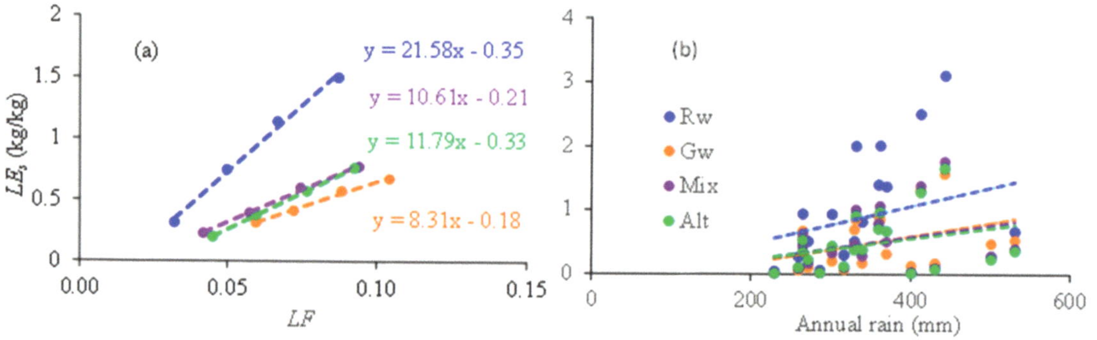


**Figure 13.** Relationship between (**a**) leaching efficiency in terms of salts leached/salts applied ($LE_s$, kg/kg) and the leaching fraction (*LF*) and (**b**) $LE_s$ and annual rainfall under river water (*Rw*), brackish water (*Gw*), mixed Rw and Gw in 1:1 proportion (*Mix*), and monthly alternate irrigation of *Rw* and *Gw* (*Alt*) at the study site.

Numerous other studies have shown a strong influence of rainfall on leaching salts out of the crop root zone. Cucci et al. [122] found that rainfall in autumn and winter produced a good salt-leaching effect in a sandy clay loam in southern Italy. Isidoro and Grattan [120] also reported that winter rainfall increased leaching compared to rainfall distributed uniformly throughout the year. Similar observations were reported for monsoonal climates [18], where annual rainfall with high-intensity events occurred during a short period (July–September).

## 4. Conclusions

An appropriate risk assessment is required for integrating brackish water (*Gw*) into crops' irrigation schedules to provide effective soil salinity management. Estimated low values of actual basal ($K_{cb\ act}$) and single ($K_{c\ act}$) crop coefficients estimated in this study over five seasons demonstrated that grapevines were irrigated under sustained deficit conditions. Under the Mediterranean-type climate, winter rainfall is considered to generate appropriate leaching to maintain a favorable environment. Simulated results confirm that rainfall reduction (20%) over the five years potentially leads to rapid soil salinization across all modes of irrigation scenarios [river water (*Rw*, 0.32 dS/m); brackish water (Gw, 3.2 dS/m); 1:1 blending of brackish and river water; and monthly alternate use of Rw and *Gw*], severely impacting the sustainable production of wine grapes. Under these conditions, it becomes quite imperative to manage the rootzone salts by applying appropriate leaching to maintain the sustainability of vineyards.

Models suggested that in the absence of favorable rainfall, leaching irrigation (*I*) with good-quality water (*Rw*) at the beginning of the growing season could best use this high-quality but low-availability water source. This strategy significantly increased the salt leaching efficiency ($LE_s$), and a threshold $LE_s$ for salinity control ($LE_s > 1$) was achieved with a leaching fraction (*LF*) of 0.07, 0.12, 0.12, and 0.15 for *Rw*, *Mix*, *Alt*, and *Gw* water quality irrigation, respectively.

Results further demonstrated that blending or alternating saline brackish water with non-saline surface water reduced the extent of salt deposition in the soil relative to the application of *Gw* only. These options can be explored further depending on the extent of salts in brackish water under projected climate change.

This study recommends applying leaching irrigation early in the season before bu burst when the soil is almost saturated with winter rain. It provides a strategic management option during drought season and periods of low freshwater allocation to develop resilience for maintaining sustainable production. However, these options need location-specific evaluation and continuous monitoring of climate, soil, and plant systems to enable long-term adaptation and resilience for irrigated viticulture and other crops.

**Author Contributions:** Conceptualization, V.P.; methodology, V.P.; software, V.P. and J.Š.; formal analysis, V.P.; investigation, V.P.; resources and project management, P.P.; data curation, V.P.; writing—original draft preparation, V.P.; writing—review and editing, V.P., J.Š., P.P., T.P. and M.C.; visualization, V.P.; supervision, T.P. All authors have read and agreed to the published version of the manuscript.

**Funding:** The authors acknowledge the financial support provided by the Department of Agriculture, Fisheries, and Forestry (DAFF), Australia as part of its Future Drought Fund—Natural Resource Management—Drought Resilience Program (4-G36U3HW).

**Institutional Review Board Statement:** Not applicable.

**Informed Consent Statement:** Not applicable.

**Data Availability Statement:** Data are available upon reasonable request to the authors.

**Acknowledgments:** The authors are grateful to Jenny Venus and Nicole Clark of Kimbolton Wines, Langhorne Creek, South Australia for providing access to the vineyard for conducting research activities and sharing site management datasets.

**Conflicts of Interest:** The authors declare no conflict of interest. The funders had no role in the study's design, in the collection, analyses, or interpretation of data, in the writing of the manuscript, or in the decision to publish the results.

## Appendix A

### *Appendix A.1. Brief Description of the Model*

This study used a numerical model (HYDRUS-2D; [53]) to simulate the water flow and salt transport in the soil under wine grapes. The governing two-dimensional water flow equation for an isothermal and isotropic medium is described as follows:

$$\frac{\partial \theta}{\partial t} = \frac{\partial}{\partial x}\left(K(h)\frac{\partial h}{\partial x}\right) + \frac{\partial}{\partial z}\left(K(h)\frac{\partial h}{\partial z} + K(h)\right) - S(h, h_s, x, z, t) \quad \text{(A1)}$$

where $\theta$ is the soil water content (L$^3$ L$^{-3}$); $t$ is the time (T); $h$ is the soil water pressure head (L); $x$ is the horizontal coordinate; $z$ is the vertical coordinate (positive upwards); $K(h)$ is the hydraulic conductivity (LT$^{-1}$); and $S(h, h_s, x, z, t)$ is the sink term accounting for water uptake by plant roots (L$^3$L$^{-3}$T$^{-1}$).

Water extraction $S(h, h_s, x, z, t)$ from the soil was computed according to the Feddes macroscopic approach [74]. In this method, the potential transpiration rate, $T_p$, is distributed over the root zone using the normalized root density distribution function, $\beta(x, z, t)$, and multiplied by the dimensionless water [$\alpha_1$ $(h)$] and salinity stress [$\alpha_1$ $(h_s)$] response functions as follows:

$$S(h, h_s, x, z, t) = \alpha_1(h, h_s, x, z, t)S_p(x, z, t) = \alpha_1(h, h_s, x, z, t)\beta(x, z, t)T_p(t) \quad \text{(A2)}$$

This model calculates plant root water uptake rates according to the local soil water pressure head, $h$, at any point in the root zone. It defines how potential transpiration (Tp) is reduced below potential when the soil can no longer supply the amount of water required by the plant under the prevailing climatic conditions. The multiplicative model for the osmotic head reduction is considered in this study as follows:

$$\alpha_1(h, h_s) = \alpha_1(h)\alpha_1(h_s) \quad \text{(A3)}$$

The reduction of root water uptake due to the water stress, $\alpha_1(h)$, is described as

$$\alpha_1(h) = \begin{cases} 0, & h > h_1 \, or \, h \leq h_4 \\ \frac{h-h_1}{h_2-h_1}, & h_2 < h \leq h_1 \\ 1, & h_3 < h \leq h_2 \\ \frac{h-h_4}{h_3-h_4}, & h_4 < h \leq h_3 \end{cases} \quad \text{(A4)}$$

where $h_1$, $h_2$, $h_3$, and $h_4$ are the threshold model parameters. Water uptake is at the potential rate when the pressure head is between $h_2$ and $h_3$, decreases linearly when $h > h_2$ or $h < h_3$, and becomes zero when $h < h_4$ or $h > h_1$. These critical values of pressure heads for grapevine were taken from previous investigations in South Australia [8,72].

Similarly, the threshold model was used to simulate the impact of osmotic (salinity) stress $\alpha_1(h_s)$, which states that water is extracted at the maximum rate below the crop threshold ($EC_e$ = 2.1 dS/m) osmotic head and the slope of the curve determines the fractional reduction of water uptake per unit increase in osmotic head (12.8%/$EC_e$ unit) beyond the threshold. These parameters for grapevine were obtained from previous regional salinity tolerance studies [75].

The spatial root distribution is defined in HYDRUS-2D according to Vrugt [123]:

$$\beta(x,z) = \left[1 - \frac{z}{z_m}\right]\left[1 - \frac{x}{x_m}\right]e^{-\left(\frac{p_z}{z_m}|z^* - z| + \frac{p_x}{x_m}|x^* - x|\right)} \quad \text{(A5)}$$

where $x_m$ and $z_m$ are the maximum width and depth of the root zone (cm), respectively, $z^*$ and $x^*$ describe the location of the maximum root water uptake from the soil surface in the vertical direction ($z^*$) and from the tree position in the horizontal direction ($x^*$), and $p_x$ and $p_z$ are empirical coefficients. These parameters for grapevine were optimized depending on the system design parameters and local soil and crop conditions [72].

*Appendix A.2. Solute Transport/Salinity Distribution in the Soil*

The governing advection-dispersion equation for the simulation of the transport of a single non-reactive ion in a homogeneous medium is described as

$$\frac{\partial(\theta c)}{\partial t} = \frac{\partial}{\partial x}\left(\theta D_{xx}\frac{\partial c}{\partial x}\right) + \frac{\partial}{\partial x}\left(\theta D_{xz}\frac{\partial c}{\partial z}\right) + \frac{\partial}{\partial z}\left(\theta D_{zz}\frac{\partial c}{\partial z}\right) + \frac{\partial}{\partial z}\left(\theta D_{zx}\frac{\partial c}{\partial x}\right) - \frac{\partial}{\partial x}(q_x c) - \frac{\partial}{\partial z}(q_z c) \quad \text{(A6)}$$

where $c$ is the concentration of the solute/salts in the liquid phase ($ML^{-3}$), $D$ is the dispersion coefficient ($L^2T$), and $q$ is the volumetric flux density ($LT^{-1}$). Soil solution salinity ($EC_{sw}$) is simulated as a non-reactive solute in the soil, as described in previous studies [57,76,85]. These studies showed good predictions of soil salinity dynamics in the soil under intensive irrigation and fertigation conditions.

The longitudinal dispersivity was assumed to be one-tenth of the size of the transport domain, with the transverse dispersivity being one-tenth of the longitudinal dispersivity Cote [77]. These values were additionally optimized during the calibration. The salinity of irrigation ($EC_{iw}$) was obtained from the water quality analysis available in the literature and from previous studies. The rainfall chemistry analysis by Cresswell and Herczeg [59] for the study region provided reliable information about rainfall salinity ($EC_{rw}$; 0.16 dS/m).

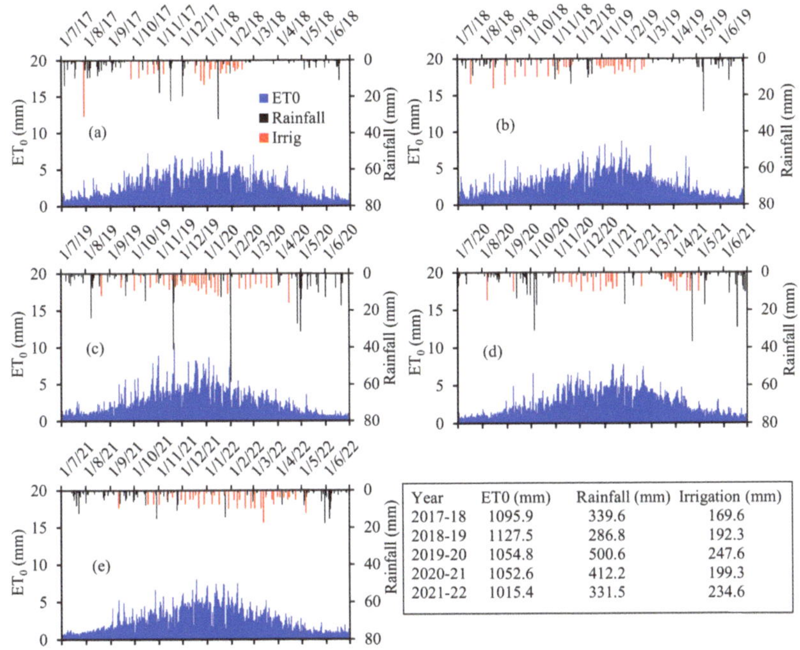


**Figure A1.** Daily values of reference crop evapotranspiration ($ET_0$), rainfall, and irrigation applied to wine grapes at the study site during (**a**) 2017–18, (**b**) 2018–19, (**c**) 2019–20, (**d**) 2020–21, and (**e**) 2021–22. Annual amounts of $ET_0$, rainfall, and irrigation are also given.

**Table A1.** Values of different parameters used to estimate daily transpiration ($T_p$) and daily evaporation ($E_s$) for wine grapes for the field site following the FAO-56 DCC approach.

| Parameter | Value | Parameter | Season | Value |
|---|---|---|---|---|
| $L_{ini}$ | 20 | Mid-season Min RH (%) | 2017–18 | 39.2 |
| $L_{dev}$ | 40 | | 2018–19 | 36.9 |
| $L_{mid}$ | 120 | | 2019–20 | 33.7 |
| $L_{late}$ | 60 | | 2020–21 | 36.9 |
| $\theta_{fc}$ (0–15 cm) | 0.23 | | 2021–22 | 41.0 |
| $\theta_{fc}$ (15–30 cm) | 0.18 | Mid-season Av. Wind speed (m/s) | 2017–18 | 1.9 |
| $\theta_{fc}$ (30–60 cm) | 0.31 | | 2018–19 | 2.1 |
| $\theta_{fc}$ (60–100 cm) | 0.29 | | 2019–20 | 2.2 |
| $f_w$ | 0.3 | | 2020–21 | 1.8 |
| $f_w$, Rain | 1 | | 2021–22 | 1.8 |
| $RAW$ (mm) | 84 | $K_{cb\ mid\ adj}$ | 2017–18 | 0.59 |
| $TAW$ (mm) | 153 | | 2018–19 | 0.61 |
| $TEW$ (mm) | 21 | | 2019–20 | 0.62 |
| $REW$ (mm) | 7.4 | | 2020–21 | 0.60 |
| Plant height $h$ (m) | 1.5 | | 2021–22 | 0.59 |
| Rooting depth $Z_r$ (m) | 1.0 | $K_{cb\ end\ adj}$ | 2017–18 | 0.46 |
| Evaporable depth $Z_e$ (m) | 0.15 | | 2018–19 | 0.48 |
| $K_{cb\ ini}$ (generic) [1] | 0.20 | | 2019–20 | 0.49 |
| $K_{cb\ mid}$ (generic) [1] | 0.65 | | 2020–21 | 0.47 |
| $K_{cb\ end}$ (generic) [1] | 0.50 | | 2021–22 | 0.46 |

[1] Basal crop coefficient values for medium-density wine grapes from Allen and Pereira [68].

## References


1. Döll, P.; Fiedler, K.; Zhang, J. Global-scale analysis of river flow alterations due to water withdrawals and reservoirs. *Hydrol. Earth Syst. Sci.* **2009**, *13*, 2413–2432. [CrossRef]
2. Huang, Z.; Yuan, X.; Liu, X. The key drivers for the changes in global water scarcity: Water withdrawal versus water availability. *J. Hydrol.* **2021**, *601*, 126658. [CrossRef]
3. Wada, Y.; Bierkens, M.F.P. Sustainability of global water use: Past reconstruction and future projections. *Environ. Res. Lett.* **2014**, *9*, 104003. [CrossRef]
4. IPCC. *Climate Change 2022: Impacts, Adaptation and Vulnerability*; IPCC: Cambridge, MA, USA, 2022; p. 3056.
5. Riediger, J.; Breckling, B.; Nuske, R.S.; Schröder, W. Will climate change increase irrigation requirements in agriculture of Central Europe? A simulation study for Northern Germany. *Environ. Sci. Eur.* **2014**, *26*, 18. [CrossRef] [PubMed]
6. Yang, H.; Luo, P.; Wang, J.; Mou, C.; Mo, L.; Wang, Z.; Fu, Y.; Lin, H.; Yang, Y.; Bhatta, L.D. Ecosystem Evapotranspiration as a Response to Climate and Vegetation Coverage Changes in Northwest Yunnan, China. *PLoS ONE* **2015**, *10*, e0134795. [CrossRef] [PubMed]
7. Phogat, V.; Cox, J.W.; Šimůnek, J. Identifying the future water and salinity risks to irrigated viticulture in the Murray-Darling Basin, South Australia. *Agric. Water Manag.* **2018**, *201*, 107–117. [CrossRef]
8. Phogat, V.; Cox, J.W.; Mallants, D.; Petrie, P.R.; Oliver, D.P.; Pitt, T.R. Historical and future trends in evapotranspiration components and irrigation requirement of winegrapes. *Aust. J. Grape Wine Res.* **2020**, *26*, 312–324. [CrossRef]
9. Vogel, E.; Donat, M.G.; Alexander, L.V.; Meinshausen, M.; Ray, D.K.; Karoly, D.; Meinshausen, N.; Frieler, K. The effects of climate extremes on global agricultural yields. *Environ. Res. Lett.* **2019**, *145*, 054010. [CrossRef]
10. Bindi, M.; Olesen, J.E. The responses of agriculture in Europe to climate change. *Reg. Environ. Chang.* **2011**, *11*, 151–158. [CrossRef]
11. Iglesias, A.; Garrote, L.; Quiroga, S.; Moneo, M. A regional comparison of the effects of climate change on agricultural crops in Europe. *Clim. Chang.* **2012**, *112*, 29–46. [CrossRef]
12. Fróna, D.; Szenderák, J.; Harangi-Rákos, M. Economic effects of climate change on global agricultural production. *Nat. Conserv.* **2021**, *44*, 117–139. [CrossRef]
13. FAO. *AQUASTAT—FAO's Global Information System on Water and Agriculture*; FAO: Rome, Italy, 2023.
14. UN. *The United Nations World Water Development Report 2022: Groundwater: Making the Invisible Visible*; UNESCO: Paris, France, 2022.
15. Assouline, S.; Narkis, K.; Gherabli, R.; Sposito, G. Combined Effect of Sodicity and Organic Matter on Soil Properties under Long-Term Irrigation with Treated Wastewater. *Vadose Zone J.* **2016**, *15*, 1–10. [CrossRef]
16. Díaz, F.J.; Grattan, S.R.; Reyes, J.A.; de la Roza-Delgado, B.; Benes, S.E.; Jiménez, C.; Dorta, M.; Tejedor, M. Using saline soil and marginal quality water to produce alfalfa in arid climates. *Agric. Water Manag.* **2018**, *199*, 11–21. [CrossRef]
17. Grattan, S.R.; Díaz, F.J.; Pedrero, F.; Vivaldi, G.A. Assessing the suitability of saline wastewaters for irrigation of Citrus spp.: Emphasis on boron and specific-ion interactions. *Agric. Water Manag.* **2015**, *157*, 48–58. [CrossRef]
18. Minhas, P.S. Saline water management for irrigation in India. *Agric. Water Manag.* **1996**, *30*, 1–24. [CrossRef]
19. Minhas, P.S.; Qadir, M.; Yadav, R.K. Groundwater irrigation induced soil sodification and response options. *Agric. Water Manag.* **2019**, *215*, 74–85. [CrossRef]
20. Muyen, Z.; Moore, G.A.; Wrigley, R.J. Soil salinity and sodicity effects of wastewater irrigation in South East Australia. *Agric. Water Manag.* **2011**, *99*, 33–41. [CrossRef]
21. Phogat, V.; Mallants, D.; Cox, J.W.; Šimůnek, J.; Oliver, D.P.; Pitt, T.; Petrie, P.R. Impact of long-term recycled water irrigation on crop yield and soil chemical properties. *Agric. Water Manag.* **2020**, *237*, 106167. [CrossRef]
22. Singh, A. Soil salinization management for sustainable development: A review. *J. Environ. Manag.* **2021**, *277*, 111383. [CrossRef] [PubMed]
23. DeGaris, K.A.; Walker, R.R.; Loveys, B.R.; Tyerman, S.D. Impact of deficit irrigation strategies in a saline environment on Shiraz yield, physiology, water use and tissue ion concentration. *Aust. J. Grape Wine Res.* **2015**, *21*, 468–478. [CrossRef]
24. Pitt, T.; Cox, J.; Phogat, V.; Fleming, N.; Grant, C. *Methods to Increase the Use of Recycled Wastewater in Irrigation by Overcoming the Constraint of Soil Salinity*; Australian Water Recycling Centre of Excellence: Brisbane, Australia, 2015.
25. Netzer, Y.; Shenker, M.; Schwartz, A. Effects of irrigation using treated wastewater on table grape vineyards: Dynamics of sodium accumulation in soil and plant. *Irrig. Sci.* **2014**, *32*, 283–294. [CrossRef]
26. Stevens, R.M.; Harvey, G.; Partington, D.L. Irrigation of grapevines with saline water at different growth stages: Effects on leaf, wood and juice composition. *Aust. J. Grape Wine Res.* **2011**, *17*, 239–248. [CrossRef]
27. Phogat, V.; Pitt, T.; Stevens, R.M.; Cox, J.W.; Šimůnek, J.; Petrie, P.R. Assessing the role of rainfall redirection techniques for arresting the land degradation under drip irrigated grapevines. *J. Hydrol.* **2020**, *587*, 125000. [CrossRef]
28. Aragüés, R.; Medina, E.T.; Clavería, I.; Martínez-Cob, A.; Faci, J. Regulated deficit irrigation, soil salinization and soil sodification in a table grape vineyard drip-irrigated with moderately saline waters. *Agric. Water Manag.* **2014**, *134*, 84–93. [CrossRef]
29. Stevens, R.M.; Pitt, T.R.; Dyson, C. Changes in vineyard floor management reduce the $Na^+$ and $Cl^-$ concentrations in wine grapes grown with saline supplementary drip irrigation. *Agric. Water Manag.* **2013**, *129*, 130–137. [CrossRef]
30. Aragüés, R.; Medina, E.T.; Zribi, W.; Clavería, I.; Álvaro-Fuentes, J.; Faci, J. Soil salinization as a threat to the sustainability of deficit irrigation under present and expected climate change scenarios. *Irrig. Sci.* **2014**, *33*, 67–79. [CrossRef]
31. Phogat, V.; Cox, J.W.; Šimůnek, J.; Hayman, P. Modeling water and salinity risks to viticulture under prolonged sustained deficit and saline water irrigation. *J. Water Clim. Chang.* **2020**, *11*, 901–915. [CrossRef]

32. Feng, D.; Zhang, J.; Cao, C.; Sun, J.; Shao, L.; Li, F.; Dang, H.; Sun, C. Soil Salt Accumulation and Crop Yield under Long-Term Irrigation with Saline Water. *J. Irrig. Drain. Eng.* **2015**, *141*, 04015025. [CrossRef]
33. Murtaza, G.; Ghafoor, A.; Qadir, M. Irrigation and soil management strategies for using saline-sodic water in a cotton–wheat rotation. *Agric. Water Manag.* **2006**, *81*, 98–114. [CrossRef]
34. Mohanavelu, A.; Naganna, S.R.; Al-Ansari, N. Irrigation Induced Salinity and Sodicity Hazards on Soil and Groundwater: An Overview of Its Causes, Impacts and Mitigation Strategies. *Agriculture* **2021**, *11*, 983. [CrossRef]
35. Laurenson, S.; Bolan, N.S.; Smith, E.; McCarthy, M. Review: Use of recycled wastewater for irrigating grapevines. *Aust. J. Grape Wine Res.* **2012**, *18*, 1–10. [CrossRef]
36. Shani, U.; Ben-Gal, A. Long-term Response of Grapevines to Salinity: Osmotic Effects and Ion Toxicity. *Am. J. Enol. Vitic.* **2005**, *56*, 148–154. [CrossRef]
37. Suarez, D.L.; Celis, N.; Anderson, R.G.; Sandhu, D. Grape Rootstock Response to Salinity, Water and Combined Salinity and Water Stresses. *Agronomy* **2019**, *9*, 321. [CrossRef]
38. Walker, R.R. *Managing Salinity in the Vineyards*; CSIRO Plant Industry: Adelaide, Australia, 2010.
39. de Loryn, L.C.; Petrie, P.R.; Hasted, A.M.; Johnson, T.E.; Collins, C.; Bastian, S.E.P. Evaluation of Sensory Thresholds and Perception of Sodium Chloride in Grape Juice and Wine. *Am. J. Enol. Vitic.* **2014**, *65*, 124–133. [CrossRef]
40. Martínez-Moreno, A.; Pérez-Álvarez, E.P.; Intrigliolo, D.S.; Mirás-Avalos, J.M.; López-Urrea, R.; Gil-Muñoz, R.; Lizama, V.; García-Esparza, M.J.; Álvarez, M.I.; Buesa, I. Effects of deficit irrigation with saline water on yield and grape composition of *Vitis vinifera* L. cv. Monastrell. *Irrig. Sci.* **2022**, *41*, 469–485. [CrossRef]
41. Martínez-Moreno, A.; Pérez-Álvarez, E.; López-Urrea, R.; Intrigliolo, D.; González-Centeno, M.R.; Teissedre, P.-L.; Gil-Muñoz, R. Is deficit irrigation with saline waters a viable alternative for winegrowers in semiarid areas? *OENO ONE* **2022**, *56*, 101–116. [CrossRef]
42. Ayers, R.S.; Westcot, D.W. *Water Quality for Agriculture*; Food and Agriculture Organisation: Rome, Italy, 1994.
43. Dudley, L.M.; Ben-Gal, A.; Shani, U. Influence of Plant, Soil, and Water on the Leaching Fraction. *Vadose Zone J.* **2008**, *7*, 420–425. [CrossRef]
44. Hanson, B.; Hopmans, J.W.; Simunek, J. Leaching with subsurface drip irrigation under saline, shallow groundwater conditions. *Vadose Zone J.* **2008**, *7*, 810–818. [CrossRef]
45. Minhas, P.S.; Dubey, S.K.; Sharma, D.R. Comparative affects of blending, intera/inter-seasonal cyclic uses of alkali and good quality waters on soil properties and yields of paddy and wheat. *Agric. Water Manag.* **2007**, *87*, 83–90. [CrossRef]
46. Phogat, V.; Mallants, D.; Šimůnek, J.; Cox, J.W.; Petrie, P.R.; Pitt, T. Modelling Salinity and Sodicity Risks of Long-Term Use of Recycled Water for Irrigation of Horticultural Crops. *Soil Syst.* **2021**, *5*, 49. [CrossRef]
47. Ben-Gal, A.; Yermiyahu, U.; Cohen, S. Fertilization and blending alternatives for irrigation with desalinated water. *J. Environ. Qual.* **2009**, *38*, 529–536. [CrossRef] [PubMed]
48. Naresh, R.K.; Minhas, P.S.; Goyal, A.K.; Chauhan, C.P.S.; Gupta, R.K. Conjunctive use of saline and non-saline waters. II. Field comparisions of cyclic uses and mixing for wheat. *Agric. Water Manag.* **1993**, *23*, 139–148. [CrossRef]
49. Phogat, V.; Mallants, D.; Cox, J.W.; Šimůnek, J.; Oliver, D.P.; Awad, J. Management of soil salinity associated with irrigation of protected crops. *Agric. Water Manag.* **2020**, *227*, 105845. [CrossRef]
50. Ben-Gal, A.; Ityel, E.; Dudley, L.; Cohen, S.; Yermiyahu, U.; Presnov, E.; Zigmond, L.; Shani, U. Effect of irrigation water salinity on transpiration and on leaching requirements: A case study for bell peppers. *Agric. Water Manag.* **2008**, *95*, 587–597. [CrossRef]
51. Singh, A. Environmental problems of salinization and poor drainage in irrigated areas: Management through the mathematical models. *J. Clean. Prod.* **2019**, *206*, 572–579. [CrossRef]
52. Wichelns, D.; Qadir, M. Achieving sustainable irrigation requires effective management of salts, soil salinity, and shallow groundwater. *Agric. Water Manag.* **2015**, *157*, 31–38. [CrossRef]
53. Šimůnek, J.; van Genuchten, M.T.; Sejna, M. Recent developments and applications of the HYDRUS computer software packages. *Vadose Zone J.* **2016**, *15*, 1–25. [CrossRef]
54. Chen, L.-J.; Feng, Q.; Li, F.-R.; Li, C.-S. Simulation of soil water and salt transfer under mulched furrow irrigation with saline water. *Geoderma* **2015**, *241–242*, 87–96. [CrossRef]
55. Phogat, V.; Skewes, M.A.; Cox, J.W.; Sanderson, G.; Alam, J.; Šimůnek, J. Seasonal simulation of water, salinity and nitrate dynamics under drip irrigated mandarin (Citrus reticulata) and assessing management options for drainage and nitrate leaching. *J. Hydrol.* **2014**, *513*, 504–516. [CrossRef]
56. Mallants, D.; Phogat, V.; Oliver, D.; Ouzman, J.; Beirgadhar, Y.; Cox, J.W. *Sustainable Expansion of Irrigated Agriculture and Horticulture in Northern Adelaide Corridor*; Technical Report Series No. 19/15; Goyder Institute for Water Research: Adelaide, SA, Australia, 2019.
57. Ramos, T.B.; Šimůnek, J.; Gonçalves, M.C.; Martins, J.C.; Prazeres, A.; Pereira, L.S. Two-dimensional modeling of water and nitrogen fate from sweet sorghum irrigated with fresh and blended saline waters. *Agric. Water Manag.* **2012**, *111*, 87–104. [CrossRef]
58. Selim, T.; Bouksila, F.; Berndtsson, R.; Persson, M. Soil Water and Salinity Distribution under Different Treatments of Drip Irrigation. *Soil Sci. Soc. Am. J.* **2013**, *77*, 1144–1156. [CrossRef]
59. Cresswell, R.G.; Herczeg, A.L. *Groundwater Recharge, Mixing and Salinity Across the Angas-Bremer Plains, South Australia: Geochemical and Isotopic Constraints*; CSIRO Land and Water: Canberra, Australia, 2004.

60. Allen, R.G.; Pereira, L.S.; Raes, D.; Smith, M. *Crop Evapotranspiration: Guidelines for Computing Crop Water Requirements*; 9251042195; FAO: Rome, Italy, 1998.
61. De Bei, R.; Fuentes, S.; Gilliham, M.; Tyerman, S.; Edwards, E.; Bianchini, N.; Smith, J.; Collins, C. VitiCanopy: A Free Computer App to Estimate Canopy Vigor and Porosity for Grapevine. *Sensors* **2016**, *16*, 585. [CrossRef] [PubMed]
62. Fuentes, S.; Bei, R.; Pozo, C.; Tyerman, S. Development of a smartphone application to characterise temporal and spatial canopy architecture and leaf area index for grapevines. *Wine Vitic. J.* **2012**, *6*, 56–60.
63. Fuentes, S.; Poblete-Echeverría, C.; Ortega-Farias, S.; Tyerman, S.; De Bei, R. Automated estimation of leaf area index from grapevine canopies using cover photography, video and computational analysis methods. *Aust. J. Grape Wine Res.* **2014**, *20*, 465–473. [CrossRef]
64. Macfarlane, C.; Hoffman, M.; Eamus, D.; Kerp, N.; Higginson, S.; McMurtrie, R.; Adams, M. Estimation of leaf area index in eucalypt forest using digital photography. *Agric. For. Meteorol.* **2007**, *143*, 176–188. [CrossRef]
65. Ben-Asher, J.; Tsuyuki, I.; Bravdo, B.-A.; Sagih, M. Irrigation of grapevines with saline water. I. Leaf area index, stomatal conductance, transpiration, and photosynthesis. *Agric. Water Manag.* **2006**, *83*, 13–21. [CrossRef]
66. Yunusa, I.A.M.; Walker, R.R.; Blackmore, D.H. Characterisation of water use by Sultana grapevines (*Vitis vinifera* L.) on their own roots or on Ramsey rootstock drip-irrigated with water of different salinities. *Irrig. Sci.* **1997**, *17*, 77–86. [CrossRef]
67. Vélez, S.; Poblete-Echeverría, C.; Rubio, J.A.; Vacas, R.; Barajas, E. Estimation of Leaf Area Index in vineyards by analysing projected shadows using UAV imagery. *OENO ONE* **2021**, *55*, 159–180. [CrossRef]
68. Allen, R.G.; Pereira, L.S. Estimating crop coefficients from fraction of ground cover and height. *Irrig. Sci.* **2009**, *28*, 17–34. [CrossRef]
69. Anzecc, A. *National Water Quality Management Strategy: Australian and New Zealand Guidelines for Fresh and Marine Water Quality: Volume 1: The Guidelines (Chapters 1–7)*; Australian and New Zealand Environment and Conservation Council and the Agriculture and Resource Management Council of Australia and New Zealand: Canberra, Austrilia, 2000.
70. Youngs, E.G. Hydraulic conductivity of saturated soils. In *Soil and Environmental Analysis: Physical Methods*; Smith, K.A., Mullins, C.E., Eds.; Marcel Dekker, Inc.: New York, NY, USA, 2001; pp. 141–181.
71. Van Genuchten, M.T. A Closed-form Equation for Predicting the Hydraulic Conductivity of Unsaturated Soils. *Soil Sci. Soc. Am. J.* **1980**, *44*, 892–898. [CrossRef]
72. Phogat, V.; Skewes, M.A.; McCarthy, M.G.; Cox, J.W.; Šimůnek, J.; Petrie, P.R. Evaluation of crop coefficients, water productivity, and water balance components for wine grapes irrigated at different deficit levels by a sub-surface drip. *Agric. Water Manag.* **2017**, *180*, 22–34. [CrossRef]
73. Skaggs, T.H.; Trout, T.J.; Šim Nek, J.; Shouse, P.J. Comparison of HYDRUS-2D Simulations of Drip Irrigation with Experimental Observations. *J. Irrig. Drain. Eng.* **2004**, *130*, 304–310. [CrossRef]
74. Feddes, R.A.; Kowalik, P.J.; Zaradny, H. *Simulation of Field Water Use and Crop Yield*; Pudoc for the Centre for Agricultural Publishing and Documentation: Wageningen, The Netherlands, 1978.
75. Zhang, X.; Walker, R.R.; Stevens, R.M.; Prior, L.D. Yield-salinity relationships of different grapevine (*Vitis vinifera* L.) scion-rootstock combinations. *Aust. J. Grape Wine Res.* **2002**, *8*, 150–156. [CrossRef]
76. Wang, J.; Bai, Z.; Yang, P. Simulation and prediction of ion transport in the reclamation of sodic soils with gypsum based on the support vector machine. *Sci. World J.* **2014**, *2014*, 805342. [CrossRef] [PubMed]
77. Cote, C.M.; Bristow, K.L.; Charlesworth, P.B.; Cook, F.J.; Thorburn, P.J. Analysis of soil wetting and solute transport in subsurface trickle irrigation. *Irrig. Sci.* **2003**, *22*, 143–156. [CrossRef]
78. Charles, S.P.; Fu, G. *Statistically Downscaled Projections for South Australia*; Technical Report No. 15/1; Goyder Institute for Water Research: Adelaide, Australia, 2015.
79. Stirzaker, R.; Thomson, T. *Fullstop at Angas Bremer: A Report on the 2002–03 Data*; CSIRO: Canberra, Australia, 2004.
80. Willmott, C.J.; Matsuura, K. Advantages of the mean absolute error (MAE) over the root mean square error (RMSE) in assessing average model performance. *Clim. Res.* **2005**, *30*, 79–82. [CrossRef]
81. Li, H.; Yi, J.; Zhang, J.; Zhao, Y.; Si, B.; Hill, R.; Cui, L.; Liu, X. Modeling of Soil Water and Salt Dynamics and Its Effects on Root Water Uptake in Heihe Arid Wetland, Gansu, China. *Water* **2015**, *7*, 2382–2401. [CrossRef]
82. Ramos, T.B.; Liu, M.; Paredes, P.; Shi, H.; Feng, Z.; Lei, H.; Pereira, L.S. Salts dynamics in maize irrigation in the Hetao plateau using static water table lysimeters and HYDRUS-1D with focus on the autumn leaching irrigation. *Agric. Water Manag.* **2023**, *283*, 108306. [CrossRef]
83. Evett, S.R.; Schwartz, R.C.; Casanova, J.J.; Heng, L.K. Soil water sensing for water balance, ET and WUE. *Agric. Water Manag.* **2012**, *104*, 1–9. [CrossRef]
84. Phogat, V.; Pitt, T.; Cox, J.W.; Šimůnek, J.; Skewes, M.A. Soil water and salinity dynamics under sprinkler irrigated almond exposed to a varied salinity stress at different growth stages. *Agric. Water Manag.* **2018**, *201*, 70–82. [CrossRef]
85. Phogat, V.; Mahalakshmi, M.; Skewes, M.; Cox, J.W. Modelling soil water and salt dynamics under pulsed and continuous surface drip irrigation of almond and implications of system design. *Irrig. Sci.* **2012**, *30*, 315–333. [CrossRef]
86. López-Urrea, R.; Montoro, A.; Mañas, F.; López-Fuster, P.; Fereres, E. Evapotranspiration and crop coefficients from lysimeter measurements of mature 'Tempranillo' wine grapes. *Agric. Water Manag.* **2012**, *112*, 13–20. [CrossRef]

87. Pereira, L.S.; Paredes, P.; Melton, F.; Johnson, L.; Wang, T.; López-Urrea, R.; Cancela, J.J.; Allen, R.G. Prediction of crop coefficients from fraction of ground cover and height. Background and validation using ground and remote sensing data. *Agric. Water Manag.* **2020**, *241*, 106197. [CrossRef]
88. Williams, L.E.; Ayars, J.E. Grapevine water use and the crop coefficient are linear functions of the shaded area measured beneath the canopy. *Agric. For. Meteorol.* **2005**, *132*, 201–211. [CrossRef]
89. Pereira, L.S.; Paredes, P.; Melton, F.; Johnson, L.; Mota, M.; Wang, T. Prediction of crop coefficients from fraction of ground cover and height: Practical application to vegetable, field and fruit crops with focus on parameterization. *Agric. Water Manag.* **2021**, *252*, 106663. [CrossRef]
90. Rallo, G.; Paço, T.A.; Paredes, P.; Puig-Sirera, À.; Massai, R.; Provenzano, G.; Pereira, L.S. Updated single and dual crop coefficients for tree and vine fruit crops. *Agric. Water Manag.* **2021**, *250*, 106645. [CrossRef]
91. Ben-Asher, J.; van Dam, J.; Feddes, R.A.; Jhorar, R.K. Irrigation of grapevines with saline water: II. Mathematical simulation of vine growth and yield. *Agric. Water Manag.* **2006**, *83*, 22–29. [CrossRef]
92. Ben-Gal, A.; Karlberg, L.; Jansson, P.-E.; Shani, U. Temporal robustness of linear relationships between production and transpiration. *Plant Soil* **2003**, *251*, 211–218. [CrossRef]
93. Shani, U.; Ben-Gal, A.; Tripler, E.; Dudley, L.M. Plant response to the soil environment: An analytical model integrating yield, water, soil type, and salinity. *Water Resour. Res.* **2007**, *43*. [CrossRef]
94. Bhantana, P.; Lazarovitch, N. Evapotranspiration, crop coefficient and growth of two young pomegranate (*Punica granatum* L.) varieties under salt stress. *Agric. Water Manag.* **2010**, *97*, 715–722. [CrossRef]
95. Skaggs, T.H.; Shouse, P.J.; Poss, J.A. Irrigating Forage Crops with Saline Waters: 2. Modeling Root Uptake and Drainage. *Vadose Zone J.* **2006**, *5*, 824–837. [CrossRef]
96. Keller, M. Chapter 3—Water Relations and Nutrient Uptake. In *The Science of Grapevines*, 2nd ed.; Keller, M., Ed.; Academic Press: San Diego, CA, USA, 2015; pp. 101–124.
97. Yunusa, I.A.M.; Walker, R.R.; Guy, J.R. Partitioning of seasonal evapotranspiration from a commercial furrow-irrigated Sultana vineyard. *Irrig. Sci.* **1997**, *18*, 45–54. [CrossRef]
98. Fandiño, M.; Cancela, J.J.; Rey, B.J.; Martínez, E.M.; Rosa, R.G.; Pereira, L.S. Using the dual-Kc approach to model evapotranspiration of Albariño vineyards (*Vitis vinifera* L. cv. Albariño) with consideration of active ground cover. *Agric. Water Manag.* **2012**, *112*, 75–87. [CrossRef]
99. Kool, D.; Agam, N.; Lazarovitch, N.; Heitman, J.L.; Sauer, T.J.; Ben-Gal, A. A review of approaches for evapotranspiration partitioning. *Agric. For. Meteorol.* **2014**, *184*, 56–70. [CrossRef]
100. Aquaterra. *Angas-Bremer and Currency Creek Flow and Solute Transport Model- AB2010*; Aquaterra: Adelaide, Australia, 2010.
101. Silva, S.P.; Valín, M.I.; Mendes, S.; Araujo-Paredes, C.; Cancela, J.J. Dual Crop Coefficient Approach in *Vitis vinifera* L. cv. Loureiro. *Agronomy* **2021**, *11*, 2062. [CrossRef]
102. Williams, L.E.; Phene, C.J.; Grimes, D.W.; Trout, T.J. Water use of mature Thompson Seedless grapevines in California. *Irrig. Sci.* **2003**, *22*, 11–18. [CrossRef]
103. Cancela, J.J.; Fandiño, M.; Rey, B.J.; Martínez, E.M. Automatic irrigation system based on dual crop coefficient, soil and plant water status for *Vitis vinifera* (cv Godello and cv Mencía). *Agric. Water Manag.* **2015**, *151*, 52–63. [CrossRef]
104. Rosa, R.D.; Ramos, T.B.; Pereira, L.S. The dual Kc approach to assess maize and sweet sorghum transpiration and soil evaporation under saline conditions: Application of the SIMDualKc model. *Agric. Water Manag.* **2016**, *177*, 77–94. [CrossRef]
105. Morgani, M.B.; Peña, J.E.P.; Fanzone, M.; Prieto, J.A. Pruning after budburst delays phenology and affects yield components, crop coefficient and total evapotranspiration in *Vitis vinífera* L. cv. 'Malbec' in Mendoza, Argentina. *Sci. Hortic.* **2022**, *296*, 110886. [CrossRef]
106. Poblete-Echeverría, C.; Sepúlveda-Reyes, D.; Zúñiga, M.; Ortega-Farías, S. Grapevine crop coefficient (Kc) determined by surface renewal method at different phenological periods. *Acta Hortic.* **2017**, *1150*, 61–66. [CrossRef]
107. Darouich, H.; Ramos, T.B.; Pereira, L.S.; Rabino, D.; Bagagiolo, G.; Capello, G.; Simionesei, L.; Cavallo, E.; Biddoccu, M. Water Use and Soil Water Balance of Mediterranean Vineyards under Rainfed and Drip Irrigation Management: Evapotranspiration Partition and Soil Management Modelling for Resource Conservation. *Water* **2022**, *14*, 554. [CrossRef]
108. Munitz, S.; Schwartz, A.; Netzer, Y. Water consumption, crop coefficient and leaf area relations of a Vitis vinifera cv. 'Cabernet Sauvignon' vineyard. *Agric. Water Manag.* **2019**, *219*, 86–94. [CrossRef]
109. Wang, S.; Zhu, G.; Xia, D.; Ma, J.; Han, T.; Ma, T.; Zhang, K.; Shang, S. The characteristics of evapotranspiration and crop coefficients of an irrigated vineyard in arid Northwest China. *Agric. Water Manag.* **2019**, *212*, 388–398. [CrossRef]
110. Williams, L.E.; Levin, A.D.; Fidelibus, M.W. Crop coefficients (Kc) developed from canopy shaded area in California vineyards. *Agric. Water Manag.* **2022**, *271*, 107771. [CrossRef]
111. Pôças, I.; Paço, T.A.; Paredes, P.; Cunha, M.; Pereira, L.S. Estimation of Actual Crop Coefficients Using Remotely Sensed Vegetation Indices and Soil Water Balance Modelled Data. *Remote Sens.* **2015**, *7*, 2373–2400. [CrossRef]
112. Phogat, V.; Šimůnek, J.; Skewes, M.A.; Cox, J.W.; McCarthy, M.G. Improving the estimation of evaporation by the FAO-56 dual crop coefficient approach under subsurface drip irrigation. *Agric. Water Manag.* **2016**, *178*, 189–200. [CrossRef]
113. Chalmers, Y.M.; Downey, M.O.; Krstic, M.P.; Loveys, B.R.; Dry, P.R. Influence of sustained deficit irrigation on colour parameters of Cabernet Sauvignon and Shiraz microscale wine fermentations. *Aust. J. Grape Wine Res.* **2010**, *16*, 301–313. [CrossRef]

114. Edwards, E.J.; Clingeleffer, P.R. Interseasonal effects of regulated deficit irrigation on growth, yield, water use, berry composition and wine attributes of Cabernet Sauvignon grapevines. *Aust. J. Grape Wine Res.* **2013**, *19*, 261–276. [CrossRef]
115. Helalia, S.A.; Anderson, R.G.; Skaggs, T.H.; Jenerette, G.D.; Wang, D.; Šimůnek, J. Impact of Drought and Changing Water Sources on Water Use and Soil Salinity of Almond and Pistachio Orchards: 1. Observations. *Soil Syst.* **2021**, *5*, 50. [CrossRef]
116. Yang, T.; Šimůnek, J.; Mo, M.; McCullough-Sanden, B.; Shahrokhnia, H.; Cherchian, S.; Wu, L. Assessing salinity leaching efficiency in three soils by the HYDRUS-1D and -2D simulations. *Soil Tillage Res.* **2019**, *194*, 104342. [CrossRef]
117. Oster, J.D. Irrigation with poor quality water. *Agric. Water Manag.* **1994**, *25*, 271–297. [CrossRef]
118. Corwin, D.L.; Grattan, S.R. Are Existing Irrigation Salinity Leaching Requirement Guidelines Overly Conservative or Obsolete? *J. Irrig. Drain. Eng.* **2018**, *144*, 02518001. [CrossRef]
119. Letey, J.; Hoffman, G.J.; Hopmans, J.W.; Grattan, S.R.; Suarez, D.; Corwin, D.L.; Oster, J.D.; Wu, L.; Amrhein, C. Evaluation of soil salinity leaching requirement guidelines. *Agric. Water Manag.* **2011**, *98*, 502–506. [CrossRef]
120. Isidoro, D.; Grattan, S.R. Predicting soil salinity in response to different irrigation practices, soil types and rainfall scenarios. *Irrig. Sci.* **2011**, *29*, 197–211. [CrossRef]
121. Läuchli, A.; Epstein, E. Plant response to saline and sodic conditions. In *Agricultural Salinity Assessment and Management*; Tanji, K.K., Ed.; Manuals and Reports on Engineering Practice; ASCE: New York, NY, USA, 1990; pp. 113–137.
122. Cucci, G.; Lacolla, G.; Mastro, A.M.; Caranfa, G. Leaching effect of rainfall on soil under four-year saline water irrigation. *Soil Water Res.* **2016**, *11*, 181–189. [CrossRef]
123. Vrugt, J.A.; Hopmans, J.W.; Šimunek, J. Calibration of a Two-Dimensional Root Water Uptake Model. *Soil Sci. Soc. Am. J.* **2001**, *65*, 1027–1037. [CrossRef]

land

Article

# Modeling Irrigation of Tomatoes with Saline Water in Semi-Arid Conditions Using Hydrus-1D

Sabri Kanzari [1], Jiří Šimůnek [2,*], Issam Daghari [3], Anis Younes [4], Khouloud Ben Ali [1], Sana Ben Mariem [1] and Samir Ghannem [1,5]

1 National Institute of Rural Engineering, Waters, and Forestry, University of Carthage, Ariana 2080, Tunisia; sabri.kanzari@iresa.agrinet.tn (S.K.)
2 Department of Environmental Sciences, University of California Riverside, Riverside, CA 92507, USA
3 National Institute of Agronomy, University of Carthage, Tunis 1082, Tunisia
4 Institut Terre et Environnement de Strasbourg, Université de Strasbourg, CNRS, ENGEES, UMR 7063, 67084 Strasbourg, France
5 Faculty of Sciences of Bizerte (FSB), University of Carthage, Jarzouna 7021, Tunisia
* Correspondence: jiri.simunek@ucr.edu

**Abstract:** In arid and semi-arid regions like Tunisia, irrigation water is typically saline, posing a risk of soil and crop salinization and yield reduction. This research aims to study the combined effects of soil matric and osmotic potential stresses on tomato root water uptake. Plants were grown in pot and field experiments in loamy-clay soils and were irrigated with three different irrigation water qualities: 0, 3.5, and 7 dS/m. The Hydrus-1D model was used to simulate the combined dynamics of subsurface soil water and salts. Successful calibration and validation of the model against measured water and salt profiles enabled the examination of the combined effects of osmotic and matric potential stresses on root water uptake. Relative yields, indirectly estimated from actual and potential transpiration, indicated that the multiplicative stress response model effectively simulated the measured yields and the impact of saline water irrigation on crop yields. The experimental and modeling results provide information to aid in determining the salinity levels conducive to optimal crop growth. The findings indicate that the selected salinity levels affect tomato growth to varying degrees. Specifically, the salinity levels conducive to optimal tomato growth were between 0 and 3.5 dS/m, with a significant growth reduction above this salinity level. The gradual salinization of the root zone further evidenced this effect. The scenario considering a temperature increase of 2 °C had no significant impact on crop yields in the pot and field experiments.

**Keywords:** soil; saline water supply; irrigation; tomato; Hydrus-1D; Tunisia

**Citation:** Kanzari, S.; Šimůnek, J.; Daghari, I.; Younes, A.; Ali, K.B.; Mariem, S.B.; Ghannem, S. Modeling Irrigation of Tomatoes with Saline Water in Semi-Arid Conditions Using Hydrus-1D. *Land* **2024**, *13*, 739. https://doi.org/10.3390/land13060739

Academic Editors: Tiago Brito Ramos, Maria da Conceição Gonçalves and Mohammad Farzamian

Received: 2 May 2024
Revised: 22 May 2024
Accepted: 22 May 2024
Published: 24 May 2024



## 1. Introduction

Saline water poses a significant challenge to agricultural sustainability in Tunisia, a country characterized by arid and semi-arid climates. With limited freshwater resources, Tunisia heavily relies on alternative water sources for irrigation, including saline waters (waters containing high concentrations of dissolved salts), which presents both opportunities and challenges for agriculture in Tunisia. The use of saline water for irrigation is widespread in Tunisia, driven by the scarcity of freshwater resources and the increasing demands of agriculture [1]. Saline water sources include groundwater with elevated levels of dissolved salts and treated wastewater from various sources [2]. In some regions, desalination plants have also been established to convert seawater into usable irrigation water [3]. Brackish water requires careful management to avoid negative impacts on soil quality and crop productivity [4,5].

When saline water is applied to the soil, salts accumulate over time, increasing soil salinity levels. It can also contribute to soil erosion and degradation, further exacerbating land degradation issues in Tunisia [6]. In addition to soil salinization, saline water irrigation can directly impact crop health and productivity. Many crops are sensitive to high soil and irrigation water salt concentrations. Excessive salt levels can cause physiological stress in plants, leading to reduced growth, yield losses, and even crop failure [7]. Certain crops, such as tomatoes, are particularly sensitive to salinity stress, making them susceptible to yield reductions when irrigated with saline water [8]. Despite these drawbacks, saline water irrigation also has some benefits for agriculture in Tunisia. Waters containing high concentrations of dissolved salts are often more readily available and less expensive than freshwater sources, providing an alternative water supply for irrigation during water scarcity [9].

Modeling irrigation with saline water is crucial for understanding its effects and implementing effective mitigation strategies. Modeling can be performed using hydrological models, such as Hydrus-1D [10]. These models simulate water movement and solute transport in soils, allowing researchers and farmers to predict how saline water affects soil salinity, crop growth, and yields over time [11]. Using different data for soil hydraulic properties, irrigation practices, and water quality, these models can simulate various scenarios and assess the effectiveness of different management strategies [12]. Modeling can also aid in identifying the most suitable crops for irrigation with water of a given salinity. Indeed, some crops have higher salt tolerance than others, and modeling can help determine which crops are best suited for cultivation in saline environments. Additionally, models can evaluate the impact of different irrigation strategies, such as drip irrigation or leaching, on soil salinity levels and crop productivity [13].

Several models have been developed to simulate water flow and solute transport in soil profiles with plants, such as Hydrus-1D [14,15], SWAP [16], and SALTMOD [17]. In this study, we use the Hydrus-1D model because of its distinct advantages over the other models in modeling root water uptake under various environmental stresses. The Hydrus-1D model offers superior capabilities in simulating water and salt dynamics in the soil profiles with plants under saline irrigation conditions, allowing more accurate predictions of crop responses to varying salinity levels [18,19].

Increased temperatures due to climate change can severely affect tomato production [20]. Increased temperatures can accelerate evaporation, potentially depleting soil moisture [21]. This can negatively impact tomato plants, affecting their growth and yield [22]. Higher temperatures can also alter the phenological stages of tomato plants, affecting flowering, fruit set, and ripening [23]. As temperatures rise, evaporation rates increase, concentrating salts in the soil. This increased soil salinity can adversely affect tomato plants, reducing their ability to absorb water and nutrients, stunting growth, and reducing yields [5].

The objectives of this study are: (i) to integrate the effects of soil matric and osmotic potentials into modeling water and salts transport, (ii) to study the effects of saline water irrigation on tomatoes cultivated under semi-arid conditions to provide a framework for understanding the complex interactions between soil, water, and crops in saline environments, and (iii) to evaluate the effects of an increase in mean air temperature on soil water and salts dynamics and tomatoes' root water uptake and, ultimately, yield. Haut du formulaire.

## 2. Materials and Methods

### 2.1. Experiment Design and Measurements

The experimental trials were conducted on a land parcel (36°51′36.791″ N, 10°11′36.795″ E) in the Ariana region, a semi-arid area in north Tunisia, during the year 2018. The soil is loamy clay. The study focused on a tomato variety, Rio Grande, commonly cultivated in Tunisia. Planting was done on 16 March 2018 in the pot experiment and on 17 April 2018 in the field experiment. A spacing of 25 cm between plants and 50 cm between rows was used in the field experiment. Tomato harvesting occurred on 10 June 2018 in pot cultivation and on 3 August 2018 in field cultivation. Both pot and field experiments were conducted under natural meteorological conditions. The field and pot experimental data are not compared against each other (since they were not carried out at the same time). Instead, they are used to calibrate and validate the Hydrus-1D model and assess its ability to account for environmental stresses.

The water requirements of the crops were estimated using climate data from the past 10-year period obtained from the National Institute of Meteorology. The average values were used to estimate the reference evapotranspiration using the Penman–Monteith formula (FAO-56). For calculating the actual evapotranspiration, crop coefficient values for the three growth stages were taken from those proposed by Allen et al. (1998) [24]. These values are as follows: $K_c$ = 0.6 (initial stage), 1.15 (mid-stage), and 0.8 (final stage). The CROPWAT 8.0 software (FAO, 1998; FAO, 2009) was utilized to establish the irrigation schedule. Water requirements for tomato cultivation are approximately 720 mm.

Surface irrigation was used to deliver water to each individual tomato plant. The irrigation scheduling in the pot and field experiments, mean temperatures, and reference evapotranspiration are presented in Figure 1. No rainfall events were recorded during the tomato growing season. In the field experiments, irrigation was applied every three or four days based on 100% of estimated evapotranspiration (Figure 1). In the pot experiments, irrigation was applied every two or three days until the soil water content reached field capacity (34% in terms of the volumetric soil water content). Soil field capacity was measured using a pressure plate apparatus (for a pressure head of −300 cm). The amounts of water added were recorded for each pot (Figure 1). The average wetted area was 45% in the field experiment and 65% in the pot experiment. Irrigation water provided to the plants had three different qualities: distilled water with a salinity of 0 dS/m, saline water with a salinity of 3.5 dS/m, and saline water with a salinity of 7 dS/m. Volumetric soil water content was measured using the gravimetric method, and soil salinity was measured using the saturated paste extract method [25] with three replicates. In the pot experiment, soil samples were taken every ten days until the end of the tomato plant cycle to characterize soil water and soil salt content. In the field experiment, the soil water content was measured on days 1, 30, 60, and 109 after planting every 20 cm down to a depth of 80 cm.

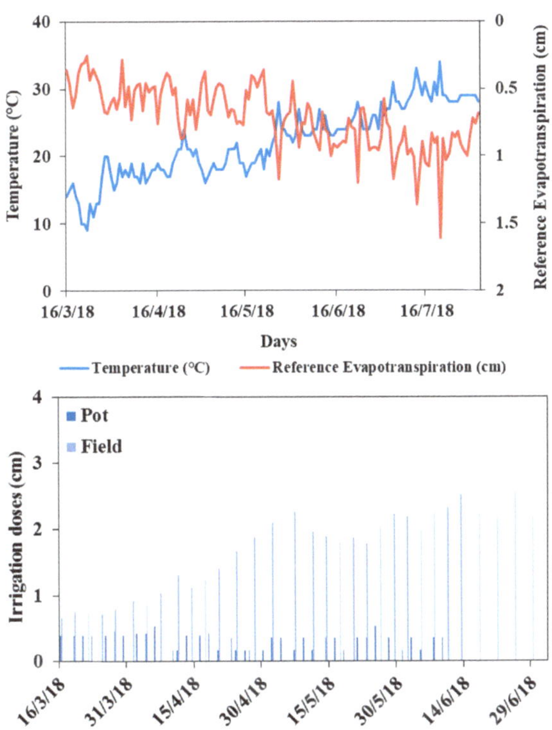


**Figure 1.** Mean temperatures, reference evapotranspiration (**top**), and tomato irrigation scheduling (**bottom**) in the pot and field experiments.

### 2.2. Hydrus-1D Model

#### 2.2.1. Theory

Hydrus-1D [14,15] is a numerical model that uses the Galerkin finite element method to solve the Richards equation governing water dynamics in the unsaturated zone:

$$\frac{\partial \theta}{\partial t} = \frac{\partial}{\partial z}\left[K\left(\frac{\partial h}{\partial z}+1\right)\right] - S \tag{1}$$

where $h$ is the water pressure head [L], $\theta$ is the soil volumetric water content [$L^3L^{-3}$], $t$ is time [T], $z$ is the depth measured from the land surface (positive downward) [L], $S$ is a sink term [$L^3L^{-3}T^{-1}$], and $K$ is the soil unsaturated hydraulic conductivity [$LT^{-1}$].

Hydrus-1D uses the van Genuchten–Mualem soil hydraulic functions to represent soil hydraulic properties. The soil water retention [26] and hydraulic conductivity [27] functions are as follows:

$$\theta(h) = \begin{cases} \theta_r + \frac{\theta_s-\theta_r}{(1+|\alpha h|^n)^m} & h < 0 \\ \theta_s & h \geq 0 \end{cases} \tag{2}$$

$$K(h) = K_s S_e^{0.5}\left[1-\left(1-S_e{}^{1/m}\right)^m\right]^2 \tag{3}$$

$$S_e = \frac{\theta-\theta_r}{\theta_s-\theta_r} \tag{4}$$

where $\theta_r$ is the residual water content [$L^3L^{-3}$]; $\theta_s$ is the saturated water content [$L^3L^{-3}$]; $\alpha$ [$L^{-1}$], $n$ ($n > 1$), and $m = 1 - 1/n$ are shape parameters; $S_e$ is the effective saturation; and $K_s$ is the saturated hydraulic conductivity [$LT^{-1}$].

Hydrus-1D uses the Galerkin finite element method to solve the advection–dispersion equations (ADE) governing solute transport:

$$\frac{\partial \theta C}{\partial t} = \frac{\partial}{\partial z}\left[\theta D \frac{\partial C}{\partial z}\right] - \frac{\partial q \theta C}{\partial z} \tag{5}$$

where $C$ is the solute concentration of the liquid phase [$ML^{-3}$], $D$ is the dispersion coefficient [$L^2T^{-1}$], and $q$ is the Darcy velocity [$LT^{-1}$]. When neglecting molecular diffusion, the dispersion coefficient is defined as:

$$D = \lambda q \tag{6}$$

where $\lambda$ is the soil longitudinal dispersivity [L].

The sink term, $S$, represents the volume of water removed from a unit volume of soil per unit time due to plant water uptake. The Feddes model, as described by [18], was employed to simulate root water uptake. Three stress response function models available in Hydrus-1D were evaluated for the simulation of salt stress: the additive model (Additive), the threshold model (T-Model), and the S-shape model (S-Model). The last two models are multiplicative models, combining the Feddes model [28], accounting for saturation stress, with either the "Threshold and Slope" model of Mass [29] or the S-shape model of van Genuchten [30], accounting for salinity stress. The theoretical underpinnings of these models are elaborated in the Hydrus-1D manual, with specific parameter values provided for tomato crops.

#### 2.2.2. Soil Hydraulic Properties and Solute Transport Parameters

The estimation of the soil hydraulic parameters, namely the van Genuchten parameters in the water content–pressure head (Equation (2)) and conductivity–saturation (Equation (3)) relationships, was based on soil column evaporation experiments [31].

The dispersion coefficient, a crucial parameter in solute transport in unsaturated soils, cannot be measured, and we thus resorted to indirect methods. Experiments on soil columns were conducted using PVC tubes with a diameter of 15 cm filled with 10 cm of soil. The columns were slowly saturated from the top with distilled water. A Mariotte bottle was placed at the top of the column to maintain a constant pressure head of about 3 cm and steady-state flow conditions. A pulse (50 mL) of a potassium chloride solution (0.8 M, 8.4 dS/m) was manually applied [32,33]. Effluent samples of approximately 100 $cm^3$ were used to measure soil electrical conductivity and determine the solute breakthrough curve (BTC). The experimental BTC values were fitted using the analytical solution of the ADE using CXTFIT 2.1 software [34] to estimate the dispersion coefficient.

#### 2.2.3. Model Calibration and Validation

The Hydrus-1D model was calibrated using measured soil water content and salinity profiles in both pots and field experiments irrigated with fresh water (i.e., 0 dS/m). The validation of the Hydrus-1D model was carried out by keeping the same input parameters and only changing the water quality in both cases of this study (pot and field experiments).

### *2.3. Statistical Evaluation of Modeling Results*

Simulation results were evaluated graphically and statistically. In the graphical method, the measured and simulated volumetric water content values were plotted as a function of soil depth. The statistical approach involved calculating the root mean square error (RMSE; in %):

$$RMSE = \sqrt{\frac{\sum_{i=1}^{n}(s_i - m_i)^2}{n}} \times \frac{1}{\overline{m}} \times 100 \quad (7)$$

where $s_i$ is the simulated value, $m_i$ is the measured value, $\overline{m}$ is the average observed value, and $n$ is the number of observations.

## 3. Results and Discussion

### *3.1. Soil Water Content Dynamics*

The variations in the soil water content in the pot experiments are presented in Figure 2. During the first month after the start of irrigation, the water content gradually increased to reach the field capacity (34%) in all three pots irrigated with different water qualities [35–37]. The water content of the soil irrigated with the saltiest water (7 dS/m) remained at the field capacity. As the soil salinity increased with frequent irrigations, osmotic pressure also increased, preventing root water uptake. Irrigations with the other two water qualities did not block water extraction by the plant roots, and the soil water content decreased over time. The soil irrigated with fresh water (0 dS/m) had the lowest water content values, as the plant roots extracted water without any osmotic potential constraints.

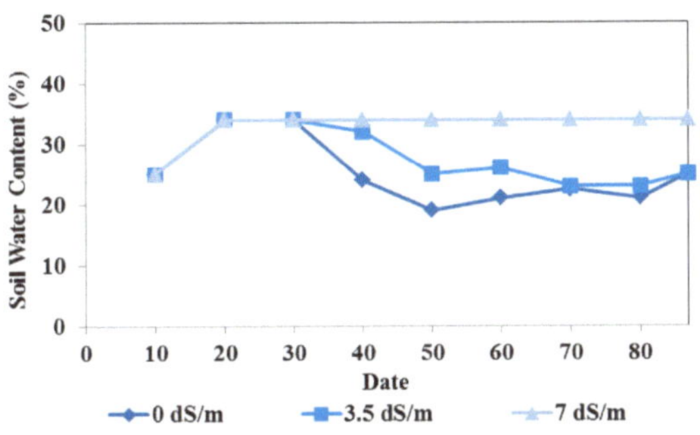


**Figure 2.** Measured soil water content (in %) in the pot experiments irrigated with different water qualities (0, 3, and 7 dS/m).

The soil water profiles, measured every 30 days in the field experiment with tomatoes irrigated with fresh water, are illustrated in Figure 3a. This figure shows the variations in soil water content without significant osmotic potential effects. Indeed, the soil water content profiles show dynamics that reflect infiltration (wetting) and evaporation (drying) episodes. After 109 irrigation days, the water content exceeded the field capacity, indicating that the roots did not undergo water stress.

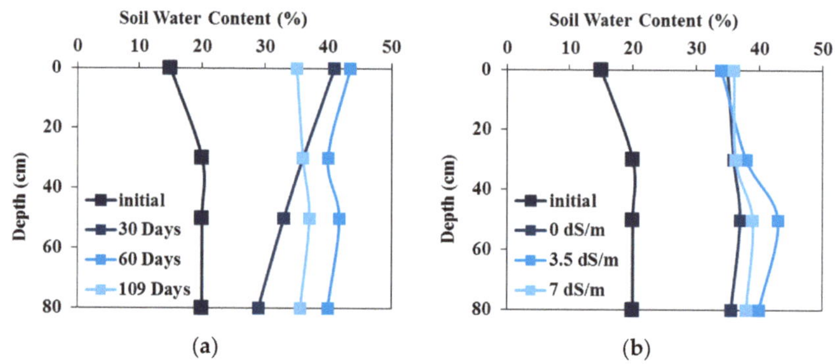


**Figure 3.** Measured soil water content (in %) in the field experiments with tomatoes irrigated with (**a**) fresh water (at 30, 60, and 109 d) and (**b**) waters of different salinities (0, 3, and 7 dS/m) (at 109 days).

Figure 3b shows the water content profiles measured after 109 irrigation days in field experiments irrigated with waters of different qualities. The water content in the surface layers is very close to field capacity (34%), while it increases considerably with depth, indicating deep water infiltration. The irrigation strategy based on calculating the water needs using climatic parameters from the previous ten years to estimate actual evapotranspiration succeeded in not subjecting the crop to water stress [38]. However, deep percolation shows that crop water needs were overestimated [39].

### 3.2. Soil Salinity Dynamics

The variations in the average soil salinity in the pot experiments irrigated with waters of different qualities are shown in Figure 4. This figure shows a nearly constant level of salinity in soils irrigated with fresh water and a gradual increase in soil salinity in soils irrigated with lower-quality waters, reaching 8 dS/m when 3.5 dS/m water was used and 12 dS/m when 7 dS/m water was used.

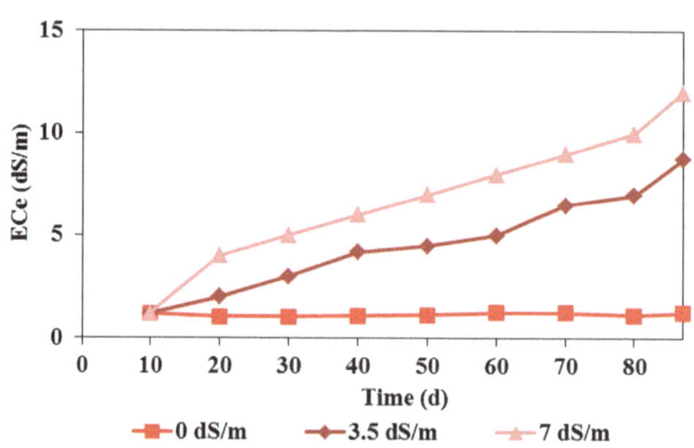


**Figure 4.** Soil salt content (ECe) in the pot experiments irrigated with waters of different salinities (0, 3, and 7 dS/m).

In the field experiment irrigated with fresh water, the variations in the soil salinity profiles measured every thirty days indicated continuous leaching of the initial salinity

(Figure 5a). In contrast, surface salinity significantly increased during irrigations with saline waters of 3.5 dS/m and 7 dS/m, reaching 7 dS/m and 12 dS/m, respectively, at the end of the crop cycle (Figure 5b). Soil salinity is higher near the surface than at depth, mainly due to evapotranspiration causing salt accumulation near the soil surface [40].

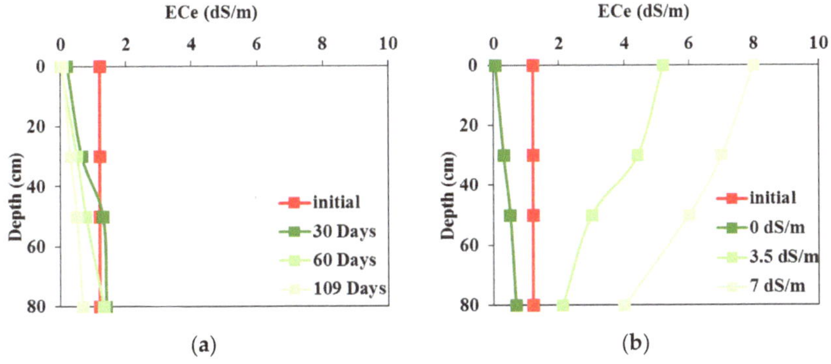


**Figure 5.** Measured soil salt content in the field experiments with tomatoes irrigated with (**a**) fresh water (at 30, 60, and 109 d) and (**b**) waters of different salinities (0, 3, and 7 dS/m) (at 109 days).

### *3.3. Numerical Modeling of Water and Salt Dynamics with Root Water Uptake*

#### 3.3.1. Inputs to Hydrus-1D

The soil was considered homogeneous in pot experiments, consisting of a single layer of 30 cm. In the field experiment, water content and salinity measurements were taken every 20 cm down to a depth of 80 cm. Four soil layers (horizons) of 20 cm thickness were considered in the simulations. The values of the van Genuchten–Mualem parameters for the four soil layers are presented in Table 1. The simulation period lasted 87 days in the pot experiments, with three output dates of 30, 60, and 87 days. On the other hand, the field trial lasted 109 days, and the output dates were 30, 60, and 109 days.

**Table 1.** The soil hydraulic parameters of different soil layers.

| Layer (cm) | $\theta_r$ (cm$^3$·cm$^{-3}$) | $\theta_s$ (cm$^3$·cm$^{-3}$) | $\alpha$ (cm$^{-1}$) | $n$ (-) | $K_s$ (cm·d$^{-1}$) |
|---|---|---|---|---|---|
| | | Pot experiment | | | |
| - | 0.1 | 0.41 | 0.27 | 1.11 | 6.41 |
| | | Field experiment | | | |
| 0–20 cm | 0.078 | 0.546 | 0.07 | 1.067 | 8.87 |
| 20–40 cm | 0.078 | 0.544 | 0.07 | 1.079 | 8.87 |
| 40–60 cm | 0.078 | 0.445 | 0.10 | 1.073 | 12.6 |
| 60–80 cm | 0.078 | 0.443 | 0.03 | 1.078 | 12.5 |

The dispersivity of the different soil layers varied around 5 cm, except for the 60–80 cm layer, where it was equal to 3 cm. The distribution coefficient of the adsorption isotherm ranged between 0.1 g/cm$^3$ and 0.3 g/cm$^3$ for all soil layers. Hydrus-1D includes a database that provides stress response function values for many crops, including tomatoes, which were considered here. The initial soil water content and soil salinity values as a function of depth in the field experiment are shown in Figure 6. In the pot experiment, the initial water content was 20%, and the initial soil salinity was 1.2 dS/m.

For the simulation of field experiments, we used the rainfall and evapotranspiration data from the National Institute of Meteorology and crop coefficients from [24].

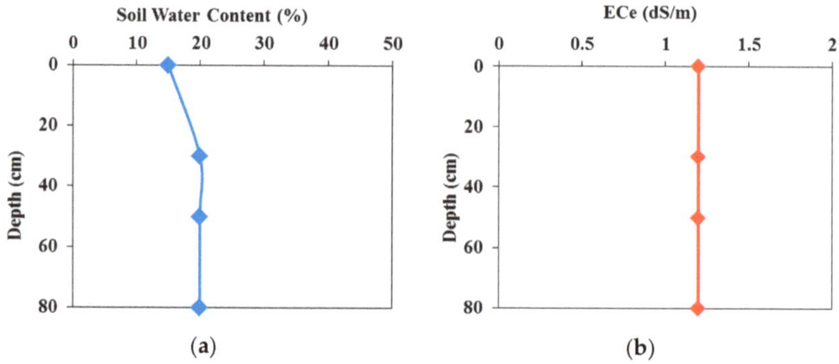


**Figure 6.** Measured initial (**a**) water content and (**b**) salinity as a function of soil depth in the field experiment.

3.3.2. Simulation Results

The Hydrus-1D model was used to simulate the movement of water and salts in both pot and field soils, considering root water uptake under the combined effects of soil matric and osmotic potentials.

Model Calibration

The Hydrus-1D model was calibrated using measured soil water content and salinity profiles in both pots and field experiments irrigated with fresh water (i.e., 0 dS/m). The measured and simulated water content and salinity values for both experiments are presented in Figures 7–9. The measured soil water content values are quite close to the simulated values. In the field experiment (Figures 7 and 8), the Hydrus-1D model underestimated the measured water content on day 30 and overestimated it on the other output dates. The model slightly overestimated the volumetric water content in the pot experiment (Figure 9).

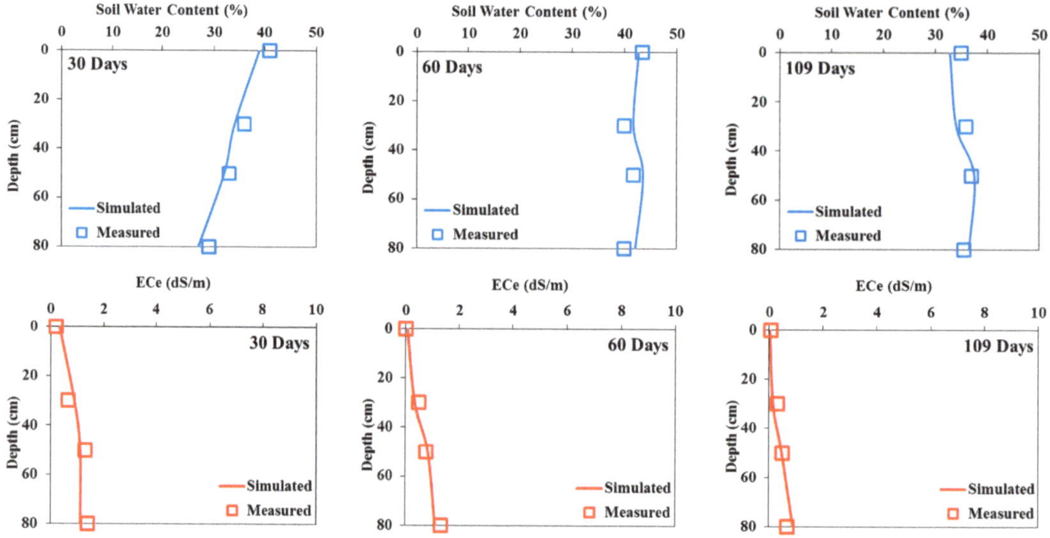


**Figure 7.** Measured and simulated (calibration) soil water content (in %) (**top**) and salinity (in dS/m) (**bottom**) in the field experiment irrigated with fresh water at 30 (**left**), 60 (**middle**), and 109 (**right**) days.

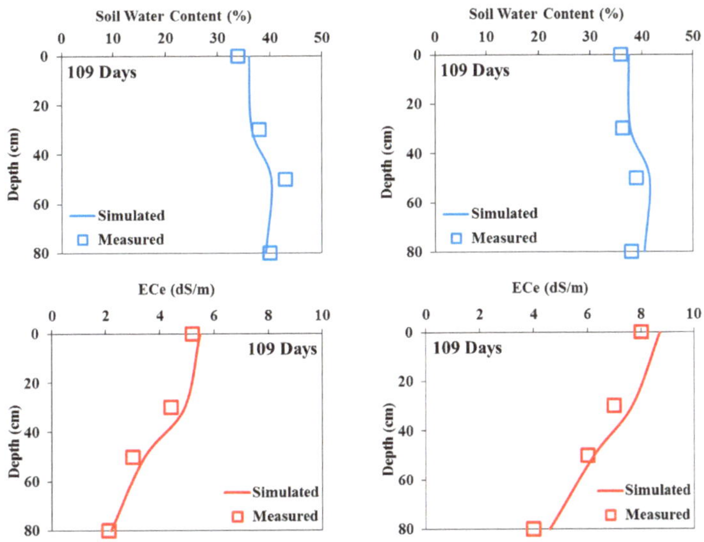


**Figure 8.** Measured and simulated (validation) soil water content (in %; (**top**)) and salinity (in dS/m; (**bottom**)) in the field experiments irrigated with saline waters of 3.5 dS/m (**left**) and 7 dS/m (**right**) at the end of the experiment.

The measured soil salinity profiles during the two pot and field experiments with tomatoes irrigated with saline waters and simulated using the Hydrus-1D model are presented in Figure 9. Overall, the measured and simulated values are similar on all three output dates. The simulated soil electrical conductivities generally overestimated the measured values in both experiments.

Model Validation

The validation of the Hydrus-1D model was carried out by keeping the same input parameters and only changing the water quality in both cases of this study (pot and field experiments). Figures 6 and 7 show that the simulated soil water content profiles agree well with the measured profiles for both pot and field experiments. Indeed, for all the output dates, the simulated profiles are very close to the measured ones.

The model's performance was evaluated qualitatively (Figures 7–9) and assessed using statistical evaluations and the RMSE values (Tables 2 and 3). These values are low (<10%) and highlight the model's reliability for simulating water and salt dynamics for both experiments and during the calibration and validation processes [41].

**Table 2.** Root mean square errors (RMSE, in %) assessing the Hydrus-1D calibration process.

| Variable | Experiment | Date 1 | Date 2 | Date 3 |
|---|---|---|---|---|
| Soil Water Profile | Field | 9.30 | 7.30 | 6.40 |
| | Pot | 10.30 | 11.10 | 8.30 |
| Soil Salinity Profile | Field | 5.30 | 2.10 | 4.60 |
| | Pot | 2.00 | 4.50 | 1.70 |

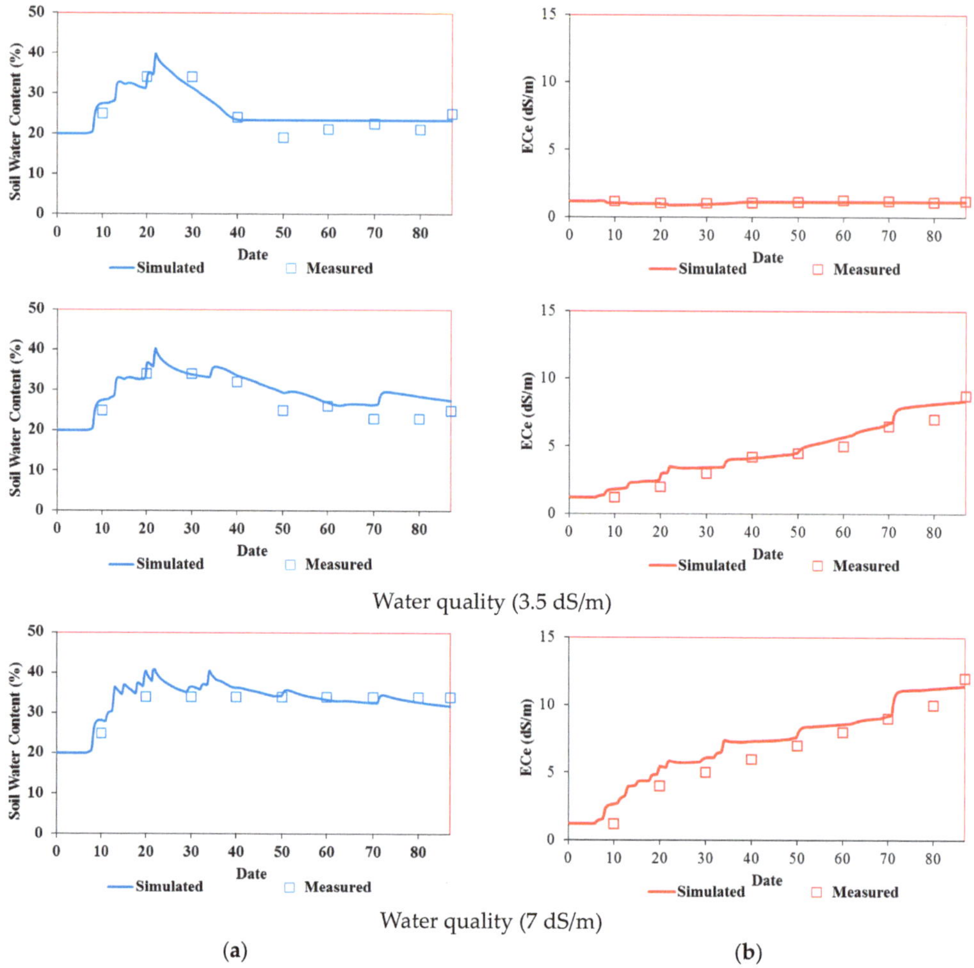


**Figure 9.** Measured and simulated time series of soil water content (%) (**left**) and salinity (dS/m) (**right**) in the pot experiments irrigated with (**a**) freshwater (**top**) (calibration) and (**b**) saline waters of 3.5 dS/m (**middle**) and 7 dS/m (**bottom**) (validation).

**Table 3.** Root mean square errors (RMSE, in %) assessing the Hydrus-1D validation process.

| Variable | Experiment | Irrigation Water Quality | RMSE (%) (on the Final Day) |
|---|---|---|---|
| Soil Water Profile | Field | 3.5 dS/m | 9.10 |
| | | 7 dS/m | 1.10 |
| Soil Water Profile | Pot | 3.5 dS/m | 3.00 |
| | | 7 dS/m | 1.10 |
| Soil Salinity Profile | Field | 3.5 dS/m | 10.20 |
| | | 7 dS/m | 5.70 |
| Soil Salinity Profile | Pot | 3.5 dS/m | 5.20 |
| | | 7 dS/m | 3.00 |

3.3.3. Crop Yield

The Hydrus-1D model cannot directly estimate the crop yield. However, according to Oster et al. (2012), the relative yield ($Y_r$) can be estimated as the ratio between the actual and potential (or maximum) transpiration, which are both provided in the model's output files [42]. Figures 10 and 11 show the relative yields for tomatoes irrigated with waters of different qualities for both experimental pot and field trials simulated using the three models mentioned above (Additive, T-Model, and S-Model) to represent the effects of osmotic and matric potentials on crop root water uptake. The simulated relative yields decreased with increased irrigation water salinity ($EC_W$) for all three stress response functions (Figure 10). The relative yield is 0.98 in the field experiment with tomatoes irrigated with fresh water (0 dS/m) for all three stress response functions. However, in the pot case, $Y_r$ equals 0.33 for freshwater irrigation. The reduced root development in pots compared to the field accounts for this reduction in relative yield [43].

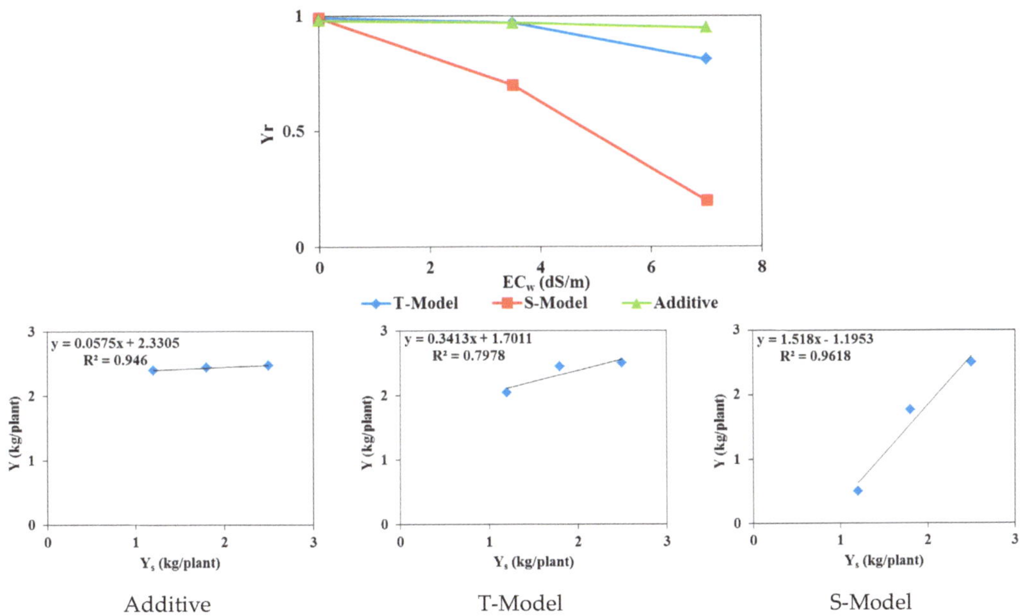


**Figure 10.** The effect of saline irrigation water on tomatoes' relative yield ($Y_r$) [−] (**top**) and the relationships between the actual ($Y$) and simulated ($Y_s$) yields (kg/plant) (**bottom**) using the three stress response models (Additive, T-Model, and S-Model) incorporated in Hydrus-1D (field experiments).

The actual yields measured in the two trials with different irrigation water qualities were compared to the simulated yields to identify the most suitable stress response function model for studying the combined effects of osmotic and matric potentials. According to FAO Note 66 (2012), $Y_r = Y/Y_{max}$, where $Y$ is the actual yield, and $Y_{max}$ is the maximum (or potential) yield. In our experiments, $Y_{max}$ is calculated for irrigations with fresh water. The $Y_{max}$ value equals 2.53 kg/plant for the field experiment and 4.16 kg/plant for the pot trial. The values of the simulated relative yields were calculated using these values [42].

Figures 10 and 11 also present linear correlations between the measured and simulated relative yields. The relative yields simulated using the additive model (Feddes) and the multiplicative T-Model (combining the Feddes et al. [28] model for saturation stress with the threshold and slope model of Mass [29] for osmotic stress) were least correlated with the measured relative yields in both pot and field trials. The multiplicative S-Model (combining the Feddes model [28] for saturation stress with the S-shape model of van Genuchten [30] for osmotic stress) is the only model capable of reproducing the measured relative yields with a correlation coefficient $R^2$ close to 1 for both pot and field experiments.

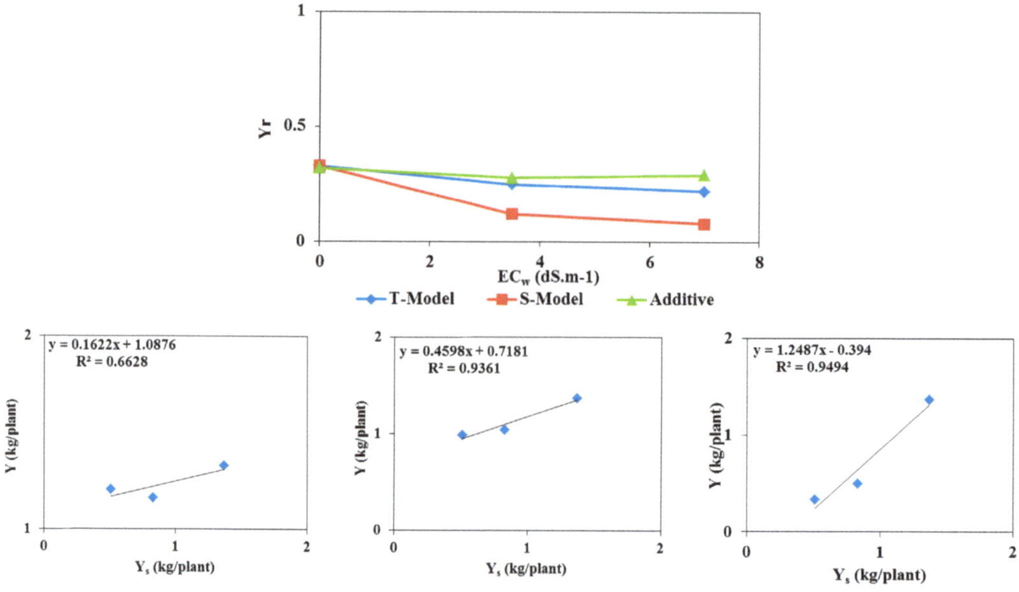


**Figure 11.** The effect of saline irrigation water on tomatoes' relative yield ($Y_r$) [−] (**top**) and the relationships between the actual ($Y$) and simulated ($Y_s$) yields (kg/plant) (**bottom**) using the three stress response models (Additive, T-Model, and S-Model) incorporated in Hydrus-1D (pot experiments).

### *3.4. Effect of Temperature Increase*

According to the IPCC (2021), temperatures in Tunisia are expected to increase by 2 °C over the coming decades. The effects of this temperature increase on tomato cultivation were studied (numerically) by increasing the temperature by 2 °C. To this aim, the reference evapotranspiration ($ET_0$) values were recalculated, and simulations were rerun using the updated $ET_0$ values while all other model parameters were kept unchanged.

The results of these simulations are presented in Figures 12 and 13. In the field experiment, soil water content exceeded field capacity in the root zone. In deeper layers between 40 and 80 cm, soil water content decreased after the 60th day, indicating infiltration beyond the root zone. On the other hand, in the pot experiment, the water content remained constant and below field capacity. Additionally, the accumulation of salts in the surface layer increased due to irrigation with saline waters and increased evaporation. The salinity in the root zone reached an average value of 6 dS/m, and the relative yields estimated using the S-Model did not change compared to the measured relative yields. An increase in temperature, thus, did not significantly affect the tomato yields in either trial [20].

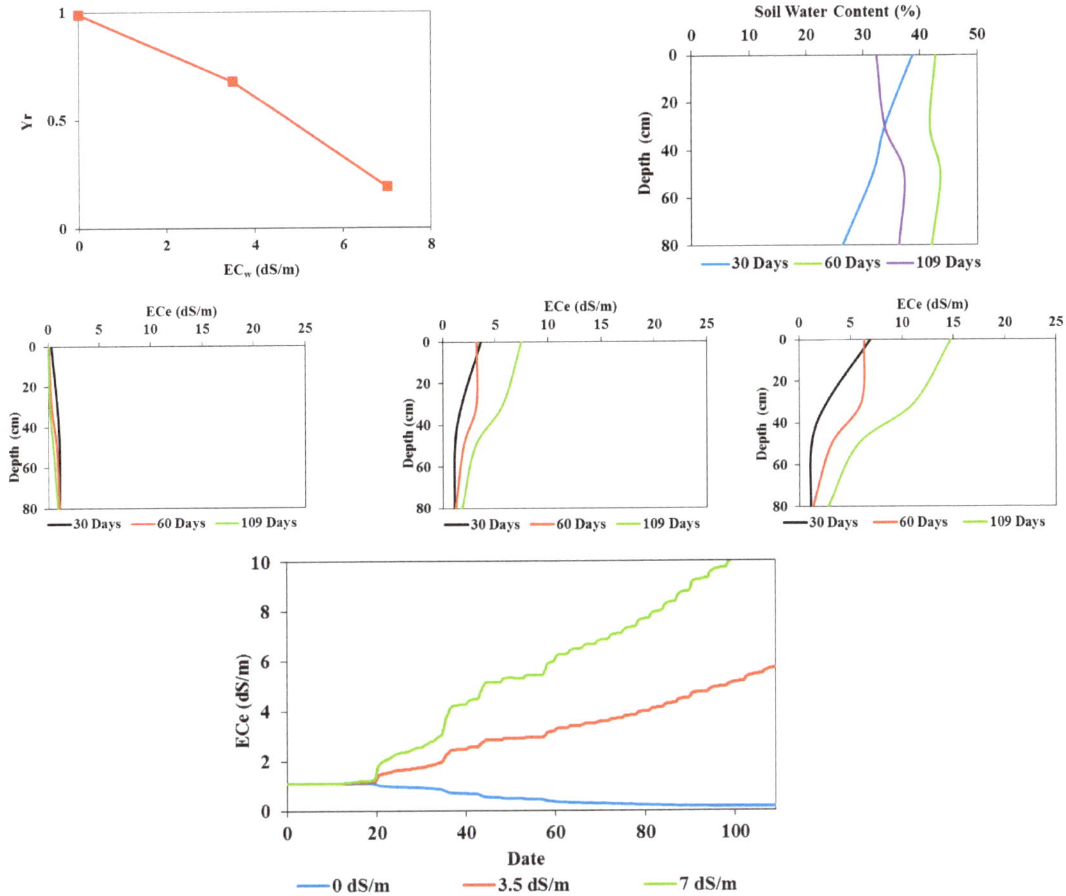


**Figure 12.** Effects of a temperature increase of 2 °C on tomatoes cultivated in field conditions: relative yield as a function of irrigation water salinity (**top left**); soil water content profiles at 30, 60, and 109 days when irrigated with freshwater (**top right**); soil salinity profiles at 30, 60, and 109 days (**middle row**) when irrigated with fresh water (**left**) and saline waters of 3.5 (**middle**) and 7 (**right**) dS/m; and average root zone salinity when irrigated with fresh water and saline waters of 3.5 and 7 dS/m (**bottom**).

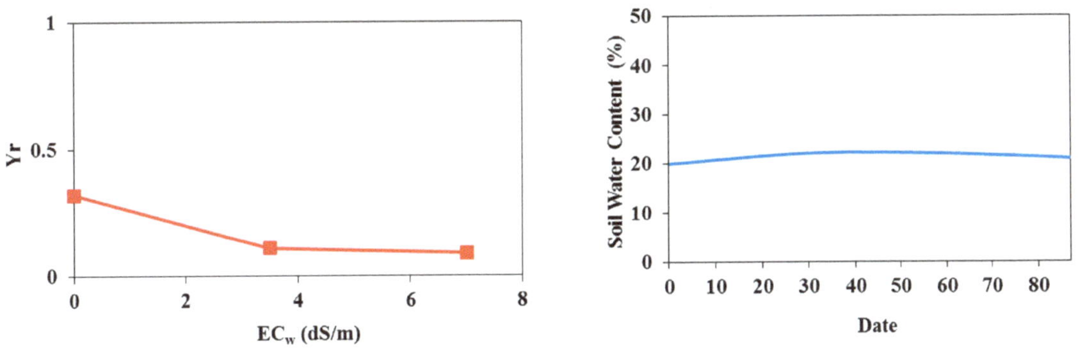


**Figure 13.** *Cont.*

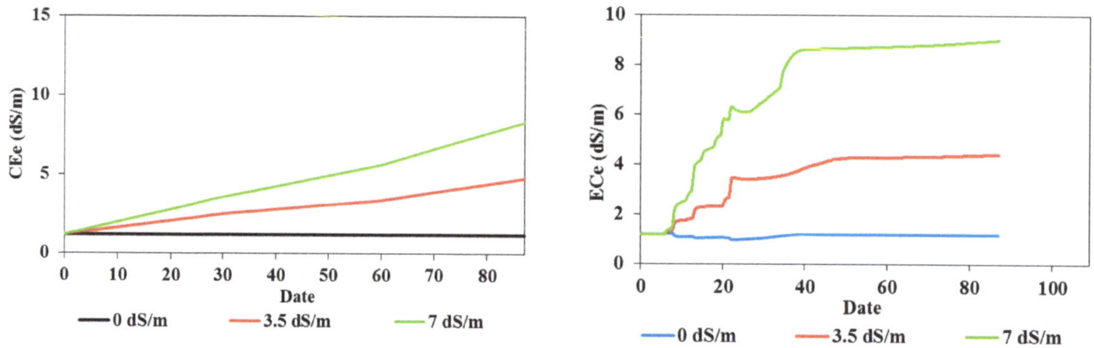


**Figure 13.** Effects of a temperature increase of 2 °C on tomatoes cultivated in pot conditions: relative yield as a function of irrigation water salinity (**top left**); average soil water content when irrigated with fresh water (**top right**); average soil salinity in the 0–30 cm soil layer (**bottom left**); and root zone salinity (at 5 cm depth) for irrigations with fresh water and saline waters of 3.5 and 7 dS/m (**bottom right**).

## 4. Conclusions

The aridity of regions like Tunisia has prompted the intensification of irrigation practices. However, the scarcity of good-quality water resources has led to the necessity of using brackish water in agriculture. The accumulation of salts due to irrigation with such waters has detrimental effects on both soil and crop yields. Experiments conducted using tomatoes grown in pots and the field and irrigated with three different water qualities ranging in salinity from 0 dS/m to 7 dS/m provided data to study the variations in soil water and salt dynamics. In both pot and field trials, the water content in the root zone (0–30 cm) remained close to field capacity throughout the crop cycle. The variations in salt profiles showed progressive soil salinization depending on irrigation water salinity, reaching 12 dS/m in the pot soil and 8 dS/m in the field soil. Crop reactions to these water qualities were evaluated using water movement and salt transfer modeling using the Hydrus-1D model. The combined effects of osmotic and matric potentials were evaluated using three stress response function models: the additive model and two multiplicative models, the T-Model and S-Model. The estimated relative crop yields showed that only the S-Model could reproduce the measured relative yields. A hypothetical 2 °C temperature increase did not significantly affect crop yields. Soil salinization slightly increased due to increased temperature and evaporation.

Adopting a modeling approach based on soil geochemical characterization and the UnsatChem module of HYDRUS [14,15], which considers transport and reactions between major ions (e.g., $Ca^{2+}$, $Mg^{2+}$, $Na^{+}$, $K^{+}$, $SO_4{}^{2-}$ $CO_3{}^{2-}$, $Cl^{-}$), would be an alternative approach to this study, where the risks of soil alkalization and salinization could be more thoroughly explored for strategic crops like tomatoes. Moreover, to obtain more accurate yield predictions, using specialized crop models, such as AquaCrop and DSSAT, would be advisable. These crop models are specifically designed to simulate and predict crop growth and yield under varying environmental conditions and provide more reliable estimates of agricultural productivity. Therefore, while Hydrus-1D excels in soil water and salinity dynamics, including crop models such as AquaCrop and DSSAT would improve the accuracy of yield predictions.

**Author Contributions:** Conceptualization, S.K., I.D. and K.B.A.; Data curation, S.K., K.B.A., S.B.M. and S.G.; Formal analysis, S.K., J.Š. and S.B.M.; Funding acquisition, S.K. and S.B.M.; Investigation, S.K., J.Š., I.D., A.Y., K.B.A., S.B.M. and S.G.; Project administration, S.K. and S.B.M.; Resources, S.K. and S.B.M.; Software, J.Š.; Supervision, S.K. and S.B.M.; Visualization, S.K., J.Š., I.D., A.Y., K.B.A., S.B.M. and S.G.; Writing—original draft, S.K. and J.Š.; Writing—review and editing, S.K.,

J.Š., I.D., A.Y., K.B.A., S.B.M. and S.G. All authors have read and agreed to the published version of the manuscript.

**Funding:** This research received no external funding.

**Data Availability Statement:** The data supporting the findings of this research are available in the manuscript.

**Conflicts of Interest:** The authors declare no conflict of interest.

## References

1. Besser, H.; Dhaouadi, L.; Hadji, R.; Hamed, Y.; Jemmali, H. Ecologic and Economic Perspectives for Sustainable Irrigated Agriculture under Arid Climate Conditions: An Analysis Based on Environmental Indicators for Southern Tunisia. *J. Afr. Earth Sci.* **2021**, *177*, 104134. [CrossRef]
2. Srivastava, A.; Parida, V.K.; Majumder, A.; Gupta, B.; Gupta, A.K. Treatment of Saline Wastewater Using Physicochemical, Biological, and Hybrid Processes: Insights into Inhibition Mechanisms, Treatment Efficiencies and Performance Enhancement. *J. Environ. Chem. Eng.* **2021**, *9*, 105775. [CrossRef]
3. Martínez-Alvarez, V.; González-Ortega, M.J.; Martin-Gorriz, B.; Soto-García, M.; Maestre-Valero, J.F. The Use of Desalinated Seawater for Crop Irrigation in the Segura River Basin (South-Eastern Spain). *Desalination* **2017**, *422*, 153–164. [CrossRef]
4. Qadir, M.; Oster, J.D. Crop and Irrigation Management Strategies for Saline-Sodic Soils and Waters Aimed at Environmentally Sustainable Agriculture. *Sci. Total Environ.* **2004**, *323*, 1–19. [CrossRef]
5. Muhammad, M.; Waheed, A.; Wahab, A.; Majeed, M.; Nazim, M.; Liu, Y.-H.; Li, L.; Li, W.-J. Soil Salinity and Drought Tolerance: An Evaluation of Plant Growth, Productivity, Microbial Diversity, and Amelioration Strategies. *Plant Stress* **2024**, *11*, 100319. [CrossRef]
6. Deeb, M.; Smagin, A.V.; Pauleit, S.; Fouché-Grobla, O.; Podwojewski, P.; Groffman, P.M. The Urgency of Building Soils for Middle Eastern and North African Countries: Economic, Environmental, and Health Solutions. *Sci. Total Environ.* **2024**, *917*, 170529. [CrossRef] [PubMed]
7. El-Ramady, H.; Prokisch, J.; Mansour, H.; Bayoumi, Y.A.; Shalaby, T.A.; Veres, S.; Brevik, E.C. Review of Crop Response to Soil Salinity Stress: Possible Approaches from Leaching to Nano-Management. *Soil Syst.* **2024**, *8*, 11. [CrossRef]
8. Li, J.; Chen, J.; He, P.; Chen, D.; Dai, X.; Jin, Q.; Su, X. The Optimal Irrigation Water Salinity and Salt Component for High-Yield and Good-Quality of Tomato in Ningxia. *Agric. Water Manag.* **2022**, *274*, 107940. [CrossRef]
9. Khondoker, M.; Mandal, S.; Gurav, R.; Hwang, S. Freshwater Shortage, Salinity Increase, and Global Food Production: A Need for Sustainable Irrigation Water Desalination—A Scoping Review. *Earth* **2023**, *4*, 223–240. [CrossRef]
10. Li, P.; Ren, L. Evaluating the Saline Water Irrigation Schemes Using a Distributed Agro-Hydrological Model. *J. Hydrol.* **2021**, *594*, 125688. [CrossRef]
11. Yu, Q.; Kang, S.; Hu, S.; Zhang, L.; Zhang, X. Modeling Soil Water-Salt Dynamics and Crop Response under Severely Saline Condition Using WAVES: Searching for a Target Irrigation Volume for Saline Water Irrigation. *Agric. Water Manag.* **2021**, *256*, 107100. [CrossRef]
12. Kanzari, S.; Jaziri, R.; Ali, K.B.; Daghari, I. Long-Term Evaluation of Soil Salinization Risks under Different Climate Change Scenarios in a Semi-Arid Region of Tunisia. *Water Supply* **2021**, *21*, 2463–2476. [CrossRef]
13. Majeed, A.; Stockle, C.O.; King, L.G. Computer Model for Managing Saline Water for Irrigation and Crop Growth: Preliminary Testing with Lysimeter Data. *Agric. Water Manag.* **1994**, *26*, 239–251. [CrossRef]
14. Šimůnek, J.; van Genuchten, M.T.; Šejna, M. Recent Developments and Applications of the HYDRUS Computer Software Packages. *Vadose Zone J.* **2016**, *15*, 1–25. [CrossRef]
15. Šimůnek, J.; Brunetti, G.; Jacques, D.; van Genuchten, M.T.; Šejna, M. Recent developments and applications of the HYDRUS computer software packages since 2016. *Vadose Zone J.* **2024**, *23*, e20310.
16. van Dam, J.C.; Groenendijk, P.; Hendriks, R.F.A.; Kroes, J.G. Advances of Modeling Water Flow in Variably Saturated Soils with SWAP. *Vadose Zone J.* **2008**, *7*, 640–653. [CrossRef]
17. Miao, Q.; Rosa, R.D.; Shi, H.; Paredes, P.; Zhu, L.; Dai, J.; Gonçalves, J.M.; Pereira, L.S. Modeling Water Use, Transpiration and Soil Evaporation of Spring Wheat–Maize and Spring Wheat–Sunflower Relay Intercropping Using the Dual Crop Coefficient Approach. *Agric. Water Manag.* **2016**, *165*, 211–229. [CrossRef]
18. Kumar, P.; Sarangi, A.; Singh, D.K.; Parihar, S.S.; Sahoo, R.N. Simulation of Salt Dynamics in the Root Zone and Yield of Wheat Crop under Irrigated Saline Regimes Using SWAP Model. *Agric. Water Manag.* **2015**, *148*, 72–83. [CrossRef]
19. Kanzari, S.; Daghari, I.; Šimůnek, J.; Younes, A.; Ilahy, R.; Ben Mariem, S.; Rezig, M.; Ben Nouna, B.; Bahrouni, H.; Ben Abdallah, M.A. Simulation of Water and Salt Dynamics in the Soil Profile in the Semi-Arid Region of Tunisia—Evaluation of the Irrigation Method for a Tomato Crop. *Water* **2020**, *12*, 1594. [CrossRef]
20. Ploeg, D.V.; Heuvelink, E. Influence of Sub-Optimal Temperature on Tomato Growth and Yield: A Review. *J. Hortic. Sci. Biotechnol.* **2005**, *80*, 652–659. [CrossRef]
21. Bhandari, R.; Neupane, N.; Adhikari, D.P. Climatic Change and Its Impact on Tomato (*Lycopersicum esculentum* L.) Production in Plain Area of Nepal. *Environ. Chall.* **2021**, *4*, 100129. [CrossRef]

22. Petrović, I.; Savić, S.; Gricourt, J.; Causse, M.; Jovanović, Z.; Stikić, R. Effect of Long-Term Drought on Tomato Leaves: The Impact on Metabolic and Antioxidative Response. *Physiol. Mol. Biol. Plants* **2021**, *27*, 2805–2817. [CrossRef]
23. Delgado-Vargas, V.A.; Ayala-Garay, O.J.; Arévalo-Galarza, M.d.L.; Gautier, H. Increased Temperature Affects Tomato Fru Physicochemical Traits at Harvest Depending on Fruit Developmental Stage and Genotype. *Horticulturae* **2023**, *9*, 212. [CrossR
24. Allen, R.G.; Pereira, L.S.; Raes, D.; Smith, M. Crop evapotranspiration: Guide-lines for computing crop water requirements. *FAO Irrigation and Drainage Paper No. 56*; FAO: Rome, Italy, 1998; 300p.
25. Hide, J.C. Diagnosis and Improvement of Saline and Alkali Soils. U.S. Salinity Laboratory Staff; L. A. Richards, Ed. U.S. Dept. o Agriculture, Washington, D.C., Rev. Ed., 1954. Vii + 160 Pp. Illus. $2. (Order from Supt. of Documents, GPO, Washington 2' D.C.). *Science* **1954**, *120*, 800. [CrossRef]
26. van Genuchten, M.T. A Closed-Form Equation for Predicting the Hydraulic Conductivity of Unsaturated Soils. *Soil Sci. Soc. Am J.* **1980**, *44*, 892–898. [CrossRef]
27. Mualem, Y. A New Model for Predicting the Hydraulic Conductivity of Unsaturated Porous Media. *Water Resour. Res.* **1976**, *1*, 513–522. [CrossRef]
28. Feddes, R.A.; Kowalik, P.J.; Zaradny, H. Simulation of Field Water Use And Crop Yield. In *Simulation Monographs*; Pudoc Wageningen, The Netherlands, 1978; p. 189.
29. Maas, E.V. Crop salt tolerance. In *Agricultural Salinity Assessment and Management*; Tanji, K.K., Ed.; ASCE Manuals and Reports on Engineering Practice: New York, NY, USA, 1990; Volume 71.
30. van Genuchten, M.T. *A Numerical Model for Water and Solute Movement in and Below the Root Zone, Unpublished Research Report*; U.S. Salinity Laboratory, USDA, ARS: Riverside, CA, USA, 1987.
31. Kanzari, S.; Rezig, M.; Ben Nouna, B. Estimating Hydraulic Properties of Unsaturated Soil Using a Single Tensiometer. *Am. J. Geophys. Geochem. Geosystems* **2017**, *3*, 1–4.
32. Vanclooster, M.; Mallants, D.; Diels, J.; Feyen, J. Determining Local-Scale Solute Transport Parameters Using Time Domain Reflectometry (TDR). *J. Hydrol.* **1993**, *148*, 93–107. [CrossRef]
33. Mallants, D.; Vanclooster, M.; Meddahi, M.; Feyen, J. Estimating Solute Transport in Undisturbed Soil Columns Using Time-Domain Reflectometry. *J. Contam. Hydrol.* **1994**, *17*, 91–109. [CrossRef]
34. Toride, N.; Leij, F.J.; van Genuchten, M.T. *The CXTFIT Code for Estimating Transport Parameters from Laboratory or Field Tracer Experiment. Research Report N°137*; US Salinity Laboratory: Riverside, CA, USA, 1999; p. 119.
35. Carucci, F.; Gagliardi, A.; Giuliani, M.M.; Gatta, G. Irrigation Scheduling in Processing Tomato to Save Water: A Smart Approach Combining Plant and Soil Monitoring. *Appl. Sci.* **2023**, *13*, 7625. [CrossRef]
36. Zhang, J.; Xiang, L.; Zhu, C.; Li, W.; Jing, D.; Zhang, L.; Liu, Y.; Li, T.; Li, J. Evaluating the Irrigation Schedules of Greenhouse Tomato by Simulating Soil Water Balance under Drip Irrigation. *Agric. Water Manag.* **2023**, *283*, 108323. [CrossRef]
37. Shan, G.; Sun, Y.; Cheng, Q.; Wang, Z.; Zhou, H.; Wang, L.; Xue, X.; Chen, B.; Jones, S.B.; Lammers, P.S.; et al. Monitoring Tomato Root Zone Water Content Variation and Partitioning Evapotranspiration with a Novel Horizontally-Oriented Mobile Dielectric Sensor. *Agric. For. Meteorol.* **2016**, *228–229*, 85–94. [CrossRef]
38. Gassmann, M.; Gardiol, J.; Serio, L. Performance Evaluation of Evapotranspiration Estimations in a Model of Soil Water Balance. *Meteorol. Appl.* **2011**, *18*, 211–222. [CrossRef]
39. Abou Ali, A.; Bouchaou, L.; Er-Raki, S.; Hssaissoune, M.; Brouziyne, Y.; Ezzahar, J.; Khabba, S.; Chakir, A.; Labbaci, A.; Chehbouni, A. Assessment of Crop Evapotranspiration and Deep Percolation in a Commercial Irrigated Citrus Orchard under Semi-Arid Climate: Combined Eddy-Covariance Measurement and Soil Water Balance-Based Approach. *Agric. Water Manag.* **2023**, *275*, 107997. [CrossRef]
40. Liu, S.; Huang, Q.; Ren, D.; Xu, X.; Xiong, Y.; Huang, G. Soil Evaporation and Its Impact on Salt Accumulation in Different Landscapes under Freeze–Thaw Conditions in an Arid Seasonal Frozen Region. *Vadose Zone J.* **2021**, *20*, e20098. [CrossRef]
41. Mai, J. Ten Strategies towards Successful Calibration of Environmental Models. *J. Hydrol.* **2023**, *620*, 129414. [CrossRef]
42. Oster, J.D.; Letey, J.; Vaughan, P.; Wu, L.; Qadir, M. Comparison of Transient State Models That Include Salinity and Matric Stress Effects on Plant Yield. *Agric. Water Manag.* **2012**, *103*, 167–175. [CrossRef]
43. Poorter, H.; Bühler, J.; van Dusschoten, D.; Climent, J.; Postma, J.A. Pot Size Matters: A Meta-Analysis of the Effects of Rooting Volume on Plant Growth. *Funct. Plant Biol.* **2012**, *39*, 839–850. [CrossRef]

MDPI AG
Grosspeteranlage 5
4052 Basel
Switzerland
Tel.: +41 61 683 77 34

*Land* Editorial Office
E-mail: land@mdpi.com
www.mdpi.com/journal/land